HERMANN EHRHARDT
THE MAN HITLER WASN'T

JOHN KOSTER

 Idle Winter Press
Portland, Oregon

Idle Winter Press
Portland, Oregon
http://IdleWinter.com

This edition published 2018
Printed in the United States of America
The text of this book is in Adobe Caslon Pro

ISBN-13: 978-1945687051 (Idle Winter Press)
ISBN-10: 1945687053

Hermann Ehrhardt
The Man Hitler Wasn't

By John Koster

*"After conquering the Huns, the tigers of the world,
I will not be beaten by the baboons."*

Winston Spencer Churchill

Hermann Ehrhardt...

He destroyed the Bolshevik takeovers of Wilhelmshaven, Braunschweig, Berlin and Munich in 1919 and saved Germany and Western Europe from Communism...

He sent his soldiers to rescue law-abiding German Jews from Fascist thugs...

He led or backed four attempts to kill Hitler—two attempts drew blood...

He married a genuine princess and lived like a prince
in a Renaissance castle until he was almost 90...

Hollywood cast Fredric March in "So Ends Our Night" and John Wayne
in "The Sea Chase" to play Hermann Ehrhardt characters—
John Wayne portrayed "Captain Ehr*lich*"...

Why haven't most people ever heard of him?

Readers Comment on *Hermann Ehrhardt*

"Hermann Ehrhardt deserves to be remembered as a heroic early opponent of Hitler and Nazism."

Thomas Fleming
Internationally famous historian, World War II naval veteran
*Author of **Loyalties**, **The Illusion of Victory**, and **The New Dealers War***

"John Koster's biography of *Korvettenkapitän* Hermann Ehrhardt is truly a page-turner and gives fresh new support to the position... which holds that Nazism was especially evil and not just a form of anti-Bolshevism... I read it twice... Koster documents the heroic struggles of a traditionalist Imperial German Naval officer to redress the egregious misjudgments of German Army Intelligence by fighting hard against both Communism and Nazism in the name of Western values and the best traditions of Germany."

John Czop
Ph.M. in Modern European History, Columbia University, NYC
Associate Editor of THE POST EAGLE, Clifton, New Jersey
Correspondent TELEWIZJA REPUBLIKA, Warsaw, Poland

"John Koster's ***Hermann Ehrhardt: The Man Hitler Wasn't*** is a fascinating biography of a man still wedded to the old world philosophy of honor. Had he succeeded in killing Hitler, we would be living in a different world. Had he failed in his anti-Bolshevik efforts, we would be living in a different and more unpleasant world. The book was fascinating reading. I had trouble putting it down. It is almost impossible to write an objective account of important individuals who lived during the Weimar and Hitler period. I believe that John Koster has accomplished this. Koster points out that the importance of Ehrhardt can be seen in the characters representing him in Hollywood movies—Fredric March and John Wayne."

John Dietrich
U.S. Army veteran, Defense Intelligence Agency
MA in International Relations, fluent in German
*Expert on post-war Europe, author of **The Morgenthau Plan***

"A re-appraisal in unprecedented detail and nuance of one of those German figures of the years following World War I who might have steered his country in a somewhat different direction than the one it took—if only a contemporary of his, one Adolf Hitler, had not prevailed. In the course of documenting their differences, the author will undoubtedly leave some readers speculating on what might have been."

Jon Guttman
MA from NYU in History, deployed U.S. Army veteran in Bosnia
Research Director, HistoryNet
*Author of **Fighter Aircraft Combat Debuts***
*and **The Origins of Fighter Aircraft**.*

"Hermann Ehrhardt was a leading German anti-Communist in 1919, but he was no proto-Nazi. For starters, he didn't hate Jews. He tried to block Hitler's rise to power politically, and four times between 1923 and 1944 he plotted to kill Hitler. John Koster, making good use of primary sources in German and French for this long overdue biography, reveals new details about an extraordinary German and explodes many of the myths about Hermann Ehrhardt's rise to notoriety in a highly troubled nation. At the end of World War II Ehrhardt was set free, and Koster leaves no doubt that this was true justice."

Greg Lalire
BA in History from University of Arizona, Senior Editor, HistoryNet,
*Author of **Captured***

Contents

Introduction

Too many books have been written about Adolf Hitler. No American book in English has ever been written about Hermann Ehrhardt, the man Hitler wasn't. *Korvettenkapitän* Hermann Ehrhardt destroyed the Bolshevik attempts to seize control of the port cities of Germany, of the capital at Berlin, and of the Bavarian capital at Munich while Hitler did nothing but rant and rave against Communists and Jews. Ehrhardt was a hard-core anti-Bolshevik and a lethal terrorist. He was also a hard-core anti-Nazi who despised Adolf Hitler. He attempted to block Hitler's accession to power. He was sentenced to death by Hitler and narrowly escaped. He was tangentially involved in four plots to overthrow or assassinate Hitler for 20 years. He described Hitler as "absolutely unacceptable," "a psychopath," and "an idiot." The fact that the bomb that almost killed Hitler on July 20, 1944 detonated on the birthday of Hermann Ehrhardt's wife is almost certainly a sheer coincidence. Yet Ehrhardt attached great importance to birthdays.

Hermann Ehrhardt's autobiography, "Adventure and Fate," was published in 1924, when he was 43, when most of his anti-Nazi exploits were still ahead of him. His anti-Bolshevik exploits had already become the stuff of legend. He lived to be two months short of 90. He deserves a full-length biography to the degree that authentic sources make this possible. Hitler has been covered more than adequately.

The contemporary and controversial German historian Lothar Machtan counted 138,000 Hitler books at the beginning of the 21st Century and added that none of the recent ones offered anything new or revealing. Professor Machtan himself offered something that was extremely revealing in "The Hidden Hitler:" the strongest circumstantial case ever printed for the concept that Hitler was a homosexual who neutered out for purely political reasons because German society of his time (outside decadent Berlin) was staunchly homophobic. The rumor had been circulated during Hitler's lifetime, mostly by his Leftist enemies. Both Ernst Hanfstaengl, once Hitler's close friend, and psychiatrist Dr. Walter Langer discussed it in the 1940s. Hitler biographer Robert G.L. Waite followed up in the 1970s. But Lothar Machtan provided a comprehensive case study that the rumor was entirely plausible, if not entirely proven.

A decade after Lothar Machtan opened one closet with "The Hidden Hitler," Jean-Paul Mulders and Marc Vermeeren published an article in the Flemish magazine "*Knack*" which opened another one. The two Belgian researchers reported that, based on samples of saliva from Hitler's known relatives, Hitler was indeed of mixed Jewish ancestry—another rumor that had been around since Hitler first became famous. The Hitler family's most dominant DNA haplogroup, E1b1b1, is rare among Western Europeans but a founding lineage of the Jewish population, found in 18 to 20 percent of Ashkenazi (northern European) Jews and in as many as 30 percent of Sephardic (Middle Eastern) Jews. Eight strands of straight brown hair on the hairbrush of Eva Braun—Hitler's curly-haired blonde official mistress—were shown to be of mixed Jewish provenance by a separate DNA test.

The world-renowned German-Jewish biographer of Napoleon, Emil Ludwig—born Emil Ludwig Cohn in Breslau (then Prussia)—had pointed out as early as his 1958 Hitler biography in "World Book Encyclopedia" that Hitler, a lapsed Catholic, frequented Jewish charities in Vienna, a city where there were plenty of Catholic charities available. "...he took advantage of Jewish charity when his funds ran too low to live in even a cheap rooming house. The shelter in which he lived was founded and supported by the Jewish Baron Koenigswarter." Hitler's half-sister was a cook in a restaurant for Jewish students. Jewish charities often fed outsiders, but they generally saved full-time paid jobs for fellow Jews.

Ludwig also implied, though he did not state, that Hitler might have been homosexual. "He was uncomfortable in the presence of women, so he removed them from public affairs.... His mustache and the way he walked and gestured were the mannerisms of a man who is trying to appear stronger than he actually is. ...He was a man on the run." "World Book" was written for middle-school-aged youngsters in a more sheltered era and Emil Ludwig's biography was not explicit. Hitler's personal lawyer, Hans Frank, had reported before he was hanged for war crimes after the Nuremberg trials that he had discovered that Hitler was of mixed Jewish ancestry and had hushed it up. Hitler's nephew, the half-Irish U.S. Navy sailor William Patrick Hitler, had allegedly tried to blackmail Hitler with one or both of these embarrassing disclosures. Hitler's personal physician, the quack Dr. Theodore Morell, listed him as "*homosex*" in a recently auctioned long-lost medical report.

Morell reported that he was injecting Hitler with female hormones—possibly to control Hitler's desire to sodomize men.

The Kaiser's Germany was full of good, useful, and patriotic citizens who were of Jewish or mixed Jewish ancestry—but these worthy people did not espouse anti-Semitism as the key to their political appeal to the thuggish lowest quadrant of the German population—the angry, economically insecure people without inheritance, education, useful skills or strong religious ties. These were the same people who, frozen out by the Christian and monarchist Right, might otherwise have become Communists. The 21st Century dawned on 138,000 Hitler biographies and progressed to stumble upon a Hitler who was plausibly a homosexual, probably of mixed Jewish ancestry, and had viciously persecuted both of these groups. The Jews, some had sometimes sheltered Hitler the vagabond loser, suffered in the worst interracial atrocity of the 20th Century.

Hitler is also said to have been as much Czech as German by ancestry. Ron Rosenbaum, among others, has pointed out that the original family name, Hiedler, was more common among Austrian Czechs than Austrian Germans in the Waldviertel section of Austria-Hungary where Hitler was born and raised. Paul von Hindenburg, Weimar Germany's last elected president and a linear descendant of Martin Luther and also of Polish royalty, referred to Hitler as "the Bohemian *Gefreite*"—*Gefreite* means lance corporal, Hitler's highest rank in the Kaiser's army. *Bohemian* means Czech. The best-selling American author Vincent Sheehan also implied in "Not Peace But A Sword" in 1939 that Hitler was visibly more Czech than Germanic in ancestry.

Ralf Jahn, a modern German historian and the self-styled Sherlock Holmes of genealogy, also researched Hitler's background. Jahn believed Hitler was part Moravian—another type of Czechoslovakian—and also found a pattern of multiple ancestral marriages that verged into incest (and several Hitler relatives had hereditary mental illness). Hitler's father and mother were uncle and niece. Hitler's father, Alois—said elsewhere to be the carrier of the alleged Jewish heritage—was probably the only person in the family whose intelligence was in any way better than mediocre. Hitler had plenty of tangential relatives who were mildly retarded and at least one cousin who was schizophrenic. She died in a concentration camp and he never lifted a finger to help her. This was the man who said he wanted to promote the Master Race to control of Europe. This was also the man who never wanted children of his own.

Hermann Ehrhardt may have inherited the slant of his eyes from remote Hunnic ancestors who rode roughshod into Germany territory in late Roman times as mentioned in the medieval "*Nibelungenlied.*" Rudyard Kipling had popularized the term "Huns" to describe the continental Germans because of their ferocious Asian ancestry, unlike Kipling's own Anglo-Saxons who were also of Germanic origin but skipped the Huns, who never got their horses across the English Channel. Madison Grant, the founding father of American "scientific" racism, had argued during World War I that the Kaiser's Germans were not purely Germanic, as Madison Grant's own Anglo-Saxons obviously were. To Winston Spencer Churchill—half American and, on his American mother's side, one-sixteenth Delaware Indian—the Kaiser's Germans were impressive barbarians: they were the 'Huns' who threatened the British Empire as their Germanic and Hunnic ancestors had threatened the Roman Empire. Churchill felt that the Bolsheviks, whom he was one of the first to oppose, were a cheap imitation of the German threat of his younger days. "After conquering the Huns, the tigers of the world, I will not be beaten by the baboons," Churchill said shortly after the second Russian Revolution overthrew Alexander Kerensky and swept Lenin into power.

Lenin, son of a Chuvash Asiatic father, had slanted eyes—otherwise social poison in Russia, where the Asian guy is always the villain in melodrama. Gemans have no such disdain of Asians, and Hermann Ehrhardt's slanted eyes were blue-grey, his hair was very dark blond, and Ehrhardt's known ancestors included at least three generations of Lutheran clergymen, including his own father. Ehrhardt's sexual normality was established by certain adolescent escapades with girls, two sons by his first marriage to a respectable middle-class widow whose first husband had been a member of the nobility, and a son and a daughter by his second wife, a genuine princess with dynastic ties to the Kaiser. Ehrhardt's 'Aryan' credentials were in far better order than Hitler's or those of several other top Nazi leaders, including notable anti-Semites. Yet Hermann Ehrhardt was not a political anti-Semite. He never organized or sanctioned racial or religious attacks on Jews. The German Communists, trying to ride the rip tide of anti-Semitism from 1919 – 1922, referred to Hermann Ehrhardt as "Ahasuerus in an armored car" because he sent his men to rescue law-abiding German Jews from thugs instead of beating them up. Ehrhardt found the Poles, Hitler's other bugbears, to be frisky and cantankerous but not a racial menace. He also admitted that his own life had once been saved by his ethnic Russian orderly and his tribal African porter. He designed a children's optical toy that was manufactured in Japan, a recent enemy but a country he clearly admired. Hermann Ehrhardt seems—like a great many educated Germans of his era, the age of Thomas Mann and Oswald Spengler and Erich Maria Remarque, who was Ehrhardt's poet laureate in two novels—to have been utterly indifferent to race and racism unless war or politics were also involved. One of Ehrhardt's assault group commanders was unabashedly Jewish. No one screened out Poles, a hardy mainstay of German militarism, from the Ehrhardt Brigade or from the Prussian Guard of the same era. Rudolf Mann, Ehrhardt's quartermaster, noted that the Ehrhardt Brigade included several White Russians, a Turk, and a Chinese, all of them respected as fellow German soldiers. The 2nd Marine Brigade was not the Nazi SA or SS.

Hitler spoke only German. Ehrhardt was completely fluent in English and French. He knew The Bible and considered himself a Christian, and he could poke his finger along through classical Latin, slowly and with a dictionary. He was a crack shot, an amateur boxer and *judoka*, a strong swimmer and a reasonably capable horseman. He sometimes drank rather heavily when he was off duty, and he loved fast driving in automobiles, especially the armored cars which he himself often steered into the battle zone. Hitler was flabby, a leery teetotaler, afraid of water even in a rowboat, afraid of horses, and afraid to drive a car until he was well past the age of 30.

At the end of his string in lawful public life, Ehrhardt would test the political wind and offer to make common cause with anybody—including Nazi drop-outs and the home-grown German Communists of the KPD—who would help him block the maniacal Hitler from power. The German politicians of the Weimar Republic had sent Hermann Ehrhardt to prison, into hiding or into foreign exile once he had actually destroyed the Bolshevik take-overs of 1919, but Ehrhardt still defended Weimar Germany from both Stalinist Communism and Nazism from the inside. Hitler tried to have Ehrhardt murdered and later sent him to prison facing the death penalty. Ehrhardt was a threat to Hitler because Ehrhardt saw through Hitler and despised him—and considered Hitler extremely dangerous to Germany. Conversely, Ehrhardt also knew that many Jews had played a significant and constructive role in German history and would be needed to do so again if Germany were to regain its power and prestige in Europe. Had Hermann Ehrhardt influenced German politics to keep Hitler and his thugs out of power—or succeeded in either of the first two of his four attempts to kill Hitler—there would certainly have been no Holocaust. Neither Ehrhardt nor the men of his command ever harmed a Jewish woman or child. They violently opposed Hitlerism as they had violently and effectively opposed Bolshevism. Hermann Ehrhardt deserves a modern biography in English.

Chapter One
Creating a Frankenstein

Captain Truman Smith, United States Army, Infantry, had one more appointment to keep before he could go home to his wife Kay and their daughter Alling, a sunny little two-year-old girl who was not in the best of health. Smith was stationed in Berlin as assistant U.S. military attaché in 1922 but he had been sent to Munich, the capital of the restive former Kingdom of Bavaria, to investigate the National Socialist German Worker's Party and to evaluate its possible future importance to German and international politics. Smith doted on his wife Kay—she was more than a foot shorter than he was—and he was worried about his daughter. He would rather have been back in his temporary home in Berlin than exploring the Robbers Roost that Munich had become after the "Soviet Republic of Bavaria," a short-lived revolutionary government headed by Bolsheviks and led by Eugen Leviné-Nissen, had been crushed by the *Reichswehr*, the German regular army, ably assisted by the notorious *Korvettenkapitän* Hermann Ehrhardt, the military chieftain of the German Right. But Truman Smith was a man who did his duty.

Hermann Ehrhardt was a known quantity. He was also a wanted man, who had ducked arrest after an attempted take-over of Berlin in March of 1920—reportedly a doomed attempt to bring back Kaiser Wilhelm II—followed by the purported murders of 354 people Ehrhardt was said to have been designated as a traitor outside the reach of the lenient Weimar Republic's courts. When a journalist asked

Ehrhardt how many murders he had actually ordered, Ehrhardt mischievously shifted his slanted Hunnic blue eyes back and forth and said: "*Acht.*" It was an evasive pun: "*acht*" with a small letter a means "eight." With large A, as a noun, "*Acht*" means "respect," or even "beware." End of interview.

Truman Smith would have been the perfect man to interview Hermann Ehrhardt, now said to be hiding in Munich, if anyone knew exactly where Ehrhardt was. Hermann Ehrhardt was the son and grandson and great-grandson of respectable Lutheran clergymen from Baden in South Germany near the Swiss border. Truman Smith was the grandson of the original Truman Smith, a U.S. Congressman and U.S. Senator from Connecticut who wrote sermons as well as speeches. He made no money in Whig politics before the American Civil War but Senator Truman Smith's political career was honorable and honest in an age that was becoming more and more corrupt. He was appointed to a judgeship in recognition of his probity.

Eugen Leviné-Nissen

The Senator's son, Captain Edmund Smith, a West Pointer and veteran of an Indian campaign in the West against the Utes, later served with the 19th U.S. Infantry in the Philippines. Captain Edmund Smith was killed in an ambush on the island of Cebu fighting Filipino guerillas. His company had dropped eight stragglers on the march and on February 4, 1900, Captain Smith led 34 volunteers out to recover his lost soldiers. The captain and his soldiers saw a cluster of *nipa* palm huts. "So the captain halted the column about 300 yards away from the huts and with three men proceeded towards them to see if they could be utilized for shelter," Major John Leefe of the 19th Infantry wrote to Captain Smith's widow. "When within about 100 yards of the enemy, the enemy, who had remained concealed, suddenly opened fire on the little party, the captain receiving two bullets in the right side of the abdomen. He did not fall, but heroically bade Private Gideon, who was close to him to engage the insurgents who, to the number of sixteen, now arranged themselves in front of the huts, and then went back calling on the rest of the men to come up and open fire, which they did, driving off the enemy. The captain realized that had had been mortally wounded." Captain Edmund Smith died the next day, an authentic hero in a war that was splattered with numerous atrocities by both sides.

Truman Smith himself had graduated from Yale rather than West Point and was studying history at Columbia University in New York City when the

Mexican Incident called him to the colors. Shortly, he was serving on the Western Front after the United States declared war on Germany. He and his men of the 4th Infantry Division helped stop the German *Friedensturm*, the Peace Offensive in the summer of 1918 which almost crumpled the Allied front. Smith's colonel was wounded during the Battle of the Argonne Forest, the bloodiest engagement in American military history. The battalion commander was killed, and Truman Smith emerged as the ranking officer of his battalion—neither killed nor wounded nor subject to a nervous breakdown due to stress of concussion, then known as "shell shock." Smith was recommended for the Distinguished Service Cross, but never got it, though he would receive the Silver Star once the new decoration for valor was authorized in 1926. He lost 40 pounds in continuous combat or combat support service in the trenches, including raids that took German prisoners. Finally Smith had to be taken out of the line not for mental problems but for sheer exhaustion. He finished the war in a hospital.

In June of 1920, Captain Smith was stationed in Berlin as part of the military mission, based on the fact that he was totally fluent in German, reasonably fluent in French, and a veteran of intense combat on the Western Front, which the Germans themselves would be apt to find impressive. The Western Front was becoming something of a cult. Smith was familiar with Hermann Ehrhardt's two Nationalist invasions of Berlin: the one in January of 1919 where his political advisor, Captain Waldemar Pabst of the Guards Cavalry Division had "executed" the Spartacist leaders Rosa Luxemburg and Karl Liebknecht with the tacit approval of the Weimar cabinet, and the one in March of 1920 where Ehrhardt and Pabst were foiled by a General Strike, which ended with Captain Ehrhardt as a wanted outlaw whom nobody wanted to personally confront. Captain Truman Smith was in Berlin in June of 1922 when Walther Rathenau, the patriotic German Jewish titan of finance and diplomacy, was murdered. Rathenau had signed the mutual aid Treaty of Rapallo with Soviet Russia that temporarily ended trouble with the Bolsheviks, only to be murdered by two young officers. One of them shouted "Long Live Captain Ehrhardt" before he shot himself in the head to avoid the humiliation of surrender. Rathenau's murder had touched off vast mourning and recriminations all over Germany and increased the liberal and leftist demands that something be done about the now-invisible Hermann Ehrhardt and his quick-on-the-trigger young henchmen.

Smith, however, was dispatched from Berlin to Munich to assess the nascent National Socialist Workers Party, not the remnants of Ehrhardt's now-illegal 2nd Marine Brigade, which had twice marched 6,000 strong into Berlin led by captured British tanks, German armored cars, thundering motorcycles, a fleet of trucks, and two batteries of horse-drawn artillery.

Foreign diplomats had described the Munich-based National Social leader as "an uneducated madman," but the U.S. Ambassador, Alanson B. Houghton, was at least curious about the movement. Captain Smith was dispatched to Munich to find out, among other things, if Bavaria, which had a long history of "particularism," might actually try to secede from the rest of Germany. Was there any danger of a new Communist revolt like the one in 1918-1919? Were the National Socialists strong enough to seize power in Bavaria? Was the 7th Division of the *Reichswehr*, the new German regular army, loyal to Bavaria or Berlin? Were the almost daily incidents of hostility between Bavarians and the Allied Military Control Commission of any real importance?

Captain Smith's first look at Adolf Hitler on Saturday, November 18, 1922, was disconcerting.

"A remarkable sight," Smith wrote later,[4] "Twelve hundred of the toughest roughnecks I have ever seen in my life pass in review before Hitler at the goose step under the old Reichsflagge wearing red armbands with *Hakenkreuzen* (swastikas). Hitler, following the review, makes a speech, stating that Berlin has prevented their movement to Regensburg. Next week, however, the National Socialists will clean up the town. He then shouts 'Death to the Jews' etc. and etc. There was frantic cheering. I never saw such a sight in my life. After the review I was introduced to Hitler. He promised to talk to me on the following Monday and explain his views."

Truman Smith first spoke to a number of other significant residents of Munich, including Crown Prince Rupprecht of Bavaria, a capable World War general, who said he was not plotting a secession from Germany. Erich von Ludendorff, the former quartermaster-general of the Kaiser's army and a rightist eccentric of some prominence, was more forthcoming and expansive when he spoke to Smith. Ludendorff, whose nervous breakdown was part of Germany's military collapse at the end of the war, told Smith that "Germany is sick from Marxism," and that "If Marxist Bolshevism continues its victorious way, England, France, and America will successively fall victim to it." Ludendorff advocated an understanding with France, saw Britain as useless, and urged Smith to tell the Americans that "only a strong nationalist government in Germany can preserve the country from chaos."

On Monday, November 22, 1922, at 4 p.m., Captain Truman Smith, U.S.A. kept his appointment with Adolf Hitler at Georgenstraße 42.

"A marvelous demagogue," Smith wrote. "I have rarely listened to such a logical and fanatical man. His power over the mob must be immense."

Smith was six-feet-four, 29 years old, a trim Anglo-Saxon of Puritan stock with thick black eyebrows and perhaps a hint of Pequot or Wampanoag Indian in his ancestry: Smith always loved Indians, rated them as great soldiers, and visited Indian reservations whenever he was in the United States. The tall American noted that Hitler, six inches shorter and seven years older, was somewhat flabby with a soft, almost feminine handshake. The setting, Smith remembered later, was drab and dreary, like the back room of a New York tenement. Hitler's Austrian accent when he spoke instead of screaming was unmistakable, rather as a Southern drawl is in New York or New England. And Hitler's pale eyes were almost hypnotic. His manner was frank and confidential, as if he and Smith were fellow survivors of the

Western Front who could look the truth square in the face as they had once looked Death square in the face. Yet Hitler answered every question instantly, almost mechanically. The effect was mesmerizing. Smith was somewhere between charmed and spooked.

"What is the essence of your movement?" Smith asked Hitler in clear German that, unlike Hitler's, had a "*platt*" North German regional accent.

"We are a union of hand and brain workers to oppose Marxism." Hitler said. The idea of the National Socialist German Workers Party, Hitler said, was to embody concern for the rights and the needs of working people without falling under the sway of Moscow.

"The present abuses of capital must be done away with, if Bolshevism is to be put down," Hitler said. Playing off Smith's precise North German accent when speaking German, Hitler played down his own anti-Semitism. The Prussians were the least anti-Semitic of Germans, the Hanoverians next door largely so, and the sort of speeches that brought frantic cheers from Hitler's coterie of thugs would have been offensive to an educated Prussian or Hanoverian—as Hitler well understood.

"What about the reparations?" Smith asked. The Allies expected the Germans to pay for the cost of World War I. The Germans claimed they couldn't pay without wrecking their economy, and the whole topic had sparked an inflation which had reduced the value of paper money to absurdity. Last week's pay wouldn't buy a pair of socks this week.

"We must reduce the reparation to a possible sum, but then pay it with all the energy of the German nation," Hitler told Smith. "Only by paying reparations can Germany regain its good name in the world."

"How would you go about this?" Smith asked.

"Only a national government can carry out a task like this," Hitler said. "There must be universal service for reparations. Two million men must give two years to it." Germany was full of unemployed veterans and Hitler explained that using their manpower constructively would solve the unemployment problem and restore Germany's credit. "The printing of paper money must be stopped," Hitler added. "This is the worst crime of the present government."

"What kind of a government do you see in Germany's future?" Smith asked.

"The national government of the future must not be encumbered with either pre-war or war personalities," Hitler said, referring obliquely to the German imperial and regional royal families and to Hindenburg and Ludendorff and the other Prussian generals, once their sworn retainers. "Monarchy is an absurdity. The German royal families ruined their cause by running away, but the whole monarchical question is of fifth or sixth importance. The people themselves can decide the question of monarchy or republic after a national government has come to power. Parliament and parliamentarianism must go. No one can govern with it in Germany today. Only a dictatorship can bring Germany to its feet."

Hermann Ehrhardt

"Soldier, to soldier, do you think there will be a renewal of hostilities?" Smith asked.

Hitler smiled.

"It is much better for America and England that the decisive struggle between our civilization and Marxism be fought out on German soil rather than on American or English soil. If you Americans fail to help German Nationalism, Bolshevism will conquer Germany. Then there will be no more reparations, and Russian and German Bolshevism, out of motives of self-preservation, must attack the Western nations. I myself want an understanding with France. The idea of launching a war of revenge is absurd."

Captain Truman Smith's account of his appointment with Hitler created a Frankenstein. Hitler, with the feral instincts of a frequently abused child from a marginal family, told the tall, distinguished American officer just what conservative Americans wanted to hear—war with Communist Russia, not with France or the Anglo-Saxons. Unable to stay in Munich any longer, anxious to get back to his wife and daughter in Berlin, Smith asked a friend of his, a German-American Harvard graduate with connections in the world of art and culture and important friends, to keep an eye on Hitler for him.

The next and last time Truman Smith met Adolf Hitler was at a reception in Berlin in 1937, when Hitler was in power and had become the dictator he predicted to Smith in 1922.

"Have I not met you before?" Hitler asked.

"Yes, Mr. Chancellor, in Munich in 1922," Colonel Smith replied.

"Oh yes," Hitler replied. "You introduced me to Hanfstaengl."

Chapter Two
Catching the Tiger

A week to the day after Captain Truman Smith had designated Adolf Hitler as the man to watch among German Nationalists, Captain Hermann Ehrhardt was on his way to prison—and on his birthday.

Princess Margrethe von Hohenlohe-Öhringen—28 years old, slim, elegant in the new form-fitting fashion, considered a beauty with regular classical features that made her look almost like a teenager, a collateral relative of the former Kaiser —was strolling from the Siegestor on the Ludwigstraße in Munich when she received a greeting from a muscular man who looked like a blacksmith with possible evil intentions.

"Serene Highness!—what an unexpected pleasure to meet you again!"

Princess Margrethe ignored the sinister stranger and kept walking, glancing around to see if a policeman might be in sight.

"Has your Serene Highness forgotten a former guest?" the sinister man in dowdy clothes asked.

"Have we been introduced?" the Princess asked primly.

"We spoke many times at the Ratibor Castle estate in Silesia in the days before the Plebiscite, and we often ate dinner at the same table," the sinister man said. "We sometimes rode together with your relatives and friends."

The Princess turned with surprise and looked at him, trying not to be frightened out of her dignified stance. She saw a badly-shaven, muscular man above average height with a slight Oriental caste to his square, regular features, a man with a large chest and broad shoulders, a thick neck and sturdy legs, formally but rather badly dressed. The man shifted his slanted blue-grey eyes from side to side, smiled mischievously and whispered: "*Hermann Ehrhardt.*"

"Impossible!" the Princess said briskly.

"Do you remember the first words Hermann Ehrhardt ever spoke to you?" Ehrhardt asked.

"What were they?" the Princess asked, now more intrigued than frightened.

"*Are you really a princess? You're so dark I thought you were Polish*! You looked at me with death in your eyes for some time afterwards."

"You *are* Hermann Ehrhardt.... No one could imagine that introduction. Captain Ehrhardt—what has become of you?"

"I'm temporarily without a command, but I still have some friends. I'm working for an optical firm here in Munich. At present, my new name is Herr Eschwege, a loyal Alsatian living in Germany. Many people seem to take an unfortunate interest in Captain Hermann Ehrhardt."

Ehrhardt had returned to Munich from the safety of exile in anti-Communist Hungary shortly after the murder of Foreign Minister Walther Rathenau. Ehrhardt himself had not authorized the assassination, but one of the killers, a former officer of Ehrhardt's 2nd Marine Brigade, had shouted out "Long live Captain Ehrhardt" before he shot himself in the head to avoid capture by the police at the midnight siege of a ruined castle in the middle of a thunderstorm. Ehrhardt, however, was widely blamed for the assassination, which was cheered by right-wing extremists and mourned and protested by the center as well as the left. The German Jewish community was traumatized and terrified.

Ehrhardt called his 1924 biography "Adventure and Fate." He had no doubt that Fate had brought him back in touch with the elegant Princess Margrethe at a time when he was living in shabby rooming-houses or snoring away with his head on the table in taverns under an assumed name, and eating at the expense of old friends "...where I had to say '*danke schoen*' for every slice of bread. I fell into disorder in terms of clothing and laundry. I left socks and shirts, shoes and suits strewn all over Munich at a time when I had no money to buy more. Every day threatened me with depression.... Every night awoke disgust, but I managed not to puke. But whoever is seaworthy is life-worthy. I gave myself orders not to get slack."

Princess Margrethe's sudden and elegant appearance delighted him, especially when she held out an offer that brought him back to the world of royalty and nobility he had once loved, even as a gate-crasher with no title of his own.

"Captain Ehrhardt, you have a standing invitation to take tea at my villa any time you wish. My lady-in-waiting and Kaete, the maid, will be delighted to welcome Herr Eschwege from Alsace to our little circle.... These are strange times. I myself am studying stenography and typing against the day when the world we knew no longer exists."

"*Durchlaucht*, if I might dare to suggest—I believe with your gracious manner and intelligence you might do better as an investment counselor rather than as a mere stenographer.... What time would it be convenient for me to call at your villa for tea?"

Ehrhardt was still leading a double life. He worked by day at the optical firm to support himself and send whatever money he could to his wife and three children in northern Germany. Money devalued so quickly during the Weimar Inflation that savings became worthless overnight and only wages paid on a day-to-day basis had any purchasing power at shops and stores. A week's wages from the week before might buy a loaf of day-old bread. But by night Ehrhardt maintained contact with his friends in the Nationalist movement opposed to Bolshevism and to what they saw as Britain's attempt to turn Germany into the next India through the Treaty of Versailles. Ehrhardt had picked up the phrase about India from the famous German historian Oswald Spengler, with whom he shared conversations and a mutual admiration. They concurred entirely that Britain was Germany's deadliest enemy for the present, and that Russia was the great threat in the future.

As another fall-back, Ehrhardt invented an optical toy based on his knowledge of gun sights and torpedo sights. "It was a box that contained mirrors, lenses, and prisms. The little children could enjoy the comical visual distortions and the spectrums of color and the older youngsters could make their own simple optical instruments. My thought was that through the pastime of play the children could learn the basic principles of optics.

"To my own astonishment I found that I had some success as a merchant. One of my acquaintances, who had founded a firm in Japan, wanted to take up the manufacture of an article. We had many orders from the Far East and opened a whole new market.... I had never dreamed of mercantile success when I was a young lieutenant."

Ehrhardt took one false step while he was signing a contract. A notary asked him for his name and, due perhaps to the new surroundings or end-of-the day weariness, he said: "*Ehrhardt*." He covered himself instantly by saying: "It's strange how one says straight-away the name he was thinking of at the moment." The notary and the office supervisor let it pass—or so he thought.

Ehrhardt as Eschwege soon became a regular guest for tea at the suburban villa that Princess Margrethe shared with her lady-in-waiting and the maid. High Tea in those circles, as much English as German due to the Kaiser's connection with Queen Victoria and the British royal family, was a regular small meal served with elegance. Ehrhardt's other meals consisted mostly of black bread, cheese, sausages, noodles with vegetables, sauerkraut and Bavaria's *flüssiges Brot*—or liquid bread—the beer from Munich's many famous breweries. He cherished the delightful treats and the company of well-bred women.

"The advantage of the soldier's profession is that a man becomes accustomed to having patience," Ehrhardt once philosophized. "Wait, spy, and spring when the moment comes."

Ehrhardt waited and the moment came. One day he arrived for tea to find the princess, her lady friend, and the Ukrainian maid nervous and apprehensive. Munich was increasingly crowded—mostly with ring-wing fugitives from Leftist Berlin and from Leftist Saxony, or, like the Princess and her lady friend, ethnic Germans from Silesia, which was demographically about 60 percent Polish and 40 percent German. The Munich Office of Rent Control had learned that the villa had a small bedroom that was not in use and had threatened to rent it to a complete stranger. "They mournfully shared their anxiety with me at having to put up with a rough total stranger as a disturbing element in their household."

Ehrhardt sprang.

"*Durchlaucht*—gracious ladies—once again a snooper has stuck his nose into my own cabin too.... What could be more natural than for each side to come to the other's help? My business activities last from eight to five, then I gobble up my evening meal in the city.... I can tell you with a good conscience that during the week you would

Hermann Ehrhardt, 1920

hardly notice my presence...." He offered to rent the vacant room himself.

"They gladly took up my proposal. The goblin of a strange compulsory tenant was lifted from them, and I found a roof over my head where betrayal had no place."

Ehrhardt accepted the smallest and draftiest of the two upstairs bedrooms, but he also got a title. As the son of a clergyman he was outside any social constraint—he had been accepted as a naval deck officer at a time when only members of the aristocracy and the sons of upper professionals such as physicians could aspire to that stratum—but the Ukrainian maid had other ideas about social standing.

"The maid, Kaete Sobezko, had been in service to her lady for many years. She was a splendid person, honest, loyal, and devoted to the Princess and she kept an eye on her like the noblest sort of hunting dog. Even though she had often seen me in Silesia at the castle of the Princess, she didn't recognize me now. I wore civilian clothes and had no more beard. Her service with royalty had given her a special perspective on the world. It was incomprehensible to her that a gentleman could live in the household of the Princess who was simply known as Eschwege. Therefore

she dubbed me Baron. I tried to talk her out of it, but she called me only by the handsome title. In the house of the Princess, it was not permissible to call a man by his own last name."

By way of gratitude, "Baron Eschwege," formerly known as Hermann Ehrhardt, took over the little garden at the villa on Sundays, trimming the trees, digging up the ground, setting out new plantings, and sparing the hired gardener any trouble. Ehrhardt had always loved Nature and enjoyed working with the peasants who were his father's parishioners, learning their Alemannic dialect in addition to the High German of his father's household and of the Luther Bible, and the French he had studied in Switzerland and elsewhere. His knowledge of English was mandatory for a German naval officer and was an added advantage in business, but speaking fluent Alemannic and good French had helped him impersonate Eschwege. Ehrhardt knew without a doubt that he could trust Princess Margrethe. Her lady friend and Kaete Sobezko both believed he was actually Eschwege. Betrayal by his own Organization Consul was almost unthinkable—the members belonged to an officer cult where lethal dueling was a recent memory and suicide was far more acceptable than disgrace. Murdering informers was also a regular occupation. Ehrhardt had set the Princess free from the fear of a compulsory tenant, and he himself was substantially free from the fear that his previous landlady would greet him with: "The Criminal Police have been inquiring after you."

Ehrhardt's rest under the roof of the Princess and her ladies, he said, was good for his worn-out nerves—perhaps too good. A few weeks after moving in, he learned from friends that a research judge from the Weimar Republic's national court system, Dr. Metz, had been questioning former junior officers of the 2nd Marine Brigade, along with some of the Bavarian officials who had turned Munich, in particular, into a safe house for the Nationalist Right. Dr. Metz, however, was an enrolled member of the German National People's Party—a moderate ring-wing group opposed to Bolshevism and resentful, as most Germans were, of the Allied Control Commission stationed in Germany and of the Treaty of Versailles. The younger officers who had spoken to Dr. Metz told Ehrhardt that Dr. Metz had been polite and even supportive, and that the whole investigation appeared to be a mere formality: "I have done my interrogation—nothing came of it."

But something came of it. When Princess Margrethe had finished her courses in typing and stenography, she asked Ehrhardt for introductions to people who might be able to find a situation for her. Ehrhardt introduced her to Walter Pruckner, a sometime Ehrhardt admirer and former landlord, who spoke to her at some length and then spoke to Ehrhardt about old times—and asked where he was living at the present time. Ehrhardt said he changed addresses and never mentioned his lodging with the Princess and her coterie. But Pruckner knew Ehrhardt by sight now, beard or no beard, even if he didn't know where Ehrhardt lived. Pruckner had also accepted rent money from Ehrhardt for the traditional three-month *terme* when Ehrhardt arrived with the 2nd Marine Brigade to overthrow the short-lived but murderous Soviet Republic of Bavaria in May of 1919. Pruckner had never returned the balance of the money. He told Ehrhardt he had invested it and lost it—not hard to believe during the catastrophic Weimar Inflation.

The next time Dr. Metz came back to Munich, he had a new lead on Hermann Ehrhardt.

On November 29, Hermann Ehrhardt was hoping to formally celebrate his birthday with a good cigar, a few drinks, and social calls from some of his friends from the former 2nd Marine Brigade. The phone rang at his office at the optical firm.

"May I speak to you—it's important!" Princess Margrethe had called, and not about his birthday.

"I'll be home in the early afternoon..." he said.

"Good... good..." she was almost sobbing.

Ehrhardt made his excuses and left work early. The Princess was in a state approaching hysteria. Her lady friend was visiting her sick father. The Princess had suffered in an agony of fear and confusion until Ehrhardt arrived.

"I received a call to appear at the Palace of Justice this morning," she said. "A kindly, polite old man with bright blue eyes greeted me with the proper terms of respect and then asked me to take a seat in a large comfortable chair with the remark 'We'll have to be quick about this—you must tell me where Captain Ehrhardt is, and what your relationship to him is!' He had a strange smile and his lips were trembling with excitement.

"I felt disgust creeping over me and I said 'I don't know where he is, and there is no relationship.'

"'You don't know where Captain Ehrhardt is? And there is no relationship with him? *Durchlaucht*, you must swear an oath about what you just said, you hear me, you must!'

"I felt faint and almost swooned, but I said, 'I want to see a clergyman!'

"'You may leave now then, but be back at one tomorrow morning,' the old man said brusquely, still trembling with excitement."

Ehrhardt took stock of the situation. The Princess was deeply religious and believed that a false oath could be a ticket to damnation. She also knew that she would be a disgrace to her family and her class if she betrayed a man who had put his life on the line to defend her estate and her life before the Silesian Plebiscite. Most German Christians and German Jews, and a considerable number of the ethnic Poles, had come out in favor of keeping Upper Silesia officially German, with a 60-40 margin in favor. Ehrhardt and the 2nd Marine Brigade has been a factor. Even the Poles had found the Ehrhardt Brigade's discipline impressive in a somewhat intimidating way.

"It's quite out of the question that you swear any oath," Ehrhardt said quickly. "You're a member of a royal house and I don't believe that he has the right to demand that you swear an oath. Call them tomorrow and tell them you're sick. We need to buy some time."

The Princess called the Palace of Justice and asked to speak to Dr. Metz. Instead, she reached Police Commissar Heldwein of the Criminal Police. Commissar Heldwein conferred with Metz, who said he could not come to the telephone, but said he would send someone to visit the Princess at home and have her swear the terrifying oath. Ehrhardt tried to console the Princess and wondered if he could better protect her by getting out of town and claiming she never knew who he was, or by simply bluffing it out. Abruptly, Police Commissar Heldwein had himself announced at the front door.

Ehrhardt, still playing Eschwege, walked out to the sitting room and took a seat. Commissar Heldwein asked him "a heap of questions. He asked where I was born, where I had lived previously, how I came to have a Hungarian passport."

"What exactly is it that you want to know?" Ehrhardt asked.

"You're suspected of having a relationship with Captain Ehrhardt," Commissar Heldwein said. "You must appear tomorrow morning at nine at the Palace of Justice to discuss the subject of Captain Ehrhardt with National Court Counselor Metz, exactly as the Princess herself must."

"I give you my word that I will be there," Ehrhardt said.

As soon as Commissar Heldwein left the villa, Ehrhardt gathered up all the letters, receipts and papers that showed he was Hermann Ehrhardt and burned them in the fireplace. He kept all the papers that showed he was Herr Eschwege.

Ehrhardt weighed his options. Metz might not know that he, as Eschwege, was actually Hermann Ehrhardt, but a great many people in Munich knew him by sight and someone was likely to identify him in court. At best, Metz might look over his false identity papers, recognize him as a fellow Nationalist, and let him slide. At worst, he was headed for a stay in prison and a formal trial—but on what charges? Ehrhardt had received orders from his superior officer, Baron General Walther von Lüttwitz, for his role in the Kapp Putsch. He had to obey the orders. This was no treason. The Leftist demonstrators his men had shot, without his order, had been asking for it....

At this moment, Lieutenant Franz-Maria Liedig, a former gunnery officer from Ehrhardt's torpedo boat Half-Flotilla 17 and later an officer from the 2nd Marine Brigade, showed up at the door to wish him a happy birthday.

"I gestured him inside, told him it had hailed on my birthday bouquet, and asked him all the questions that I was struggling with." Liedig, not directly involved and a law student in Munich, spoke with Ehrhardt at some length, spoke with the Princess at greater length, and then explained the obvious: Ehrhardt and the Princess both needed a good lawyer. Liedig took off to find one. Ehrhardt briefly wondered if Bavaria—a separate kingdom before 1871—might take up his cause to defend the Bavarian national sovereignty. The bloodshed in Munich in 1919 had started when a group of war resisters—ironically, many of them Jewish émigrés from Prussia—had taken Bavaria out of the war before the Germans themselves had signed the Armistice. In 1920, when Ehrhardt himself was leading the 2nd Marine Brigade through the northern farm country of Hanover, a separate kingdom before 1866 with strong ties to Britain, the Hanoverian monarchists had offered him the command of their own secret army to secede from the rest of Germany, Prussia in particular. The Hanoverians were reputed as the toughest soldiers in Germany but their numbers and lack of industrial power had made the offer absurd, if not treasonous: Ehrhardt had laughed bitterly and then apologized and politely refused. Prussian authority was still widely disliked both in Bavaria and in Hanover, the second and third largest states in Germany. Secession was not an option for Ehrhardt, but pressure for local sovereignty might lead to his continued safety in Bavaria. "The outlaw knows what the right of asylum is worth."

At worst, Ehrhardt thought, he would probably get a short prison term which ended in an amnesty. Two million Germans had died in the World War by battle, others by starvation due to the British blockade, a million others had been crippled; the courts granted amnesties to political offenders of both the Left and Right with a very free hand. Germany needed more German patriots, not more executed war heroes.

Liebig appeared with the lawyer. The attorney spoke first with Ehrhardt, then with the Princess at somewhat greater length.

"He was of the opinion, as I was, that the Princess could not be compelled to take an oath. Liedig, because of his youth, commanded a strong optimism that did us all a lot of good. I've asked myself in retrospect why this dumb summons held us in such hypnotic fascination.... I myself went to the hearing, even though another change of location would have been better for me, because I couldn't run out on the lady in her highly nervous state. She went because of the law of hospitality and her duty to protect her previous protector. Before everything else, however, we consoled ourselves with the expert legal opinion that under German law the Princess could not be forced to take an oath."

Chapter Three
Caging the Tiger

The next day at nine Hermann Ehrhardt, still posing as Eschwege, the loyal German fugitive from Alsace, escorted Princess Margrethe into the Palace of Justice. "I remained at her side and could cover her from any angle. In the Justice Palace we were led to a waiting room. Immediately a white-haired gentleman with a sweet facial expression came into the room and greeted us with exceptional courtesy. I quickly recognized him due to the descriptions as Dr. Metz."

Ehrhardt was adept in boxing and judo and he had to hold himself back from attacking the elderly Dr. Metz with what would have been mutually lethal consequences.

"I kept myself under control, went up to him, presented myself as Herr von Eschwege and said: 'You wanted to interrogate me, here I am.' But with a lovable little bow I shrugged the Research Judge off and said 'A moment of patience, I have something very brief to do with the Princess."

Ehrhardt still believed that under German law the Princess could not be forced to swear an oath, and he reminded her of this in the anteroom. When he returned to Dr. Metz, he was surprised to find that the Princess was not to be present at his own hearing. Dr. Metz was highly excited. His hands and lips trembled, and he nervously wiped his spectacles as thick beads of sweat stood out on his forehead, even though Ehrhardt said the hearing room in the Place of Justice was "dog cold" in late November, a peasant expression which meant that dogs were allowed to sleep inside the house.

Dr. Metz asked about "Eschwege's" birth, parents, origin, the course of his life, his military service and "a thousand other things. I had memorized the dates of the Eschwege biography so well that I could answer promptly. There were no grounds for controversy. But he tried to make me insecure. At every opportunity he said with his sweet voice: 'Your account seems very unbelievable.'"

Dr. Metz tripped him up only once when he asked about his educational background. Herr von Eschwege obviously couldn't say that he was a naval officer—*Korvettenkapitän* Hermann Ehrhard—so he made something up.

"I studied law in Straßburg for a year" 'Eschwege' responded.

"What was the name of your professor?" Metz asked.

"Dr. Wiesenbach."

"I know the name of every law professor at every university in the Empire," Dr. Metz snapped back. "Your Wiesenbach has never existed."

Dr. Metz then played his trump. He laid two photographs of Captain Hermann Ehrhardt in his trim beard and full uniform in front of 'Eschwege.'

"Do you recognize this man?"

Ehrhardt choked down a laugh at the humor of the situation. Slowly he drew out a pair of black horn-rimmed spectacles—produced at the optical firm with window-glass for lenses, since Ehrhardt's eyesight was perfect—and looked over his own photographs very slowly.

"I'm sorry, but this person is unknown to me. The gentleman must be Captain Ehrhardt, because it's on his account that I'm here."

Ehrhardt wondered how long the theatrical performance would go on. He no longer believed that Dr. Metz, as a German Nationalist, would cut him any slack. But he held out due to the sheer joy of battle, and a belief that a soldier at the Front can sometimes pull things off that General Staff officers believe to be impossible.

Dr. Metz demanded that he swear an oath that his testimony had been truthful. Ehrhardt knew enough law to believe that an oath under duress was no oath at all, so he swore that he was Eschwege—since with his passport, he actually *was* Eschwege for purposes of identification

"Where do you work?" Dr. Metz asked.

"In my firm," 'Eschwege' replied.

Dr. Metz made a dubious face and asked Police Commissar Heldwein to go with 'Eschwege' to his optical firm—to see if it actually existed. The people at the optical firm office all vouched for the fact that Herr von Eschwege was the assistant office manager and had worked there for some time. 'Eschwege,' in fact, stayed at the office and tried to get back to work. Shortly, Commissar Heldwein showed up and invited him to another hearing session that same afternoon.

"It was clear that he had ducked into the nearest tavern or cigar store and telephoned the Research Judge," Ehrhardt said. Ehrhardt knew he should get out of Munich—but he wanted to hear from the Princess first.

The Princess, he learned, had already buckled under the court pressure and sworn a non-religious oath that everything she had said in court was true. Ehrhardt made a tough choice: he hated the idea of confinement but come what might, he now had to convince the court that the Princess had not actually known that he was Hermann Ehrhardt, so that she herself could not be charged with perjury. He showed up at the afternoon hearing expecting the worst, at least for himself.

First Dr. Metz asked to see Herr von Eschwege's passport. When he had the passport in hand, Dr. Metz broke into a sweat with excitement. Then, incredibly, he gave the passport back. Ehrhardt was surprised but assumed this was some sort of tactic to give him a false sense of security.

Dr. Metz called for Frau Weiss, a rather chatty older woman who had sometimes had tea with the Princess at the villa where they both rented rooms. She knew Ehrhardt by sight.

"That's Herr Eichmann, I'll swear to it," Frau Weiss told Dr. Metz.

Ehrhardt and Judge Metz were both astounded. Eichmann was one of Ehrhardt's assumed names, and also, as it turned out, the actual gardener at the villa, employed by Professor Schloesser, the landlord.

The next witness was a pretty little waitress who had served Ehrhardt and his wife and three children when they had vacationed at Gmund on the Tegernsee for five weeks some time before.

"Can you identify this gentleman?" Dr. Metz asked.

"I can't be sure if this is Herr Ehrhardt or not," the girl said. Either she was frightened or Ehrhardt, minus his beard and fierce expression, was actually able to shed his memorable fearsome appearance.

"Now, little lady, you can swear to it, can't you?" Dr. Metz wheedled in his Lower Rhenish accent, in a smarmy bid to win the girl over.

"I just can't be sure it's him," she said.

All at once, the girl jumped up and ran out of the room "as if she were shot out of a pistol" and vanished.

The next witness was Walter Pruckner, the former landlord who said he had lost Ehrhardt's unexpired rent money to a bad investment.

"Can you recognize this man?" Dr. Metz asked theatrically.

"That man is Captain Hermann Ehrhardt," Pruckner said bluntly.

Dr. Metz stuck his face close to Ehrhardt's, whose own face was flushed with rage against Pruckner, who had betrayed him for money.

"*ARE YOU* Captain Ehrhardt?" Dr. Metz demanded

"That's right!" Ehrhardt said. "I've lost the game! I left money with this man when I came here to free Munich from the Soviet Republic of Bavaria after they murdered Countess Westarp and the Prince of Thurn und Taxis. He lost the money and wants to send me to prison so he won't have to pay me back. He's a thief and a swine. Do you know what kind of a dirt-bag you have here as a witness...?"

"That's enough," Dr. Metz said. "The issue is whether or not you are Captain Ehrhardt."

"My oath is rescinded," Ehrhardt said flatly.

"It's good that you do so," Metz said, smiling lovably now that he had made his case. "The recision of your oath will undoubtedly lead to a milder sentence."

The Princess appeared a few moments later from the hotel lobby across the street and Dr. Metz said: "Take your oath back" so he himself could claim credit. Metz was still warily impressed by royals. The Princess stood her ground. She said she had sworn the oath based on Ehrhardt's paper identity as Eschwege. Ehrhardt told her the game was up and that she should rescind the oath because he actually was Hermann Ehrhardt: he stuck out his own neck to cover her tracks. She pressed his hand in a farewell, understanding that he had once again stuck his neck out for her, and just for a moment, Dr. Metz ceased to exist for both of them.

Dr. Metz had problems of his own. Once the word got out that he was holding Captain Hermann Ehrhardt at the Palace of Justice, a rescue attempt by dozens, perhaps hundreds of heavily armed Rightists was a near certainty. These veteran fighters had hand grenades, automatic pistols, submachine guns and light machine guns and knew how to use them. "On this day there were ten nationalist groups in Munich," Ehrhardt observed. "If it became known in any way that I was in custody, the boys would have come to take me back and trampled anyone who got in their way."

Dr. Metz once again started to sweat as he realized that custody of Captain Ehrhardt just might get him killed, along with a large part of the police force and the judicial system. The battles of 1919 had killed more than 1,000 citizens of Munich: once the Ehrhardt Brigade had arrived the battles were entirely one-sided.

"With which class coach would you like to travel to Leipzig, third or second?" Dr. Metz asked Ehrhardt in his sweetest voice.

"If it's at state expense—second," Ehrhardt replied brusquely.

Ehrhardt spent the next two hours in a waiting room with Criminal Police officers keeping an eye on him. Metz, still smiling and still sweating, looked to be scared half out of his wits. Twenty minutes before the train was scheduled to arrive, three members of the Criminal Police led Ehrhardt to a waiting automobile and climbed in with him. A second-class railroad compartment had been reserved for them. One of the officers had bought a loaf of bread and a cluster of sausages and the four of them ate a hasty supper together. Ehrhardt soon learned that he had been betrayed not only by Walter Pruckner in court but off-stage by Eichmann, the gardener at the villa, and by a tax inspector who wanted any possible reward. After the bread and sausages, they all slumped over and tried to sleep. Ehrhardt woke in the middle of the night and noticed that all three policemen seemed to be sleeping.

But he could tell by the breathing that one man was actually awake, keeping him under surveillance.

Ehrhardt suddenly remembered that he still had the Hungarian passport on his pocket, so he asked to go to "the place where even the Kaiser has to be alone." The police officers escorted him to the coach's toilet and once Ehrhardt locked himself inside he shredded the passport and flushed it. He thought about jumping through the restroom window. But the train was traveling fast enough to make a broken neck or leg a near-certainty.

The train pulled in to the huge electrified chasm of the Leipzig railroad station and the police stuffed Hermann Ehrhardt into a green wooden lock-box on the horse-drawn paddywagon and headed for Beethovenstraße 2, the national prison. Ehrhardt noticed that Leipzig, unlike Berlin, Hamburg, and Munich, still used horses rather than cars and trucks. A sleepy-looking guard came out when they rang at the prison door and said in a thick Saxon accent:

"Well, you guys came at just the right time! Ya woke us up out of the best morning sleep. What did ya do, my darling? Ya must be a real baddie if it took three cops to bring ya here."

"Chief, this guy's no criminal," the top Munich policeman said in a mild Bavarian accent. "He's a political prisoner an' from the better circles, as you'll soon find out."

The Saxon guards shook Ehrhardt down for everything he had on him, including his tie-pin. Then Ehrhardt shook hands with all three Bavarian Criminal Policemen and praised their good manners and efficiency. "I thank you, gentlemen. You've shown an impeccable, first-class manner toward me. I hardly had the feeling that I was a prisoner while traveled with you."

"You mustn't think ill of us, Captain, because we had to bring you here," the senior Bavarian Criminal Policeman said. "You know about duty. We had our orders and instructions from the Research Judge. It scared us half to death just transporting you. But hopefully, Captain, things won't be this bad for long. Bad sometimes turns to good."

The Bavarian policemen left and the grumbling Saxon warders took Hermann Ehrhardt up the dark, twisting, smelly corridors to his new lodgings. He later called the prison "a symphony of stenches." He also said he felt as if he had been thrown into a heap of dirty laundry with no way to wash it. Ehrhardt found that, as a political prisoner, his assigned cell was on the third floor—to prevent escape. The cell was gloomy, with no electricity. One wall backed up into one of the prison's stone chimneys that radiated excessive heat and burned his hand when he touched that part of the wall. The cell was so tight that he bumped into something every time he turned around. The iron frame bed, close to the floor, was the only furniture.

"The feeling of sitting in a cage overcame me and immediately awaked the instinct—how quickly can I get out of this cage?"

The single window was barred and the lower sill was six feet above the floor, with a two-inch opening for ventilation. When he jumped and did a pull-up on the bars, all he could see outside was the night sky.

Chapter Four
The Lausbub

Fate was a palpable reality to Hermann Ehrhardt. He was an instinctive Christian and he knew The Bible quite well. He also knew some Latin from his years at the *Gymnasium*, the classics-driven secondary school he attended before his temper earned him an early dismissal without a diploma. The years of his youth were the years when Richard Wagner's stupendous cycle of four music dramas, "The Ring of the Nibelung," with its tragic fate-driven heroes and villains, was incontrovertibly Germany's national epic. The *Leitmotiv* of "Fate" thundered from the orchestra, concealed beneath the stage, and heroes and villains died their spectacular deaths atop the stage as Fate had demanded. Ehrhardt's biographical collaborator, the novelist and silent screenplay writer Friedrich Freksa, in the introduction to "Adventure and Fate" called Ehrhardt a figure from the Sagas. Hermann Ehrhardt himself had a bit of modesty and a wry sense of humor. He called himself a *Lausbub*.

Georg Hellmuth Hermann Ehrhard—the final T came later—was born the son of Pastor Georg Ehrhard and his wife, the former Marie Weissler, in Diersburg in Baden, near the border with Switzerland, among the people known as Swabians —the Romans had called them *Suebi*, and to the Swiss peasantry, any German national who crossed the border was a "*Swobe*."

"From my great-grandfather all the Ehrhardts have been pastors, and father and mother couldn't think otherwise but that I too would one day stand in the pulpit. But the Dear Lord God gave me too much of a *Lausbub* sensibility. From the time I was a little boy I loved anything that could shoot, and I bought my first pistol, which wasn't worth much, by saving up from my school lunch money."

"I didn't have much to do with my two sisters because to me, as a little boy, girls were pretty much worthless.... My mother taught me respect for the feminine gender. She had a quick hand and she punished any dumbness or rudeness with a quick box to the ear. But she never held a grudge for long."

Ehrhardt enjoyed working side by side with the peasant farm family who lived near the parsonage in Diersburg. He enjoyed playing with the farm animals and helping out with the farm work far more than he enjoyed sitting around the house.

"Once my mother got me a handsome velvet suit with a pointed white collar, that I wore for the first time on a Sunday morning. My mother warned me not to get this holiday costume dirty. I promised her. Then I walked proudly over to the neighbors' house to show myself off in my new finery. Their yard was wrapped in Sunday stillness. I looked around and in the middle of the yard I saw a barrel with a very thick stopper next to the farm wagon. I knocked on it. The barrel was full. So I began to play with the stopper. Suddenly the thick wooden stopper shot out against my chest and I fell over backwards. A stream of yellow liquid manure as thick as my arm shot out of barrel. It was all so sudden and scared me so much that I lay there for a moment in spite of the violently stinking stream. Finally I got to my feet and ran home to my mother as fast as I could. I left a long stream of smelly filth behind me right into the clean Sunday parlor. As soon as my mother saw that, she gave me a box on the ear. She only gave me one—but that was enough.

"She forgave me in a minute, but after she washed the new Sunday suit, the stink wouldn't come out of the velvet. My father refused to let her throw the clothing away. He said: 'Punishment must follow. He must wear the suit as long as it's in one piece.' For a long, long time the suit smelled like liquid manure. Wherever I went, people rumpled up their noses. For years, the neighbor would call after me at a distance: 'Little man, how does it smell now?'"

Though he was a self-admitted *Lausbub*, Hermann remained fastidious about cleanliness and personal hygiene. The expression—*Lausbub* means "louse boy"—refers not to dirtiness with lice but to constant irritation. Hermann more than lived up to the description. Lice would also play a part in his fate.

He remembered that his mother, the family disciplinarian, had punished him for tracking wet snow into the parlor by making him stand outside buckled over in the freezing night to ponder his wrong-doing. Then, afraid that he might catch cold, she forced him to drink hot herbal tea that always made him crack a sweat.

One day a letter reached the parsonage with a foreign postmark and stamp. The whole family was excited and Ehrhardt's sister, who collected stamps in an album, was particularly delighted. The postmark read "India."

The sealed letter was reserved as an after-dinner treat and Pastor Ehrhardt opened it and read an account of the adventures of an Ehrhardt cousin who described his life in India in terms that Hermann, even as a young boy, found rather self-important. But the letter stirred him. "For the first time, I saw far-away lands outside a book or a map. A fellow from our own family was out there in the wide world, beyond the ocean, in India, the land of sun and stories."

Ehrhardt talked about the letter with "my blond friend Berti"—Hermann's own atavistic Hunnic blood may have been brought to his attention as a possible source of his reckless mischief—and Berti was also fascinated.

"We decided when our school days were over to go on this wandering path ourselves," Hermann wrote. "During the school day I thought of nothing but India, and so it was no wonder that I didn't get especially good grades from my teacher.... My teacher persecuted me with contempt and mockery, to shake my self-respect, but with a kid like me, this had the reverse effect. I had already assigned myself the role of a man who hunted tigers, defended himself against the murderous Thug stranglers, and climbed the pointed peaks of the Himalayas.... My teacher told me to stay after school so he could have a talk with me."

"It's not so much the lack of intellect I observe in you as it is the failing of any character," the teacher said. He told Ehrhardt that he was not so much stupid as he was lazy, and that he was a disgrace to a family of clergymen and professors, including Ehrhardt's own father, whom the teacher deeply respected, a father who has enjoyed reading the Odes of the Roman poet Horace with the teacher in the garden of the parsonage. The mathematical laws of heredity were little understood in the late 1890s and the teacher may have mistaken Hermann's atavistic Hunnic features for Down Syndrome, which was not well understood either. Ehrhardt sulked and wondered what would happen if he beat the teacher up, but he thought better of it. He hoped none of his friends heard any of this. The teacher assigned extra homework and told Hermann that if he didn't improve his attitude he would never be of any use to his father or to his Fatherland. Then he slammed Hermann's exercise book on the desk and dismissed him.

The Fate theme thundered in Ehrhardt's soul.

"Berti, I'm not sitting this one out! I'm going to India *now*! Are you coming with me?"

"You mean we're ready so soon?"

"How much do we need? We need weapons and money. If we want to, we can do it."

"Right, money..." Berti said. "I've got a *Christaller* and I've got a Mark my uncle gave me, but that's all I have."

"I have twelve Marks in my savings box, and we can sell the stamp album in Basel," Ehrhardt said.

"But doesn't that belong to your sister?" Berti asked.

"Half of it's mine, and when I find a diamond mine in India or get a job as the prime minister of a maharajah, I'll pay her back ten times over."

"Has your cousin written again?"

"...A fine letter. He was on a tiger hunt with the maharajah and some other lords on an elephant. Straightaway the tigers broke out of a thicket. He lost his rifle in the excitement but his elephant, Murad, strangled one tiger with his trunk. That one was dead and the Maharajah shot the other one. The tigers killed a couple of the beaters, but in India that's no big deal."

"After that they had a part in the Maharajah's castle. They bathed and they put on fresh silk robes. A hundred Indians pulled the rope ceiling fans so they hardly noticed the heat."

"Where the air is hot, the wind is hot," Berti said dryly, suggesting he found the cousin's hunting story somewhat improbable.

"But Berti, the Maharajah has two ice-cold fountains in the palace."

"What else does your cousin do in India?"

"My cousin bet the maharajah that he could build a steam plow that was faster than the prince's elephant plow. My cousin bet 500 pounds and the maharajah bet the big brooch on his turban. My cousin won, of course, but as clever as he is, he gave the maharajah back his brooch and the maharajah brought two of his steam plows."

"But is this true?"

"Of course it's true, and he wrote me that all we have to do is come over. I think we should go to him right away."

Berti quickly acquired a couple of hunting knives from his cousin, a forester. He also cut several maps of Switzerland out of his cousin's atlas. Hermann brought some "plaids," woolen blankets for sleeping or for covering in case of rain. He also took two shirts, two pairs of socks, and some extra collars. The boys planned to stay for a few days with Hermann's uncle in Geneva, and then walk through Switzerland and northern Italy until they reached the fabled seaport of Brindisi.

"How are we going to get the passage money for the ship?" Berti asked. His longest trip so far, Hermann noted with a certain condescending attitude, had been a visit to Munich and the Bavarian mountains.

"Same as everybody does who doesn't have passage money," Hermann said. "We'll hide ourselves in a couple of trunks, and when we're too far from land for them to turn back, we'll come out and tell them we want to pay for our passage as cabin boys. Maybe we won't even have to worry about it. Once we're on the ship, everything will go just fine."

"But food..." Berti asked. His mother had always told Hermann that Berti was a huge eater.

"There's always something on a ship," Hermann said, as he knew from reading adventure stories instead of doing his Latin homework. "They've got barrels of salt pork and there's always a food locker."

"What if they catch us?"

"First, that's just food snitching and we won't be punished, and in the second place we won't actually be stealing. As soon as we get to India and find something, we'll send the captain whatever money we cost him and something extra, and we'll send him a keepsake, like a tie-pin or a handsome ring."

Hermann and Berti set their escape for a certain summer Friday night. Hermann slipped through the quiet parsonage while the family was asleep, lifted his sister's stamp album, dropped his bundle of clothing out the window and jumped after it. He reached Berti's window and gave the cry of a crow.

"I think they've noticed something," Berti said as he slipped out of the house.

"And if they've noticed something, we're outside, we're off. Nobody has anything to say to us. We have our fate in our hands."

The boys slipped past the hated school in the bright moonlight and headed for the Swiss border. Hermann called a halt at some sheds on the outskirts of town while it was still dark.

"From here it's only an hour to Basel. What do we want with the city at night? We can spare the rent of a room. In the early morning we'll wake up, sell the album, buy some pistols, and get back on the road by ten. There's no point in going there now."

The boys rolled up in their blankets and conked out in the summer warmth. They awoke to the crack of a whip. A freight wagon came clattering up the road with a jolly-looking young driver.

"Boys, do you want a ride?"

"Going to Basel?" Berti asked.

"Yeah," the freighter said. The boys clambered up into the wagon.

"What are you up to in Basel?" the freighter asked.

"Business," Berti said briskly.

"Business?" the freighter asked, and winked his left eye. "What kind of business could you two be up to?"

"We want to sell a stamp album," Berti said.

"Stamp album? Lemme see it."

Hermann held out the stamp album and the freighter looked it over.

"What might it be worth?" the freighter asked.

"A thousand francs," Hermann said.

The freighter shoved the stamp album between the barrels in his wagon, out of their reach.

"...And the thousand francs?" Hermann demanded.

"The police will see about the thousand francs," the freighter said. "You can't convince me you came by this stamp album honestly."

"Just look in there, my sister's name is inside," Hermann said.

"Anybody can say that," the freighter said and whipped up his horses. The two boy boys slid to the back of the freight wagon like two sacks of potatoes.

"Do you think he's really going to take us to the police?" Berti whispered. Hermann thought that the freighter simply meant to steal the album for himself.

"*Now I know what I'll do,*" Hermann whispered to Berti.

Hermann whipped out his trusty bean-flipper, took a sharp stone out of his pocket, and shot the wagon's right-hand horse on the buttock. The horse jolted and broke into a run with the other horse keeping time. The freighter had to jump off holding the reins to restrain the horses. Hermann lurched forward, tugged the album free and threw it out of the wagon down the slope. He and Berti jumped out of the back of the wagon.

"Thanks a lot for the ride!" Hermann shouted sarcastically.

The freighter laughed. He had been struggling to hold the horses when Hermann retrieved the stamp album: he thought the thousand-franc album was still in the wagon. Hermann and Berti peered from behind the bushes like hostile Indians in a harum-scarum novel. The young freighter's smirk turned to a grimace when he discovered the stamp album was missing. When the freighter drove away, cursing under his breath, the two boys found the stamp album, broken in half at the binding but with all the stamps still stuck in place.

"There are bad people in this world," Berti mused. "He looked so happy I never mistrusted him."

"Let's eat," Hermann said. "It will give us strength."

"If we had a revolver, he never would have tried it," he added.

The boys enjoyed their lunch—white bread with butter, sausages and dried fruit—and drank some clear cool water from a stream, with the mountains in the distance. Then they walked on to the border station where the customs guard knew Hermann by sight and waved them through without an inspection. At the first shop they encountered, Berti asked after a stamp dealer. The shop-keeper sent them to Füssli, whose shop was down by the Rhine.

"I'm after American stamps here," Herr Füssli said. "You've got nothing but European stamps here, and I'll trade some American, Egyptian or Japanese stamps with you."

"My sister said these stamps are worth more than foreign stamps and she must know," Hermann said. "She's been collecting for a long time."

Füssli shook his head.

"What do you want for the whole collection?" he asked.

"A thousand francs," Hermann said flatly.

Herr Füssli snapped the book shut, set it on a chair, stretched out his felt slippers and laughed.

"A thousand francs! Show it to me and I'll give you the shop and the whole collection, along with the furniture!"

Hermann tried to argue and Herr Füssli took out a catalog—Hermann said it looked as if Füssli had written the catalog himself—and offered to go over the prices.

"I'll show you what the whole collection is worth," he said. After he counted the 922 stamps at a rough estimate of one centime each, he offered out of the goodness of his heart to add a few francs in case there were any more valuable stamps he had missed.

"I'll round it out and give you twenty francs," he said.

"We stopped breathing," Hermann said. "Twenty francs was a lot of money for two people who had none, but we had both believed that the stamp album was worth a thousand."

"If you don't want it, then try your luck with Farina," Füssli said. "He's not far from here, next street down, then go left on the crooked alley. Take him the best regards of Herr Füssli."

Hermann and Berti followed Füssli's directions and found "a dark, black-bearded Italian who looked like Napoleon III on the gold coins, reading an Italian newspaper in sight of his front window."

"Herr Farina," Berti said. "We've been sent by Herr Füssli. Herr Füssli looked at our collection and only wanted to give us twenty francs, and said you would pay more."

Hermann bumped Berti and said: "No. Two hundred francs!"

"Two hundred francs!" Berti agreed.

Signor Farina took the halves of the book, looked it over, and hit Berti on the head with it.

"Get out, brat, get out!" he shouted in Italian.

Hermann snatched up the book, dodged out of the shop, and caught Berti on the way out.

"Go to Füssli and sell it for twenty," Hermann the Hun told blond Berti. Hermann's eyes were slanted *up*—the attack signal. "I'm going to have revenge on this loser."

Hermann ripped a blank page out of the stamp album. He wrapped it around a clot of horse manure he had spotted in the street. With deadly aim, he shot through the open window and caught Farina right in the face. Hermann was half-way up the street before Farina could recover from his shock. He caught up with Berti on the way back to Herr Füssli's shop.

"What did Farina say?" Herr Füssli asked.

"Farina is a lump," Hermann said. "We gave him your greeting but he hit poor Berti over the head with the book. But I took care of him. I sent a horse noodle through his window and caught him right in the nose."

"You did right," Herr Füssli said. "But show me the book again. If I consider it right, twenty francs may have been a bit too much. Farina seems to have understood this. Take fifteen."

Hermann got angry. "No! I'd rather throw it in the Rhine!"

"We've got a long way to go, all the way to Geneva," Berti said.

"Good," Herr Füssli said. "I'll let myself be softened up. I'll pay you twenty."

"Where can one buy revolvers here?" Hermann asked as he counted the money.

Füssli left the boys in the shop, walked out for a stroll and came back with a revolver loaded with blank cartridges and a rusty pocket pistol. The two guns cost thirteen francs.

The boys left Basel behind them and marched on toward Geneva, buying flour and eggs and sleeping in haystacks. On the second day, Hermann aimed the rusty pocket pistol at a tree. The first shot jammed. Hermann reloaded. The second shot blew the pistol apart and tore a hole in his hand. Berti wrapped a cloth around the blood. The boys met a forester who put a proper bandage on the wounded hand —the wound was not serious—and asked where they were going. He also gave them a place to sleep near the stove and his wife made them breakfast with fresh bread, milk and coffee. The boys kept heading for Geneva and Hermann's uncle. Herman observed that each of them felt they would like to go home—and neither of them had the courage to say so. They encountered a tourist who asked them where they were going. They told him.

"I know your uncle!" the tourist said. "Come with me, boys, and I'll take care of your expenses and get the money back from your uncle."

"No, we'd rather go alone," Hermann said.

Berti bumped Hermann gently. "My legs hurt."

"The expense won't be too much for your uncle and leaving you here wouldn't be a good idea," the tourist said.

The boys and the tourist got on a train and the two boys, happy to be riding for a change, felt half asleep. The tourist took them to Hermann's uncle's house in the middle of the day. The housemaid spoke to the tourist and promised to pay him for their tickets.

The uncle came home in the evening.

"It's good that you're here," the uncle said. "Your father had telegraphed that you'd be coming."

Hermann was shocked. "I always had a great respect for my father, but I couldn't grasp exactly how he knew that we would show up in Geneva."

After a good dinner Berti entertained the uncle by describing their adventures: In his story the freight wagon driver turned into a highway robber and the tree Hermann had shot turned into a menacing vagabond. The uncle smoked his pipe and laughed.

The next day a Swiss policeman showed up and took them to the train.

"Your parents will be waiting in Basel," the Swiss policeman said. "Now don't do anything dumb." The Swiss policeman stayed at the railroad station until the train left. The boys felt like convicts but were glad that they didn't have to walk all the way back home.

Berti shouted with joy when he saw his parents. Hermann slipped out the opposite side of the train rather than face his father in public. He walked all the way home, wondering what his father would do to him—and still wondering how his father knew he would go to Geneva. The housemaid let him in. He slipped into his own bed and lay waiting. He heard the wagon roll up outside.

"Is he here?" his father's voice asked.

"Yep, he is," the maid replied.

Hermann shut his eyes and played dead.

He heard his parents walk into the bedroom.

"He's already asleep," his mother said.

"That won't help him a bit, he must get up," Pastor Ehrhard said. "He must see what he's done!"

"But the poor baby has already had enough anxiety and fear, otherwise he wouldn't be here lying in bed. Let him sleep out. In the morning he'll see what a dumb trick he's played on us. Believe me, he'll never do it again. Just look how pale his face is."

His mother and father gently shut the door and let him sleep.

Hermann's father, in fact, later decided to send him back to a little village near Geneva for French lessons, since there was no really qualified French teacher in the local school system around Diersburg. The Swabian people were proud of their Alemannic dialect—"A Swabian can do anything but speak textbook German," was their watchword. The Swiss were proud to be tri-lingual in German, French, and Italian. Hermann was now an amateur photographer, so he took his camera and his portable darkroom to his uncle's house.

"In this village there was a really obnoxious old schoolmaster that I took an instant dislike to. In order to scare the girls and show the boys how funny he really was, I fired my new pistol when I was near him. I wanted them to hear the crack of the bullets. One of these bullets bounced off a stone and hit a girl. Fortunately it struck a corset stay and the corset stay saved the poor terrified creature from any injury.

"I myself was shaken enough. In order to comfort the girl, I told her of the wonders of photography, took her photograph and promised to show her and the girls the wonders of photography in the darkroom along with the boys.

"The whole event was naturally tattled on, and the schoolmaster, who was shocked that we boys had been in a darkroom with the girls, sent a report of the whole thing to the schoolteacher at the *Gymnasium*."

Back in the hometown school, Hermann was confronted by his hometown enemy, the *Gymnasium* teacher. The teacher asked him to read out a passage of

Latin as a sort of final evaluation. "I stammered a bit, because I naturally wasn't prepared for this particular passage, and the fellow said rudely to me: 'Of course! You can't do anything right. Work is naturally a little different that sitting in the dark with little girls.'"

Hermann jumped up, struck the teacher in the face and knocked his glasses off. He stormed out of the classroom.

When he got home and told Pastor Ehrhard what had happened, he was grateful for the amazing fact that his father seemed to understand him completely.

Hermann was expelled from the *Gymnasium* without the *Abitur* certificate that qualified graduates for the universities or to the professional schools for subjects such a medicine or law. His father had already decided, in the kindliest possible way, that Hermann was not cut out to be a pastor.

"What should become of this young man, all my relatives wondered, once I was kicked out and no *Gymnasium* in my tight little land of Baden dared to take in the criminal Hermann Ehrhardt."

Chapter Five
Sea Cadet Ehrhardt

'The Criminal Hermann Ehrhardt" was saved from a life of rustic obscurity by the worst blunder that Kaiser Wilhelm II ever made.

"The Kaiser needed sea cadets," Ehrhardt wrote. "All the newspapers had advertisements for enlistment. There were only three or four days left in the opportunity for enlistment in the sea school. I quickly decided to head for Kiel and was able to pass all the preliminary tests. But my mother cried and said 'Now I've lost my son!' For her, the waters had no bottom."

The opportunity for enlistment came in the aftermath of the murder of two Catholic German missionaries, Father Richard Henle and Father Franz-Xavier Nies in Shandong, China on November 1, 1897. The murderers, 20 to 30 Chinese members of the Big Sword Society, were looking for Father George Maria Stenz, whom they had been told had referred to the Chinese family gods as demons. Father Stenz had been humbly sleeping in the Chinese servants' quarters of the mission at Juye in Shandong when the killers arrived. He lived another 30 years.

Wilhelm II himself, a devout Lutheran, had Jewish friends, notably the shipping magnate Albert Ballin. But he saw a clear chance to win some loyalty from his sometimes unruly Catholic subjects, most of them acquired when Germany was unified in 1871 and the Hohenzollerns of Protestant Prussia became the imperial family. The new German state needed overseas colonies as markets just as Britain needed India, Burma, and large parts of Africa and China for the sale of manufactured goods. Perhaps not coincidentally, just the year before the murders, in 1896, Rear Admiral Alfred von Tirpitz had visited China and pointed out that Kiautschou Bay on the Shandong Peninsula would be the ideal location for a German naval base on the Asian mainland. Tirpitz's successor with the tiny German Pacific cruiser fleet, Otto von Diederichs, sent a telegram to Berlin just a week after the murders: "May incidents be exploited in pursuit of further goals?" The Kaiser and told Diederichs to proceed to Kiautschou with the entire cruiser squadron and exploit away. Diederichs had only four cruisers not laid up for repairs but he reached Kiautschou two weeks after the murder of the missionaries. With no actual fighting—Manchu-ruled China had recently lost a very one-sided war with Japan, including a naval catastrophe—the Germans forced the Chinese to sign a 99-year lease on the Kiautschou Bay region of Shandong, just as the British had earlier done at Hong Kong and at Shanghai. The Germans then plunked an estimated $100 million into harbor facilities, safe water and sewer systems, and schools that accepted Chinese academic and vocational students.

Wilhelm was Queen Victoria's first and favorite grandchild through her eldest daughter and his mother, Princess Victoria. Wilhelm himself had echoed Bismarcks's "land rat—water rat" policy of keeping Germany strong on land and leaving the high seas to Britain until a few years before Victoria died in 1902, cradled by Wilhelm's one good arm. But

Albert Ballin with the Kaiser

even before Victoria's death, Wilhelm decided to build up a navy second only to Britain's rather than maintaining Otto von Bismarck's vision of a clear understanding with Britain: peaceful German dominance of the Eurasian continent through largely German-Jewish banking, commerce and industry and the mere potential of military power, while Britannia ruled the waves. Kiautschou became a 99-year German leasehold in 1898. The call went out for even more naval cadets for a greatly expanded Imperial German Navy even after Hermann Ehrhardt answered the call with an immense sense of relief in 1899.

The Germans based their naval cadet training on the British Royal Navy—so much so that learning English was mandatory: commands were habitually given in English. The training took place at sea on tall-masted windjammers, long out-moded for sea battles but a great place to harden the cadet's body and study the ocean in all her moods.

"We slept only in hammocks," Ehrhardt remembered. "At the first grey of morning the boatswain's whistle and the order *"Rise! Rise!"* woke us from a deep and well-earned sleep…. Seven minutes after the wake-up call we had to have all our sleeping gear hung up and be standing on deck. Then most of the time, just to get our arms and legs going, we were chased three times 'over the top,' which meant we had to climb the shrouds up to the tip of the mast and then clamber down the other side. Breakfast tasted especially good after this, but we always had some appetite left over."

In the morning, the sea cadets studied academic subjects related to the sea such as navigation and mechanical engineering. The afternoons included boat drill, sail-shifting exercises, and infantry drill in case of land action. The Germans had their own small Marine Corps, known as *Seesoldaten*, sea soldiers, but armed sailors were sometimes expected to take part in landing parties, as they had at Kiautschou. The sea cadets also handled the maintenance of the ship and quarters. Each cadet stood a two-hour watch at night and helped run the ship under sail whenever opportunities arose or emergencies threatened. Sea Cadet Ehrhardt said it was a toughening experience—storms on the Atlantic seemed to threaten immediate extinction—but he clearly loved just about every minute of it.

Sea Cadet Ehrhardt wryly admitted that he had trouble at first making himself understood because his everyday speech was in the South German Alemannic dialect and the Kaiser's sea cadets spoke *"platt"*—North German, the closest existing language to Anglo-Saxon English. For instance, a South German would ask *"Was ist das?"* and a North German *"Wat is dat?"* when they saw something they didn't recognize on the sea or on the dinner table. Sea Cadet Ehrhardt claimed that he missed a lot of good food when his North German messmates pretended they couldn't understand what foods he was asking for in his Alemannic accent at the cadet's table and refused to pass the dish.

"By nature I'm inclined to be defiant," he wrote. "But the hard service at sea put me through a wringer so that the only part of my defiance that remained was the tough will to maintain my own individuality."

Individuality could be troublesome. He remembered the strong surf at Tenerife—so tempting that the young sea cadets swam through the waves despite the danger. The heat in the Canary Islands in the summer was such that they wore their British-style straw hats in the water to shield their heads from potential heat stroke even while they were swimming.

"I had the bad luck to lose my hat. The surf quickly took it away and I couldn't get it back. I asked our boatswain to get me a new hat, but the one I got had a round Spanish shape, miles away from the British mode. This hat made me immediately recognizable. Even when I was couple of miles from the school ship on boat drill the duty officer would spot all my mistakes and shout out 'Sea Cadet Ehrhardt!' and have a thunderstorm ready to fall on me when I got on board. Sometimes he shouted 'Sea Cadet Ehrhardt!' when I wasn't even doing anything wrong. That's how I learned how important it was to always be in proper uniform."

When the school ship arrived in Rio Di Janeiro, Ehrhardt called the bay and the roadstead the most beautiful on Earth. The only detriment about Rio Di Janeiro, he said, was that the summer heat discouraged mountain climbing of the magnificent peaks that overlooked the bay. The heat was also palpable when he was part of a team in a boat race with another whaleboat in front of spectators. He accidentally dropped an oar.

"The petty officer, who saw this, sent the cutter back to the ship and reported me to the watch officer. As a punishment he sent me 'over the top' ten times. This was an excessive punishment but in the heat of Brazil it was an atrocity. As I was going up for the last time I sensed myself drifting into unconsciousness, made it down the shrouds almost without knowing where I was, and clenched my teeth until I reached the washroom under the deck. I fell on the cool stone slab, took a couple more breaths, and passed out."

Ehrhardt, once he recovered, gloried in the toughness required of sea cadets. As a first-year cadet he reported a potentially cancerous growth on his neck. The ships' medic cut the lump out of Ehrhardt's neck with a pocket knife—no anesthetic, plenty of bleeding followed only by iodine. Ehrhardt took the extreme pain without flinching and the wound bled out and healed by itself. "I can still see the scar every morning but it's not important."

Sea Cadet Ehrhardt was promoted to *Fähnrich* in 1900 after a satisfactory term as a sea cadet. A *Fähnrich*, literally ensign, was the junior grade of commissioned officer in the Imperial Navy, but the rank gave Ehrhardt a great feeling of satisfaction. He credited the fact that he was instructed by sturdy naval officers who served as role models rather than pedantic old schoolteachers teaching subjects that had no interest for him—and with the excitement of storms as sea, exotic destinations, and the pretty much constant exhaustion of shroud climbing and boat drill. The rowing and climbing the shrouds 'over the top' developed his muscular physique to the point where, though he was five-foot-ten, then well above average height, he looked squat when photographed by himself because of his big shoulders and chest and powerful legs. He looked every inch the Hun—and Huns were about to become a staple of conversation in Germany and all over Europe.

In the summer of 1900, Chinese peasant nationalists known as the Boxers for their martial arts skills, humiliated and enraged by European and Japanese predation on Chinese territory, turned their displaced aggression on the thousands of Chinese Christian converts and on the missionaries who tried to protect them. In the city of Peking *(modern Beijing)*, the Boxers mutilated Chinese converts and raped and murdered screaming Chinese Christian girls within harrowing earshot of the foreign embassies.

On June 19, Baron Clemens von Ketteler, the German ambassador, asked the official Chinese government troops to meet him for a conference to have the Boxers removed from the area around the embassy compound. The baron refused an escort of a naval petty officer and four German Marines because he thought it would look cowardly or mistrustful. On the morning of June 20 he set out in a sedan chair, with his interpreter, Heinrich Cordes, following in a second sedan chair. Halfway to the proposed meeting, Manchu imperial soldiers—not Boxer rebels—fired on the two unarmed German diplomats. Baron von Ketteler was hit multiple times and Cordes, seriously wounded, ran for it, holding his hand over his wounds until he reached the shelter of the American mission. "The murderers themselves were Chinese Imperial banner troops in full uniform," said Heinrich Cordes, who understood Chinese. "I heard them shout—there's a foreigner who got what he deserved."

Baron von Ketteler's murder and the murder of the Christian missionaries reached Germany and Kaiser Wilhelm II went predictably ballistic. Germany declared war on China—as did six other western nations and Japan. The Japanese had also lost a diplomat to street murder. The Germans received 2,000 serving military volunteers for the first "East Asian Expeditionary Corps"—German Marines and armed sailors had already been sent from Shandong. The soldiers soon to head for China from the military seaport at Bremerhaven heard an address by Kaiser Wilhelm II on July 27 that gave the world a new nickname for the Germans. No authenticated version exists but a journalist reconstructed the best complete version of the speech from what he remembered and from his notes. The version that reached the public contained many of the same elements:

"For the first time since the reconstruction of the German Empire, a great undertaking overseas is upon you," The Kaiser began. "...You know it well, you will be fighting against a cunning, brave, well-armed and cruel enemy. When you come upon them, understand: pardon will not be given. Prisoners will not be taken. Use your weapons so that a thousand years from now, no Chinese will so much as dare to squint at a German..." The unofficial version, also widely circulated, added: "When you come upon the enemy, he is to be killed. Pardon will not be given. Prisoners will not be taken. As the Huns made a name for themselves a thousand years ago under their king Attila, make sure that the name of Germany will be known in the same way, so that no Chinese will dare to so much as squint at a German."

The foreign press in England and France lapped it up at the time and even people who thought Wilhelm was a bit of a show-off believed that the Boxers and Wilhelm deserved one another. The *Hunnenrede*, or "Hun Oration" first became notorious in Germany itself.

Kaiser Wilhelm II

The Boxers ultimately murdered 20,000 to 30,000 Chinese Christians, 156 Protestant missionaries with 84 of their children, and 41 Catholic priests and nuns. Europeans and Americans considered the Boxers dangerous savages. But later that summer, letters began to appear in German Socialist newspapers, letters said to be from German soldiers who claimed to have murdered Chinese prisoners and raped and bayoneted Chinese women under the influence of the Kaiser's Hun Oration. August Bebel, the leader of the Social Democrats, argued that Wilhelm had infused his soldiers with Hunnic barbarism. The *Hunnenbriefe*, or Hun Letters, were used by both the Socialists and the Polish separatists to attack the Kaiser and his militarists.

"As we won the first battle, you should have seen how we broke into the city," said one letter the Socialist August Bebel read in the *Reichstag*. "All that came in our way, whether man, woman, or child, all were slaughtered. Well, how the women screamed! But the Kaiser's order was louder: give no pardon! And we had sworn to be loyal and obedient and we held to it."

Cynics noted that not one of the *Hunnenbriefe* was signed with the writer's name—The letter about the screaming women was simply signed *Kamerad*, or "comrade." Some of the *Hunnenbriefe* described atrocities at the capture of the Taku Forts on the Peiho River approach to Peking—but the Taku Forts were captured by the British and the Japanese, not the Germans: the capture took place on June 17, more than a month *before* Wilhelm II made his notorious Hun Oration in Bremerhaven, and three days before Baron von Ketteler was murdered. Modern consensus attributes the *Hunnenbriefe* to a mixture of scanty facts, adolescent bravado, and deliberate political fakery. The concoction of fake atrocities and the inflation of numbers for real atrocities was to become a Leftist staple during the years when Hermann Ehrhardt achieved his greatest notoriety.

The Boxer Rebellion took place just about the time that three separate scientists rediscovered the mathematic rules of genetics first demonstrated by Gregor Mendel in 1868. Full-blooded Germans who had straight dark hair, slanted eyes, flat faces, and an absence of adolescent freckles—each individual feature was reasonably common in Germany, and in some parts of eastern France—were now seen as an atavism: though genetic lottery they were the sons of the Huns. British propagandists, taking up from August Bebel's Socialist reading of the unsigned and dubious *Hunnenbriefe*, could now explain why people who had the same bloodlines as the British royal family, many of the world's great composers and scientists, and most of the English, the Lowland Scots, and the American upper classes, could revert to savagery at the drop of a hat and the hat's replacement with a spiked Oriental-looking helmet. In his fierce outward appearance, and his penchant for dangerous mischief involving pistols and girls in darkrooms, not to mention punching his schoolteacher, Hermann Ehrhardt—now *Fähnrich* Hermann Ehrhardt—was the archetype of the atavism.

Chapter Six

Heroics and Horrors

Ensign Ehrhardt, to his obvious regret, was too junior an officer to be sent to China, though he would have gladly taken on the Boxers to save the missionaries and all those Chinese girls. He wrote obliquely of his chagrin at not serving in China:

"No one is glad to die. But the highest part of the soldier's calling is that men lose their anxiety about death through discipline. If humans ever lost their willingness to risk their lives for one another, nursling babies and helpless women would die in burning heaps. Hundreds would drown if the rescuers feared for their own deaths. The whole society of mankind would collapse if the readiness to sacrifice one's own life for one's neighbors disappeared.... The great moral idea of war for a man is not that he kills, but that he can die for his own people."

Ensign Ehrhardt, now all of 20 years old, had spent the Boxer Rebellion training recruits in Germany, which he tacitly found somewhat embarrassing. He was good at it, however, and was promoted to *Leutnant zur See*, naval lieutenant, in 1902. He admitted that he was also personally training himself for service on land and sea even as he trained the recruits. He enjoyed slipping away in the ship's boat in civilian clothes for a glass of cognac or poaching wild game for the officer's mess. He said he saw this as training for future adventures in the service of the Kaiser and the Fatherland. "The main thing was not to get caught. But soldierly virtues were also exercised, namely agility and decisiveness.... My life was like that of every sea officer, full of all kinds of work, because a modern warship was a work of art without comparison, and the officers of the first line had to master it. We were seaman, technician, artilleryman, infantryman in one person. We had to be good in mathematics and in the basics of astronomy. The calling was not limited, the sea itself expanded it."

Leutnant zur See Ehrhardt was elated when volunteers were requested for an expedition to German South West Africa where a tribal uprising that started at the end of 1903 had claimed the lives of about a hundred German settlers, including four or five women, and 13 German soldiers.

"I volunteered immediately, was sent to South West Afrika and came under the command of *Oberstleutnant* [Ludwig] von Estorff. My section commander was *Kapitän-Leutnant* Mansholt, my battery chief *Oberleutnant* Stempel. The war made us into good comrades. The Hereros were tough, brave, cruel and experienced fighters. They waged war according to barbaric custom, mutilated any wounded man they found, and the women tortured him to death. This was a hard war but it hardened us to war as if it were hunting."

The Herero, a Bantu tribe of cattle breeders and herdsmen, had long waged was on the smaller Proto-Mongoloid Nama, the so-called Hottentots, who sometimes picked off Herero cattle as if they were hunting wild game. When the Herero caught a smaller Nama hunting their cattle, the Herero captors would recite a formula: "This will teach you not to listen for the Herero cattle"—and cut off the ears; "This will teach you not to smell the Herero cattle"—and cut off the nose; "This will teach you not to look at the Herero cattle"—and gouge out the eyes; "This will teach you not to eat the Herero cattle"—and cut the throat. The Herero were so devoted to cattle that the Herero women sometimes wore headdresses which looked like the horns of a cow. The Herero men often had multiple wives, but Herero women could have only one husband—so they sometimes took lovers on the side.

German missionaries arrived and peacefully converted many of the Herero in a matter of a few years. The horrible ritual mutilation of the Nama was stopped and the Herero herdsmen and the Nama hunters and gatherers, now both mostly Christian, learned to share the sparse grasslands and scarce waters of Namibia in a way that was tolerant if not exactly friendly. The first years of German administration were years of benign neglect. Dr. Ernst Göring—Hermann Göring's father—convinced the Herero to sign some treaties giving away large swaths of land, but Göring let the missionaries control most of the trade and the missionaries were widely respected, even loved, by the Africans of both tribes.

Unfortunately, the German settlers who came to the pacified country included a number of ruffian cattle traders and crooked trading post owners who were not the sort of people the missionaries were. Younger Herero women were quite often beautiful, and were quite often coerced into brief affairs with German settlers, sometimes actually raped. Revenge murders took place by Herero husbands, but unequal justice prevailed: Herero men who killed German settlers were generally hanged or sent to prison for life. German settlers who killed Herero men generally got short prison terms.

The Herero found the cattle swindling by some of the German settlers even worse that the seduction or rape of the Herero women. When an epidemic of rinderpest wiped out many of the native cattle, the cattle traders monopolized the vaccine and charged the Herero exorbitant numbers of unvaccinated cows taken in trade for those which had been vaccinated. The cattle traders also insulted the small Nama tribesmen as well as the proud Herero by calling the Nama "baboons" or referring to them as *kaffir*—Arabic for pagans or savages—even when most of the Herero and Nama had been baptized and some of them knew how to read. The German governor in 1903, Colonel Theodor Leutwein, was rather less racist than the German cattle traders, but he must have been asleep at the switch not to notice that the Herero gradually turned down trade offers of beads and cloth and traded only for guns.

Colonel Theodor Leutwein

Samuel Maharero was somewhat shaky in his position as chief, and needed to reinforce his prestige. Maharero was an intelligent man and understood that the Herero, with or without the Nama, were not about to win a war with Germany. He devised a strategy: Herero warriors were told that when the fighting broke out they were not to harm missionaries, European women or children, Boers, or Englishmen. Chief Maharero knew that the British had fought an exceptionally brutal war against the Boers to the south just two years before. Unable to subdue the commandos of Boer mounted riflemen, the British had put the Boer women and children and their loyal African farm hands into concentration camps, a term borrowed from the Spaniards in Cuba before 1898. Bad nutrition, dysentery and typhus had killed 18,000 to 28,000 white Boer civilians, most of them women and children, and another 12,000 African farmhands. The Boers, on their side, routinely shot any Africans who had been armed by the British and at least some of the unarmed laborers who worked for the British. Maharero's probable intent was to touch off a war between the British and the Boers against the Germans, as well as to win over world opinion to the Herero. He knew that many adventurous French and German officers had volunteered to fight as comrades-in-arms of the Boers against the British and may have hoped that the whites would fight to exhaustion and then leave the Africans alone. He waited until half the Herero men owned new Mauser rifles—obtained from the German trading post owners.

Samuel Maharero

The opening feint made a mockery of the cattle traders who thought the Africans were stupid: Maharero convinced his some-time friend Hendrik Witbooi, a principal chief of the Nama 'Hottentots', to start some trouble down near the border with British South Africa. The Nama had a long-standing grievance against the Germans: the governor before Leutwein had shelled a Nama camp to force the Hottentots to sign a treaty some years before and had killed a number of women and children. The Nama were hunters, gatherers, and stalkers, good at escape and evasion. Leutwein's Germans sent troops into Nama country but had trouble finding the Nama 'Bushmen.' At this moment, Samuel Maharero—once Colonel Leutwein's endorsed candidate for Herero chief—struck at the cattle traders and slaughtered about 100 German men and teenage boys with the traditional mut-ilations. German rapists or illicit lovers also had their private parts amputated. The Her-ero generally followed the agreement and spared the women and girls, the little boys, and the missionaries. But the four or five white women who were killed and 13 murdered German soldiers—some of them Africans in German uniform—were an ample pretext to the decision to neutralize the Herero and the Nama, if anybody could find them.

Leutnant Ehrhardt arrived with the first group of German volunteers from Europe, while the fighting was still an issue in doubt. He had no knowledge of the Herero or Nama side of the story. He found Africa utterly amazing.

"This broad grassland has its own magic. On a reconnaissance ride in the plains I suddenly saw thousands of animals grazing. I thought they were a giant herd of cattle. A Boer who was with us told us they were antelope.

"I've always been a passionate hunter. From my European background it was unfathomable that thousands of game animals stood on this one piece of land. I took my binoculars, recognized the horns, and thought that if we ever ran out of fresh meat, especially for the sick men, this would be a great place to get some. I took a patrol with me and went after the herd but they were flighty and ran for it. We had to shoot from a great distance. About five guns went off at the same time, two of them mine, and unfortunately we shot more animals than we needed."

Ehrhardt found this waste personally embarrassing. "I often experienced senseless shooting at wild game and at any animal people saw out there. But an officer has to set an example before he can give orders to his men." From then on, he shot only what he needed for the mess table.

"On Easter Sunday [*April 3, 1904*] I experienced my first heavy fighting," Ehrhardt wrote. "Our detachment of about 200 men was marching on the grassland. I was the column leader of the battery and commanded two new 3.3-centimeter automatic cannon. To reach the open flatland we had to move through thick underbrush. As we came out of the thicket, the Hereros

Herero child soldier. The Herero were, as Ehrhardt himself noted, great warriors.

hit us with a burst of rifle fire. A heavy one-hour fight took place. I opened up with the automatic cannons immediately. They were decisive, but out of our 200 men we lost 40 killed and wounded.

"During a pause in the fighting I lay near the battery and suddenly noticed that some of the shots were coming not from ahead, but from above. I discovered a 150-foot tree, with an African in the treetop directing his fire at us.

"I took a rifle from a man lying nearby, put the front sight on the guy, and pulled the trigger. First the rifle fell from the tree, then the Herero. I later picked the gun up as a trophy. It was a K-98 [*state-of-the-art Mauser*] with special Herero decorations on the stock.

The Herero wore European clothing as shelter from the intense sun. The Germans had read about slavery in the United States—Germans found it appalling and Prussia, by royal order, had played no role at all in the African Slave Trade after

1704. The volunteers may have felt as if they were shooting down the devout, lovable "darkies" from "Uncle Tom's Cabin." But the Herero were also capable marksmen, as they proved in many engagements.

The first Herero Ehrhardt personally killed indicated the one failing of the reliable long-range Mausers used by both sides: the high-velocity bullet passed through the victim so cleanly that unless the marksman hit the heart or the brain, the man who was shot still had a lot of fight left in him. Americans fighting the Moro tribes in the Philippines a few years before had found that a Moro warrior could be shot multiple times and still rush up to the American and take his head off with a bolo knife before the Moro died of internal bleeding. The Americans replaced their new .38-caliber revolvers with the old .45 Colt revolver from the Indian Wars, which delivered a lethal shock that knocked a man over backwards. The 7.63-mm cartridge used in the reliable C96 "broomhandle" Mauser automatic pistol had the same problem as the K-98 Mauser rifle—the bullets punched small neat holes without much stopping power. Winston Churchill grudgingly admitted that his German-made C96 Mauser automatic had saved his life at the battle of Omdurman in 1898 because the automatic pistol had never jammed. But Churchill was on a horse at the time, immune to close contact with swordsmen. As soon as he could, Hermann Ehrhardt swapped his issue automatic for a Browning, the generic name for most American-made .45 automatics. It would become his signature weapon.

Ten days after Ehrhardt and the automatic cannons took on the Herero and survived with heavy losses, Colonel Leutwein took on the Herero and lost. Leutwein attempted a counter-attack. Leutwein's under-sized army was surrounded by 3,000 Herero warriors armed with Mausers at Oviumbo on April 13. The ragamuffins proved effective warriors and some were crack shots. The Herero forced the Germans into a humiliating retreat. Leutwein's military replacement, General Lothar von Trotha, had already been dispatched from Germany to take over the military forces. Leutwein remained governor. Leutwein, who knew about the cattle swindling and the coercive prostitution and rape, wanted to win one decisive victory and offer the Herero a peace treaty. Lothar von Trotha had other ideas. Known as an Iron Man during the Boxer Rebellion, decorated with the Iron Cross, Second Class, as a young officer during the Franco-Prussian War of 1870-71, Trotha wanted to destroy the power of the Herero completely. He arrived in Africa on June 11—fighting had been going on for five months—and he soon found out that the Herero were tough adversaries. The climate was also tough.

"We came deeper into the country and learned the effectiveness of the desert," "Ehrhardt wrote. "Once I was sent out with a strong cavalry patrol into the grasslands. We were on the march for two days without finding a single drop of water. As the sun rose high at midday and burned us up with its heat, we could barely keep moving forward. We searched for wild game in vain, so we could drink their blood to quell our thirst. Not a living thing as far as we could see. Even the birds seemed to stay out of the sky because of the heat. The first of my boys collapsed. He had dragged himself this far but couldn't go any farther. That broke the ban on collapsing. Nobody had wanted to be first. One after another we fell out of our saddles. Devoid of any energy, completely dried out, we remaining lying where we fell. We weren't actually tortured, because even the fantasies that had shown us mirages dried up. We slipped into sheer stupefaction. But at evening, when the sun lost its power, we got up and reached a waterhole. The waterhole had been contaminated with the carcasses of dead animals. We dived to the bottom and lapped up the mud, which seemed delicious to us.

"Shortly after that the detachment leader, [*Kapitänleutnant*] Mansholt, came down with typhus. I sat beside his bed for three days, but no care seemed to help. He died as a sacrifice for the Fatherland. I closed his eyes and experienced the worst part of war through this death by sickness. With three sacks of cement that I had dragged away from an abandoned farm with my own hands, we made a tombstone for him and left him to rest in African soil."

Trotha cornered the Herero at Waterberg—the last water before the Omaheke arm of the Kalahari Desert—on August 11 of 1904. The Herero fought back with a night of hand-to-hand fighting so fierce that some German officers, driven from their assigned positions, reportedly committed suicide to avoid disgrace. The issue was in doubt all night. When the sun came up the German artillery broke all resistance and the Herero covered the withdrawal of their women, children, and cattle into the same sort of sandy wastes that had almost wiped out Ehrhardt's cavalry patrol with collective heat stroke.

Lothar von Trotha

"The whole Herero people were driven to a death by thirst," Ehrhardt wrote with a degree of dismay that verged on horror. "As we drove after them, we found their dead cattle, and nearby, only dead or half-dead women and children. It hit me very hard: a whole people had been driven to a death by thirst."

Some Germans wrote back to their families that they gave the women and children water when they found them. The fact that the Herero women often wore stitched-up versions of German peasant costumes made them sympathetic figures. Three farm boys in tropical uniform were touched when they found a younger woman carrying a baby and leading her blind mother or mother-in-law on a sort of tether. To their delight, the three volunteers found a goat that had some life left in it and they gave the goat to the two women and told them to keep moving. Their covert act of mercy may have placed them at risk.

Marine-Feldkompanie, 1904

Typhus had killed Germans and Herero alike from the start of the campaign. At the Waterberg and afterward, the disease became an epidemic. The Herero wore European clothing they had no way of washing, which turned their bodies into incubators for lice. The Germans had to cut short the pursuit after three days because of heat exhaustion and because of the menace of typhus contracted from the dead or dying Herero fugitives. Spread by body lice and fostered by malnutrition, the disease struck both sides impartially with deadly results,

"The typhus hit me on the return march from the sandy wastes," Ehrhardt remembered. "They pulled me out of line like a mangy sheep and the wagon that I lay in had to be kept at a distance from the rest of the command. Nevertheless my comrades found me. Once they found me the doctor gave me sour cherries to eat. I didn't want the whole portion. My comrades were so thirsty they ate the rest of them with my own spoon."

"Soon after that I was completely unconscious. I have only the loyalty of my orderly Kerschowsky to thank for my life. He and an African porter stayed with me for four long weeks, during which I hardly had an eye-blink to show I was alive. Only a shot that was once fired at us brought me back for a second of clarity."

Troops in German Southwest Africa during the 1904 Herero Revolt

The Germans, who ultimately had about 20,000 soldiers and sailors in the field, reported 676 men killed in action, 907 wounded in action, and 689 dead from disease, primarily typhus and dysentery. The Herero casualties are impossible to calculate but they obviously suffered greater losses in battle and far greater losses from typhus and thirst. The African catastrophe was potentially compounded when the Herero survivors, unwanted anywhere else, began to attempt to return to their old herding grounds in German South West Africa, still armed with Mausers, though obviously not eager for a re-match with the German artillery. Ehrhardt had already shipped back to Germany for convalescent leave, to the delight and relief of his mother, but Leutwein and Trotha were stuck with the menace of an expanded typhus epidemic, as well as vengeful feelings for their own casualties and the mutilation of the German dead and wounded.

On October 11, Trotha had made a speech that put Kaiser Wilhelm II's *Hunnenrede* in deep shade for arrogance as well as for brutality.

"I, the great general of the German soldiers, send this letter to the Hereros. The Hereros are German subjects no longer. They have killed, stolen, cut off the ears and other parts of the body of wounded soldiers, and are now too cowardly to want to fight any longer. I announce to the people that whoever hands me one of the chiefs shall receive 1,000 marks, and 5,000 marks for Samuel Maharero. The Herero nation must now leave the country. If it refuses I shall do it with the long tube [*cannon*]. Any Herero found inside the German frontier, with or without a gun or cattle, will be executed. I shall spare neither women nor children. I shall give the order to drive them away and fire on them. Such are my words to the Herero people."

Trotha seemed to have realized almost immediately that his orders were an outrage to humanity and the next day he amended the order.

"This proclamation is to be read to the troops at roll-call, with the addition that the shooting at women and children is to be understood as shooting over their heads, so as to force them to run [*away*]. I assume absolutely that this proclamation will result in taking no more male prisoners, but I will not degenerate into atrocities against women and children. They will run away if one shoots at them a couple of times. The troops will remain conscious of the good reputation of the German soldier."

When news of the first order got back to Germany, the churchmen were horrified and the Socialists had a field day printing letters from Germans who claimed they took no prisoners and bayoneted the Africans. The fact that Herero warriors seldom if ever surrendered, and that the Herero were far deadlier in hand-to-hand fighting than the average German recruit, suggested a replay of the *Hunnenbriefe* from the Boxer Rebellion.

Chancellor Bernhard von Bülow, Kaiser Wilhelm's friend and confidant, told Wilhelm II that Trotha's *Vernichtungsbefehl*, or extinction order, was a crime against humanity, was impractical, would deprive the colony of potentially useful cattle herders, and would be a blot of Germany's reputation around the world. Kaiser Wilhelm was not yet the Beast of Berlin he later became in British war-time propaganda. After a two-week delay, Wilhelm told Bülow to cancel the order to shoot the Herero men on sight and drive the women and children back into the desert. The Germans were ordered to accept the surrender of any Herero man who was not a known murderer or a ring-leader, and of all women and children. The order arrived as the Nama—some of whom had scouted for the Germans against their old enemies, the Herero—touched off a revolt of their own.

The multi-barreled pom pom gun on a pedestal is the weapon described as used fighting the Herero in Southwest Africa. These guns were designed to fight off torpedo boats from the decks of naval ships; their use against people was somewhat questionable.

As the exhausted warriors of both sides gradually came in and gave up their weapons—about half of the 900 Nama warriors also had Mausers—they were lodged behind barbed wire and fed an unfamiliar diet, mostly boiled rice. The typhus struck the camps and about 7,000 Africans died before they could be signed out to German settlers as contract labor. Subsequent British propaganda attempted to turn a war that had been rampantly brutal on both sides into an attempted extermination. In fact the Germans, even the shady cattle traders, wanted the Herero and the Nama alive as long as they could exploit the tribesmen's labor and forestall any future uprisings. German missionaries, largely sympathetic to the Herero and the Nama, argued that the whole uprising could have been prevented if two or three of the worst German offenders had been hanged for murdering Africans or raping their women. Instead, two African leaders were hanged. Samuel Maharero and Hendrick Witbooi were both spared, and essentially reinstated as chiefs. The survivors of both tribes—as many as half of the Herero may have died due to bullets, shells, typhus and thirst—returned to a subsistence economy that maintained German South West Africa as a financial liability for the rest of its history. Hermann Göring's bureaucrat father, the former governor Dr. Ernst Göring, once a considerate administrator, turned into a hopeless drunkard in chagrin and his wife moved in with a titled, remotely Jewish baron so that little Hermann Göring could enjoy the good things of life.

But *Leutnant zur See* Hermann Ehrhardt had had his first test of battle, and while he deplored seeing the Herero driven into the desert for the rest of his life, he had proven himself in armed fight against a capable and merciless enemy. He also kept the captured Herero K-98 Mauser as a souvenir.

Chapter Seven
Beard, Boat, and Bride

While Hermann Ehrhardt was experiencing colonial war and its horrifying aftermath in German South West Africa in 1904, the world first experienced modern naval war as the Japanese attacked the Russians at Port Arthur in Manchuria in February of 1904. Ehrhardt was convalescing from his near-fatal bout with typhus when the Japanese, using their torpedo boats with daring and skill, sank most of the Russian battle fleet at Tsushima in May of 1905, and the world would never be the same as it had been.

Americans of the era saw Asians—first the Chinese, then the Japanese—as "cheap Oriental labor" and adopted legislation first to ban Chinese from America and then—after the upset victory over Russia—to limit Japanese visitors to students who would return home after their studies without inflicting an outright ban: this was the so-called "gentlemen's agreement." The Germans generally thought of Asians as elegant and exotic—the port city of Hamburg had long had its own quaint and popular Chinatown. The Germans used Russians or Russian Poles for cheap labor rather than Asians. Itō Hirobumi, the peasant-to-samurai-to-noblemen who was the architect of modern Japan, was a drinking buddy of Otto von Bismarck. Itō based Japan's constitution on the Prussian constitution. Kitasato Shibasaburō helped the German scientist Emil von Behring discover the cure for of diphtheria in 1901, just as Sahachirō Hata would help the German Jewish scientist

Paul Ehrlich discover a chemical remedy for syphilis in 1908. But Kaiser Wilhelm, while he greatly honored Paul Ehrlich and asked him to seek out a cure for cancer, the scourge of the Hohenzollerns, had invented the term *"Die Gelbe Gefahr"*—the Yellow Peril—to warn that the insidious modernized Asians could someday take over the West as they almost had taken over Central Europe in the days for the Huns in late Roman times and the Mongols during the High Middle Ages. Kaiser Wilhelm—startled by Japan's victory over China in 1895—sketched out a painting, later completed by Hermann Knackfuß, a professional artist, showing the nations of Europe, personified as apprehensive warrior women while the Archangel Michael with wings and a sword warns the European beauties to protect themselves and their culture against the Asian beasts, personified by the Japanese Buddha of Kamakura glowering with menace from the East. Wilhelm imposed upon his good friend Albert Ballin, the Jewish shipping magnate, to display this artistic warning in the lounges of his Atlantic liners, which

Itō Hirobumi

took affluent travelers on Mediterranean cruises during the stormy winter months. Field Marshal Alfred von Waldersee—a parlor anti-Semite like his American-born wife, Mary—dryly asked Albert Ballin if he had ever noticed the Kaiser's omnivorous appetite for flattery. Ballin, a cultivated and reasonable man, glumly told Waldersee he had no choice but to cater to it. Then they had lunch together and savored a brief moment of commiseration.

The Russian debacle at Tsushima—the Russians lost eight battleships, several other ships, and more than 5,000 sailors, the Japanese lost three torpedo boats and 116 sailors for a devastating total victory—led to a drastic shift in world politics. The once-mighty Russians had long been hated by the Finns due to political oppression, by Polish Catholics due to religious persecution, and by the Jews because the Tsarist Russians tolerated *pogroms* in which dozens of Jewish men were murdered and their women were raped by looters who also stole their valuables. Russia received no sympathy from Asians, who had also experienced rampant Russian brutality up to and including outright massacres. The Japanese, who had left a clean reputation behind them after the Boxer Rebellion was crushed—no rape, no murder, fastidious and appreciative looting—were the heroes of the hour. Britain gloried in its 1902 alliance with the "gallant little Japs," which had helped make the war possible, since the British forbade third parties from entering the conflict under threat of facing the Royal Navy. Jack London, the most popular American author of the day, extolled Japanese courage and their iron disciple and predicted that the Japanese "brown man" would take over the management of the hard-working Chinese "yellow man," not a coward either when well led. London and Homer Lea, a blue-eyed, red-headed Virginian who served as a Chinese general, predicted that the "brown man" and the "yellow man" would shortly join forces to kick the white man out of Asia. President Theodore Roosevelt covertly cut two deals with Japan: the Taft-Katsura Agreement recognized Japan's right to control Korea, and the subsequent Root-Takahira agreement recognized Japan's right to economic control of Manchuria. Fighting Japan or Russia to maintain the Open Door policy where all sides had equal rights in Manchuria, Roosevelt argued "would require a fleet as good as that of England and, plus an army as good as that of Germany." In return for Korea, the Japanese looked the other way at the American conquest of the Philippines. In return for Manchuria, Japan became the leading anti-Russian power in the Pacific. There was also a huge public fascination with Japanese art and culture all over the Western world. Sessue Hayakawa and his wife, Tsuru Aoki, played leading roles in Hollywood films opposite whites in the same era and nearly at the same level as Douglas Fairbanks, Mary Pickford, and Charlie Chaplin. "Sometimes I got the girl, but in the end I always died, so it was okay," Sessue Hayakawa said in fond nostalgia.

Peoples of Europe, Guard Your Most Sacred Possessions (1895), by Hermann Knackfuß

Young Ehrhardt was a troublemaker who failed to follow in the footsteps of his father, a pastor. Instead, he joined the Imperial German Navy in 1899, advancing to the rank of commander.

Hermann Ehrhardt made the most of his vaguely Asiatic appearance, which had raised some eyebrows when he seemed to constantly get into trouble in his *Lausbub* days. Now he cultivated the Asian look. Wilhelmine German admirals and ships' captains usually grew big mustaches, often waxed to a point, and full bushy beards like buccaneers: Ehrhardt cultivated a trim dark mustache that traced his upper lip and a small dark goatee on the tip of his chin. He actually looked something like the Japanese naval officers who turned up in the newspapers illustrations. The resemblance was deliberate.

More to the point, Ehrhardt had read the message of Tsushima: He knew his chance of commanding a battleship in the caste-conscious Imperial German Navy without an aristocratic title or an influential father or uncle was miniscule. What Hermann Ehrhardt wanted to command was a torpedo boat, the battleship-killer of Tsushima.

The first German warship launched after Germany federated following the Franco-Prussian War was essentially an upgrade of the U.S.S. *Monitor*, which had changed naval design forever a decade before during the battle with the Confederate States Ship *Virginia* (formerly U.S.S. *Merrimack*) at Hampton Roads on March 8, 1862. Monitors were powerful in coastal defense—the world's navies still used monitors for shore bombardment in World War I—but their low freeboard made them death traps on the high seas.

Pre-war torpedo boats moving at speed under Imperial Battle Flag

Naval design improved greatly. The year Ehrhardt was promoted from sea cadet to full-fledged ensign, 1901, the SMS *Braunschweig* was laid down in the naval shipyard at Kiel, the first of her class. The *Braunschweig* (Brunswick in Anglo-Saxon, after the North German city) was launched on 20 December 1902 and commissioned into the fleet on October 15, 1904, when Ehrhardt was in Kiel recovering from the typhus he had contracted in Africa. He may have attended the commissioning ceremony. The ship was 419 feet long, 73 feet wide and drew 27 feet of water. The three propellers driven by 16,000 horsepower from 14 coal-powered boilers gave the ship a top speed of 18 knots. The *Braunschweig* had four 11-inch guns in two armored turrets mounted fore and aft, 14 6.7-inch guns—four in single turrets, the other ten mounted broadside sticking out of the hull—and a battery of six 88-mm quick-firing guns mounted on the front or rear superstructure with coverage of all angles of torpedo-boat approach. The ship also carried six 18-inch torpedo tubes mounted below the waterline.

Aft torpedo tubes

The SMS *Braunschweig* was a formidable ship for two years—and then turned into a floating anachronism. In 1906 the British launched HMS *Dreadnought*, which mounted ten 12-inch guns in five armored turrets, backed by quick-firing guns of smaller caliber and machine guns, and driven by oil-fired steam turbines that gave the Dreadnought a top speed of 21 knots. One extra inch of gun caliber—due to the range and impact of the main battery—and three extra knots of speed meant that the SMS *Braunschweig* and similar coal-fired battleships were obsolete. Germany, Britain, the United States, and Japan all began to build battleships with 12-inch main batteries and stream turbine engines at a frantic pace.

The stopgap—Ehrhardt realized this—was the torpedo boat. Operating at night or in fog, torpedo boats could close with any battleship, as the Japanese had at Tsushima, and sink or cripple them with a single torpedo hit. The original torpedo boats were little more than speedy little ships with torpedo tubes—the coal-fueled killers of the Russian battleships at Tsushima had only a single 6-pounder gun and two 3-pounders (about the size of World War I tank guns) to augment their three 14-inch torpedo tubes. But the Japanese boats could travel at 28 knots, far faster than any battleship. The torpedo boats Hermann Ehrhardt would command were ranked by the British as "destroyers" because they were fitted out to fight enemy torpedo boats as well as for daring attacks on battleships. The Krupp-Germania boats weighted 400 tons, and were driven by coal-fired boilers to a top speed of 26 knots. The boats carried three four-pounder cannons—one on either side of the bridge and one on the fantail facing aft—along with two machine guns to repel boarders. Each of the three 17.7-inch torpedo tubes was mounted on a separate turntable, usually with one tube forward, one tube between the two smokestacks and one tube behind the second stack near the fantail. The single full mast was behind the second stack, and was used only for signal flag hoists, like the half-mast on the bridge. The boats had a crew of 56 to 80 officers and sailors and carried 110 tons of coal. The size and reliability of the German torpedoes made the series of boats—T 108 to T 113, and their successors of the T 132 to T 136 series, with upgraded gun power and with the two aft tubes between the smokestacks—some of the deadliest small ships afloat. All German torpedo boats had a forward rudder just behind the bow, with the main rudder behind the propellers under the fantail. The double rudders meant that German torpedo boats could turn with astonishing speed—and also meant that the deck crews were hurled around like potatoes and often developed chronic rheumatism due to the stress on their joints and the cold North Sea or Baltic weather. The torpedo boat service was in its infancy when Ehrhardt returned to active service. Ultimately, the Germans would build 178 torpedo boats.

"As a special recognition on my return I was given the position of watch officer on His Majesty's Light Cruiser *Berlin*, which in this year served as the escort for the Royal Yacht *Hohenzollern* on the Kaiser's sea voyages. These were beautiful, diversified voyages in the northern waters. The only ominous part for me was my passion for hunting."

Ehrhardt indulged his passion for hunting wild game all over Scandinavia, with mixed results—especially in Denmark, which had lost territory to Prussia in two wars in 1864 and in 1866. Prussians and Germans were not admired in Denmark. One night Ehrhardt and a sailor took a dinghy ashore on a small Danish island looking for game for the officer's mess. Ehrhardt shot at a hare, which vanished into the grass. As he searched for it, he bumped into a Danish forester.

"What are you doing with the gun?"

"I wanted to shoot shore birds,"

"I already heard a shot..."

The Danish forester sent his dog out and the dog shortly returned with the dead hare in its mouth.

"The look the forester threw me penetrated to my kidneys," Ehrhardt wrote. "I said nothing but 'Good Day' and went away feeling sick. I let myself be rowed back to the ship.

"After a few days I thought to myself 'Praise God, that went well.' But sure enough, the next morning, as I was still in bed, my old hunting buddy, *Kapitänleutnant* Grube came in and shouted: 'Ehrhardt, it was *you!*'

"What's the matter then?" I asked, gagging to myself.

He opened up the '*New Tidings of Kiel*' and read to me merrily: "A Danish newspaper writes under the title—'A German Officer As Poacher.' And the whole story, very much decorated, followed."

"I made use of my rights as the accused and denied everything.

"But in the afternoon *Kapitän [Friedrich von]* Ingenohl of the *Hohenzollern* also read the article and sent a message by blinker to the commander of our cruiser to the imperial yacht. The 'old man' came back, called us together in the mess room, and laid the blue-penciled copy of the 'New Tidings Of Kiel' on the table.

"I stepped forward immediately and said, '*Herr Kapitän*, it was me."

"The blinkers flashed between the cruiser and the *Hohenzollern* and Grube, the old sinner, said sharply: '*Leutnant* Ehrhardt must report to the *Hohenzollern*."

Ehrhardt was a potential victim of royal politics. Kaiser Wilhelm II had married his wife, Augusta Viktoria von Schleswig-Holstein, at least in part because her family members were the heirs to Schleswig-Holstein, which Wilhelm's grandfather and Otto von Bismarck had lifted from Denmark—with the approval of most of the German residents and some of the more progressive Danes—in 1864. The annexation was confirmed after Prussia's defeat of Austria in 1866. Wilhelm II came to love Dona, as she was called in the family, but he also wanted to patch things up with Denmark, whose royal family had domestic ties to most of Europe. The Kaiser later approved his one daughter Viktoria Luise's love-match to the heir to the (then non-existent) throne of the Kingdom of Hannover for the same reason. Wilhelm was also known to have sired several illegitimate children with courtesans he met on maneuvers before his marriage, and the strong-willed Dona was a notable prude who hated that sort of thing, especially when her husband had done it. The Kaiser's affairs were so notorious that Arthur Conan Doyle's first short story about Sherlock Holmes, "A Scandal in Bohemia," (1891) featured a wayward Germanic monarch whose six-foot-six physique was that of Bismarck but whose outrageous wardrobe was that of Wilhelm II. The plot revolves around recovering a compromising photograph and love letters exchanged with an American opera singer, Irene Adler, to appease a "Scandinavian bride of... strict principles... the very soul of delicacy." This bride sounds like Dona. *Leutnant zur See* Ehrhardt—whose favorite English author was Arthur Conan Doyle—considered the possibility that he might become an interesting human sacrifice to Dona as he stepped aboard the *Hohenzollern*.

"I announced myself about the *Hohenzollern* strictly according to regulations, and didn't get a bad reception from *Kapitän* Ingenohl." Ehrhardt wrote."Then I was sent on to His Majesty and in that moment my whole career was at risk. I could rightly and conveniently, as one might say, be beheaded."

"You service on the escort vessel for the Hohenzollern would be an honor for any of my naval officers," Wilhelm told Ehrhardt. "I really have to wonder that you didn't have this honor before your eyes at all times. But in regard to your youth and your fine service in South West Africa: three days arrest. I thank you."

Ehrhardt heaved a sigh of relief as soon as he backed away from the imperial presence to his three days arrest in his cabin. But no more than two weeks later, he was back at it, poaching for black grouse in Sweden with his old hunting buddy Grube.

"Our navigation officer in the *Berlin* was a thundering moralist and because he was of senior rank I had to hold my tongue through many abusive lectures. But God is just. We came around the point of a small island in a little sailboat and we spotted the moralist in a small boat near the coast having a good time with a Swedish girl. Now he too was famous aboard ship. Some have a passion for one kind of hunting, and some for another."

Kapitän Ehrhardt's unusual interview with Kaiser Wilhelm, shortly led to a wonderful result. Wilhelm II had sometimes been a naughty boy with naughty women and he too was a voracious hunter who, unlike Hermann Ehrhardt, shot far more game than he needed for the table. Both exploits may have been a compensation for his withered left arm, a birth defect that made it difficult for Wilhelm to manage a rifle or ride a spirited horse. After years of therapy, Wilhelm could grip objects with his left hand but could not lift anything heavy. He exercised his good right arm with dumbbells, constantly sawed and chopped wood single-handed, and gave people crusher handshakes with his strong right hand, reportedly wearing multiple rings to increase the victims' agony. Wilhelm had been denied the dueling field of honor as a member of a reigning house: dueling with a disguised fencing expert could be a great cover for an assassination. An officer once reportedly challenged Wilhelm to a duel while drunk. Wilhelm refused, as was his duty. The drunken officer socked Wilhelm in the eye. The drunken officer was given a loaded pistol, told to go to his cabin and do the right thing, did it, and was quietly slipped over the side. But Wilhelm almost routinely pardoned young officers who killed or wounded their adversaries in duels over a lady's honor. Wilhelm also appeared to have pardoned *Leutnant zur See* Ehrhardt. A few months after his brief dressing-down on the *Hohenzollern*, Ehrhardt got his fondest wish: he was assigned to command one of the newer torpedo boats in the Kaiser's Navy.

Emperor Wilhelm II with his wife Auguste Viktoria during World War I

"I experienced my happiest time with the black weapon." Ehrhardt wrote, meaning the underwater torpedo and the need for night attacks. "For a torpedo boat commander, I stood on my own deck, on a boat that cost 30,000 pounds sterling, with a crew of 70 men, and I had to take full responsibility for the crew and the boat. In the torpedo boat service the prerequisites of existence were a clear eye and a quick decision.... Grand Admiral [Hans von] Koester's night exercises were at a level nobody at the time had ever considered in England... The tactical necessity for the torpedo boat is to break smoothly through a line of battleships while they are moving... More men and boats were to be lost through failure of nerve rather than through recklessness."

The favored tactic was for the faster torpedo boats to move parallel to the slower German battleships and then turn quickly and slip through the intervals between the racing capital ships to charge forward and torpedo the enemy battleships as suddenly as possible.

"We used to have a custom of letting an empty champagne bottle hang from the fantails of the battle cruisers and battleships on a spar," Ehrhardt wrote. "The torpedo boat commander had to bring his boat in so tight on the big ships' fantails that he ripped the bottle off. Full bottles would then be emptied during the critique of the exercise. This sort of exercise cost us a lot of dented bows. But it gave us an undoubted security, even a superiority, over the English torpedo boat commanders."

While preparing for daring exploits in case of war, Ehrhardt was promoted to *Kapitänleutnant*. Since officers in peacetime often languished at a lieutenant's rank until they retired, Wilhelm may have been keeping a friendly eye on the *Lausbub*. On his promotion, Hermann Ehrhardt promptly got married.

The bride was Anna Friederike von Gilsa, the daughter of a merchant, Edwin Dieckmann, who had himself served as a soldier in the 4th Battalion when the Free City of Hamburg had its own army before the German Unification of 1871. Anna Friederike was the recent widow of Lieutenant Julius Bernhard Wilhelm von Gilsa, a member of the Gilsa family, a baronial house with their own coat of arms and a notable history in German military circles. The witness to their marriage was Major Louis von Löwenstein. The couple had produced one child, a daughter, before the husband's untimely death. A century before, Captain Frederick von Gilsa, though a Prussian, served with the Hanoverians of the Kings German Legion of the British Army at La Haye Sainte, the North German Thermopylae that was the key to the Battle of Waterloo. Captain Gilsa's unit, the First Light Battalion, along with the Second Light Battalion and the Fifth Line Battalion of the KGL were all but exterminated buying several hours of time for the Duke of Wellington—the Eighth Line Battalion was also mauled. When the Prussians arrived they caught Napoleon from behind—ending his career.

Less fortunate was Colonel Leopold von Gilsa, veteran of the first Schleswig War of 1848-49 and later commander of the 41st New York Volunteer Infantry

Regiment, a German-speaking unit formed in New York and New Jersey during the American Civil War. Gilsa, once wounded leading his men in battle against the Confederates, became a scapegoat when his regiment and the tandem 45th New York were the first two federal units struck by General Thomas (Stonewall) Jackson's famous surprise flank attack at the Battle of Chancellorsville in May of 1863. Leopold von Gilsa, in fact, had tried to warn Union General Joseph Hooker that his regiments were in the process of being outflanked through the heavy woods. Hooker ignored Gilsa. Most of Gilsa's men in the 41st and 45th New York Volunteers made a fighting withdrawal until their supports collapsed. Then they ran for it like everyone else. A month later, five of Leopold von Gilsa's 41st New York companies rallied from an initial flight at Gettysburg, returned to action with heavy losses, and helped stabilize the Union line. The other four companies just kept running. Gilsa never made general, retired in disgust, and died working at a clerk's job to support his childless wife in 1870, the same year that the Prussians and other Germans defeated Napoleon III and re-established the military reputation that the Germans had lost in the United States during the Civil War. For the next three decades, U.S. soldiers would wear felt imitations of the leather Prussian spiked helmets to try to emulate the Prussians' prowess.

Ehrhardt's marriage seems to have been a Prussian classic—a respectable wife with an ample dowry, widow or otherwise, was seen as an important asset for career stability. These were the years of "The Merry Widow"—the operetta where a profligate but charming nobleman stalks the widow of a possible Jewish court banker. Prussian officers sometimes married Jewish girls—Bismarck himself had approved because he said that the upper classes could use some added brain power. The Gilsa marriage was obviously not a romantic rapture but Ehrhardt must have had at least some time for married life because he and his wife had two sons within the first few years of their marriage.

"Since I belonged to the North Sea station, I reckoned that it would be best to be stationed in Wilhelmshaven. I was delighted that my father-in-law, who came from Hamburg, had built a little house for us in preparation for the marriage.

"I was the first to be settled in the cluster of townhouses that were under construction. Many of my comrades shook their heads and asked 'How can anybody with any sense build a villa in Wilhelmshaven?' But I said to myself, 'If I have to live in the middle of a dump, at least I'll have a nice house.

"But the nemesis followed quickly. After a year I was assigned to a very interesting but labor-intensive command in Kiel, as a consultant on torpedo research. Eight hours a day on the job was like nothing. I spent the whole day working with ballistic and explosives research. When I got back home after supper I found a thick mat of paper work on my desk and stayed up at it until midnight or one o'clock. My wife lamented this and said: 'All I ever see of my husband is his back.'

"However, the life for my wife and children was merrier and more open in Kiel than it had been in Wilhelmshaven, the actual home of the German Navy."

Kapitän Ehrhardt's naval career got an inadvertent boost from an unexpected series of events. In November of 1908, shortly after he first took notice of Hermann Ehrhardt, Wilhelm II was vacationing at an estate in the Black Forest in southern Germany. Shortly after dinner, Dietrich von Hülsen-Haeseler, Wilhelm's military secretary, appeared out of the closet, as it were, in a girl's ballet *tutu*, offered the Kaiser and friends a stylized dance performance, and dropped dead of a heart attack. The whole thing may have been a he-man stunt to cheer Wilhelm up and make him smile, or it may have been the beginning of a homosexual orgy—no one knows for sure because the *première danseuse* died on the spot before anything else happened. The incident allowed Maximilian Harden, a critic and journalist, to ramp up his previous charges that Wilhelm was surrounded by a *camarilla* of secretly homosexual or bisexual men who had an unnatural influence on his policies. Aside from the scandal, homosexual contact between men was an official crime under the constitution of the German Empire. Harden, strikingly handsome, fond of silly costumes, a man of the theatre and a first-generation convert from Judaism—his birth name was Felix Ernst Witkowski—was not hostile to homosexuals as a group. His intent was to foster a more liberal cabinet, not dominated by older military officers. His prime target, Prince Philipp von Eulenburg, was a decorated veteran of the Franco-Prussian War, the father of eight children, and Wilhelm's dearest friend and surrogate father or uncle. Eulenburg loved to sing, to play the piano, and to write children's stories. He hated hunting—and this was something of an anomaly among German aristocrats. Prince Eulenburg's wife denied any evidence of his homosexuality, but police files indicated a number of long-past contacts, mostly with lower-class men who sometimes "borrowed" money from him. Prince Eulenburg was so mortified by the moral and criminal implications that he collapsed at his first trial and never actually stood trial at all due to poor health. He was never convicted and never acquitted. The case was still open when he died in 1921. Some of the other people Harden accused also appear to have been actively homosexual. Some of them committed suicide, pretty much mandatory for a serving officer found to be a sodomite. The Kaiser suffered such a huge blow to his popularity with the German people when he first stood by Eulenburg that Wilhelm quietly dropped him as a confidant.

"Why did he do this to us," Wilhelm asked. "Why did he not tell anyone about it?"

Eulenburg, whatever his private life may have been, was a tragic loss for Wilhelm II but an unwitting benefactor to Hermann Ehrhardt. Eulenburg had long advised Wilhelm to pursue the policies of Otto von Bismarck—build the strongest possible German Army but leave control of the global oceans to Britain, Prussia's ally in the previous wars against Louis XV and Napoleon, and the homeland of Wilhelm's beloved grandmother, Queen Victoria. Wilhelm himself had felt that way while Victoria was still alive and Eulenburg advised him to keep to that policy. But Germany verged into a greatly expanded naval building program after Eulenburg fell from influence.

Long out of favor, Eulenburg summed up the ideas he had long before whispered in Wilhelm's ear in an essay published in 1912: "Our standing German Army would become almost fantastically strong if even one quarter of the billions spent on the Navy were used to expand it... the countries of Europe would bow to the peaceful leadership of such a mighty Germany in the same way that the federal states bow to the leadership of a mighty Prussia."

Eulenburg's fall from favor conversely led to an expansion of the German Navy at the expense of the German Army. While agitation by August Bebel and the Social Democrats limited military conscription to 55 percent of eligible young man, as opposed to universal military service for all healthy males, Germany kept building warships of all types—including torpedo boats. The pre-war Germans shortly had 102 of them. The torpedo boats improved in size and power—some of the newer boats built between 1908 and 1910 could do 32 knots and had two 3.4-four inch guns along with four machine guns, and four big 19.7-inch torpedo tubes, which amply qualified for the British designation of German torpedo boats as "destroyers." *Kapitänleutnant* Hermann Ehrhardt could look forward to steady advancement—especially if war with England broke out.

Chapter Eight
Nibelungenlied

In 1909, Albert Ballin, Kaiser Wilhelm's Jewish friend and trusted advisor, met with Sir Ernest Cassell, a major power in British banking, to try to head off what they both knew would be a disastrous war from Europe, for Germany and for the Jews, whose admiration for German tolerance, culture and progress transcended national borders. The British Jewish banking magnate Sir Ernest Cassell told the German Jewish shipping magnate Albert Ballin that any understanding between Britain and Germany was now impossible, partly due to the Kaiser's ship-building but mostly due to Britain's economic envy for the Kaiser's rising Germany.

"Under the rule of Emperor Wilhelm II, with a consciousness of goals which cannot be more highly praised, Germany has been taken into the world market and Germany's industry and merchant marine have developed to a previously unimaginable flowering.... England has suffered immense losses in overseas trade. English trade is receding and it is undoubted that in the long run England will not be in a position to maintain free trade."

Theobald von Bethmann-Hollweg had replaced Bernhard von Bülow as German chancellor shortly before the conversation, and encouraged by the Kaiser, with Ballin's advice and Cassell's support, Bethmann moved to smooth over relations with Britain. Economics intervened: The Germans of all social classes, including the Social Democrats, supported a certain degree of German militarism due to fear of Russia, but the Social Democrats and other liberals kept conscription to 55 percent and denied the Prussian generals the army they felt they needed to fight Russia and France at the same time. But Britons of all social classes became politically disenchanted with Germany because they hated not only the challenge of the new German Navy but the challenge to Britain's world trade that the German Navy represented. The Anglo-Japanese Alliance of 1902 had been signed not just to keep the Russians out of India but to keep the Germans from expanding out of the Shandong Peninsula into the rest of China, where the British controlled Shanghai and Hong Kong and a vast trade network with the people who had invented the word "*cash*." Russia's alliance with France, a former enemy, against Prussianized Germany, a former British ally in Frederick's and Napoleon's times, wrote the doom large. Britain's growing hostility to Germany made it explicit: the next general European war would be a German catastrophe.

Albert Ballin with the Kaiser

The German national epic, the *Nibelungenlied*—a tangential source of Richard Wagner's "Ring of the Nibelungs," the modern German national epic—begins with a quarrel between two betrayed but powerful women and ends as the doomed Germanic heroes, honor-bound to celebrate the wedding of a woman they know wants them all dead, ride to the wedding anyway, and die. Theodore Roosevelt, once a great friend of the Kaiser, ominously called the second part of the "*Nibelungenlied*" his favorite epic.

The last line is "*Das war der Nibelungen tod*"—"That was the death of the Nibelungs." The metaphor was apt. The German diplomats and the Kaiser himself, angry at the assassination of Austrian Archduke Franz Ferdinand by Serbian terrorists, felt honor-bound or anger-bound to give the Austrians and their dubious multi-ethnic army a blank check to beat up on Serbia. Russia—whose generals had long advocated a strike against Germany, partly because German-Jewish bankers and their American and British friends had funded the humiliating Russo-Japanese War—gave the frightened Serbs a blank check to resist the Austrians. Austria invaded Serbia. Russian declared war on Austria. Germany declared war on Russia. France mobilized to support Russia and Germany declared war on France—though the French mobilization was a *de facto* war declaration of sorts. The Germans knew that they had to knock rapidly mobilizing France out of the war before gigantic but slow-moving Russia could mobilize, so they offered Belgium, ruled by a German dynasty, a return of sovereignty after the war for neutrality and free passage. The Germans were rebuffed, and they invaded Belgium on a temporary basis. The British then declared war on Germany to defend Belgian neutrality from the Kaiser and his Huns.

The Kaiser with Teddy Roosevelt

"This is our reward for Waterloo!" the Kaiser shouted melodramatically, remembering the battle where the sacrifice of the Hanoverians—then British subjects —and the arrival of the Prussians had saved Wellington from defeat and finished off Napoleon for good.

Japan, Britain's ally since 1902, shortly prepared to declare war on Germany and annex Shandong in China with British approval as they had annexed Korea in 1910 with American approval. *Die Gelbe Gefahr*—the Yellow Peril—confirmed Wilhelm's darkest forebodings, prompted, perhaps, by his most lurid fulminations. Things fell into place as they were intended to everywhere but in Germany.

Hermann Ehrhardt had not followed the politics but he was up on the latest tactics and was more than ready for battle.

"Suddenly the day of the declaration of the impending war arrived and surprised me in my work as a researcher," he wrote. "Sitting at a writing desk while the shells were bursting at the front was worse to me than prison." He immediately had himself announced at the admiral's office. Ehrhardt told the staff that his co-researcher, who was invalidated for a sea command for reasons of age and health, would be able to do the job far better than he could himself. He asked to be relieved as a researcher and returned to active command with the torpedo boat squadrons. Within a few days of Britain's declaration of war on Germany, Ehrhardt was patrolling the Baltic with two torpedo boats, guarding sea traffic from neutral Denmark to Germany. International traffic with Germany had collapsed within a few days of the declaration as the British clapped a naval blockade around Germany, but Hermann Ehrhardt's first few days at war proved shocking.

"Toward morning we anchored and I went down from the bridge to my cabin… straightaway I heard a deafening explosion. The whole boat shuddered. I jumped up the stairwell onto the deck and saw everything burning. My second boat had torpedoed itself! The launchers had understandably been loaded with armed torpedoes. As the boat swung, a man on deck had tried to steady himself and had grabbed the launching handle. The torpedo launcher had swung inboard and the torpedo launched into one of the smokestacks and detonated. The monstrous power of the explosives blew the deck right off the other boat, while my own boat was barely damaged. The other boat burned like a torch. I rescued as many of the crew as could be rescued. Sadly, thirty-two brave seamen lost their lives."

Ehrhardt's first important wartime mission was to bring the burned and injured survivors back to the hospital and to give those dead sailors whose bodies had been recovered a burial on land. Ironically, the German civilians had been told that the deaths and injuries were the result of an early naval victory. Ehrhardt and his men hung their heads in embarrassment as the delighted civilians cheered and waved flags at them. Ehrhardt said that he felt somewhat responsible for the design failure that led to the self-torpedoing: "Such thoughts have to arise to a human who has not yet turned into a machine…. How much of the responsibility did I have to the proper readiness of the weapon? Bite your teeth together and look ahead! It's better to go to bed with courage than with cowardice… The Lord God knows better than men."

Ehrhardt remained on the Baltic patrol for five months. He had already experienced the pattern that had been established and would last through the war and into the aftermath. While German civilians were being treated to a "victory" that was actually a lethal accident—the torpedo explosion—and told that their side could never lose, British and American civilians were being treated to a German atrocity that was rooted in reality but expanded with extravagant lies as a warning that a German victory would mean the end of European civilization.

Ehrhardt had soldiered and nearly died in Africa, and his favorite author in English was Arthur Conan Doyle, whose 1909 book "The Crime of the Congo" had documented Belgian brutality in Africa, including the lopping off of African children's hands, that was numerically a hundred times worse than anything either the Germans or the British had done to the Africans. British propaganda would shortly blame the Germans for lopping the hands off Belgian children, which never took place, or with the mass public rape of Belgian virgins and nuns, which never happened either. The mass shooting of Belgian civilians in retaliation for sniper attacks on German troops was a dreadful fact. The advancing armies reportedly killed 5,281 civilians in Belgium and 896 in Northern France. At least some few of the hapless victims were women and children, and this was exploited in the British press as an example of Prussian *Schrecklichkeit*, or frightfulness. The troops who did the actual shooting were mostly panicky Saxon reservists, not Prussians, and they told stories of being shot at and seeing their friends shot while standing in chow lines. The Saxons reservists also told of bursting into houses to find civilian men firing and women reloading for them. Most of the snipers were probably Belgian reservists out of uniform, or outright civilians with hunting weapons. More than 100 of the 500-plus German dead were killed by shotguns—use of shotguns was illegal in warfare. Post-war investigation showed that not one German, either Prussian or Saxon, had lopped the hands off a Belgian child. The public rapes of Belgian virgins and nuns never happened either. A team of eight American correspondents traveling with the German Army telegrammed,

"WE UNITE IN DECLARING GERMAN ATROCITIES GROUNDLESS. AFTER SPENDING TWO WEEKS WITH GERMAN ARMY ACCOMPANYING TROOPS UPWARD TWO HUNDRED MILES WE UNABLE REPORT SINGLE INCIDENT UNPROVOKED REPRISAL. ALSO UNABLE CONFIRM RUMORS MISTREATMENT PRISONERS OR NON-COMBATANTS... NUMEROUS INVESTIGATED RUMORS PROVED GROUNDLESS. TO TRUTH THESE STATEMENTS WE PLEDGE PROFESSIONAL WORD."

Nobody had had any trouble documenting Belgian atrocities in the Congo a decade before. Conan Doyle's 1909 best-seller showed actual photographs of hand-maimed Bantu and Pygmy children from Belgium's Congo Free State, where an estimated two million Africans were murdered. President Theodore Roosevelt, Kaiser Wilhelm II, and Winston Churchill, who had soldiered in Africa, all protested the Belgian brutality in the Congo. Conan Doyle had never visited the Congo himself. He had relied on the reports of Edmund D. Morel and Sir Roger Casement. Conan Doyle shortly became an anti-German propagandist. Morel became a pacifist during the war and advocated a moderate peace afterwards. Casement was hanged for supporting the German-supported Easter Uprising of Irish Republican patriots in 1916. The Kaiser himself had been the first major rule to recognize Irish independence.

While Ehrhardt was still patrolling the Baltic, the German Navy was being routed from Kiautschou, Germany's largest overseas naval base, by Britain's Japanese allies. A German pilot and a Japanese pilot fought the world's first air battle, circling each other in struggling aircraft with motorcycle-sized engines and shooting at one another with pistols. Both missed. Each side lost a destroyer. When the Germans gave up, the Japanese marched in, treated the Chinese decently, and virtually adopted the Germans, some of whom decided to open businesses with Japanese partners or colleagues, like the one that later manufactured Ehrhardt's optical toy. The Japanese anatomist Ryuhei Adachi had written, during the Kaiser's "Yellow Peril" rampages, that the Germans were the best looking but the worst smelling of all Europeans, and the novelist Mori Rintaro Ōgai had written a cautionary tale called "The Dancing Girl" in which a Japanese student disgraces himself when he forms a misalliance with the original Dumb Blonde—a chorus girl named *Eritsu*, which is to say, Elise—while studying in Berlin. Mori Rintaro Ōgai also wrote a satire called "Vita Sexualis," an allusion to Dr. Richard von Krafft-Ebing's world-famous "Psychopathia Sexualis," in which Japanese are mostly normal and Germans are very strange. Ōgai was probably also inspired by the Eulenberg scandal. Both sides mellowed after the Japanese grabbed Shandong and hoped to keep it.

The German cruiser squadron, expelled from their Chinese base and trying to return to Germany, fought the British twice off South America: the first clash was a clear-cut German victory. The second was a massacre of outnumbered and outgunned German ship who fought as a pont of honor

Mori Rintaro Ōgai

—and because they were too slow to escape. The British themselves were appalled by their own poor marksmanship and made a foreboding decision: they decided on trying for the highest possible rate of fire and authorized keeping bags of guncotton in the turrets of their battleships. Turret doors to the magazines would now be left open. The S.M.S. "*Emden*," a light cruiser and last survivor of the German Pacific Fleet, became a famous commerce raider and sank a Russian cruiser and a French destroyer along with a number of merchant ships before being sunk herself. Some of the crew escaped back to Germany in a maritime adventure of small-boat handling and great navigational skills.

A few months into hostilities, Ehrhardt was ordered to take his half-flotilla —five torpedo boats—to the harbor at Libau (modern Liepāja, Latvia), where English submarines were active. The Germans had decided to block the harbor entrance with outmoded steamers that were filled with cement and scuttled as an underwater barrier.

"The blockship steamers were several hours ahead of us and steaming wearily forward. Hardly had we reached them and made ourselves known with blinker signals than a thick snowstorm came up at night and reduced visibility to almost nothing. We steered by map and compass with a watch in hand. We could hardly see the outline of the boat next to us. We had to cut down to half speed so as not to lose contact with the slower blockships. So we steered quietly into Russian territory, until by map and compass and watch I had the presentiment that we must by close to Libau. A sudden gust blew a hole in the snowstorm and just for a moment we could see the shadows of the buildings against the night sky. We were on the right course.

"While I was collecting my boats, an orderly brought me a radio telegraph message: Our covering ship, the armored cruiser '*Prinz Friedrich Karl*' a few miles west of us had run onto a mine and was in shaky condition. What to do? Give up the mission and call them back and pick up our people, or to carry the order through?

"I made my decision in the blink of an eye. Such a night would never come again. '*Prinz Friedrich Karl*' would have to help herself or rely on some other ship. For us the higher mission was to destroy the enemy."

SMS Friedrich Karl midsection (Miniature housed in the Deutsches Museum)

Ehrhardt slipped into the harbor. He found no English submarines. He stood by with his torpedo boats as the blockships were positioned and sunk with scuttling charges. The sunken blockships jammed the mouth of the port. "The snow was so thick and the sleep of the Russians was so good that the harbor defenses never saw us..." For good measure, he shelled some oil tanks he had spotted and touched off an enormous series of explosions that appeared to have driven the coastal battery gun crews away from their guns. With his mission accomplished, Ehrhardt was headed to try to rescue the crew of the *'Prinz Friedrich Karl'* when he received a second radio telegraph message: The *'Prinz Friedrich Karl'* was on the bottom, but the light cruiser *'Augsburg'* had arrived and taken off the crew. "It would have been too bitter if my success was crowned with the loss of the crew as well as the loss of the cruiser," he wrote with immense relief. *Kapitänleutnant* Hermann Ehrhardt himself received the Iron Cross, second class, for this exploit, which he noted dryly was still worth something in 1914 before medal proliferation took over. His crews were applauded through the streets of Danzig by cheering crowds and feted at the *Rathskeller* where approving citizens sent them many bottles of wine to celebrate with.

The next adventure began with a perilous embarrassment. Ehrhardt's torpedo boats were escorting the *'Augsburg'* and the light cruiser *'Magdeburg'* off the coast of Finland in a heavy fog when the *'Magdeburg'* ran aground. Ehrhardt and some of his men boarded the cruiser and, working "like slaves," off-loaded ammunition and threw the other gear overboard in a frantic effort to lighten ship.

"Shortly after the sun came up the fog vanished. A squadron of Russian armored cruisers appeared a few miles away and hit right and left in the water with heavy shells. Now we at least had the opportunity to save the crew of our cruiser. We took the men aboard, and a couple of scuttling charges blew holes in her hull. She sank slowly beneath the surface."

Running a ship aground was something of a disgrace and the Baltic Squadron wanted revenge. Ehrhardt and the officers of the "*Augsburg*" headed back to the coast of Finland a few weeks later. This time they brought the U-26 with them. The plan was to lure the Russian armored cruiser squadron out so that the U-boat could have a field day.

"We sailed out as bait, perhaps rather rudely. The Russian cruisers came out as if by right, and we zigzagged under their shellfire and tried to lure then into the vicinity of the U-boat.

"As the exploding shells got closer and the water spouts began to fly up, we didn't worry much. But as the shells came closer so we could feel the heat, we began to ask one another—is the U-boat ever going to shoot? Only the hunter's instinct kept me calm—it's stronger in action than the instinct for self-preservation.

"Suddenly a giant water column leaped up from the side of the leading Russian cruiser, the *'Pallada.'* Her hull turned over steeply and then she smoothly sank.

This was the main ship that had disturbed our rescue efforts on the '*Magdeburg.*' The loss of the '*Magdeburg*' was expiated. The other Russian cruisers turned away immediately and left the crewmen to their fate." The '*Pallada,*' launched in 1906, was an oddly formidable ship which mounted two 8-inch guns and eight 6-inch guns with 22 12-pounders. Her sinking on October 11, 1914 and the loss of most of the crew more than made up for the '*Magdeburg,*' whose crew had been rescued.

Russian cruiser Pallada

Ehrhardt never condemned the Russians for leaving their sailors in the water. He pointed out that on September 22, two months before, the U-9 had sunk three British cruisers, the '*Hogue,*' '*Aboukir,*' and '*Cressy,*' one after the other. The three older cruisers were patrolling the North Sea blockading Germany. As the first and second cruisers stopped to pick up survivors, they inadvertently gave the lone U-boat a chance to re-load. The triple hit against the hated blockade led to celebrations in Germany and deep concern in Britain.

"It may sound like a paradox, but it would have been more humane to let a few hundred men drown than to attempt their rescue and give the U-9 a chance to cost a greater number of men their lives." Ehrhardt wrote. "That is war."

The drastic weather in the Baltic led to complications that Ehrhardt referred to as "highly humorous." He was leading his half-flotilla near Riga when he suddenly realized that the five torpedo boats were surrounded in the darkness and fog by a number of bigger ships he was unable to identify. "I wanted to see what sort of uncles these were," he said. "They made smoke and ran for it and we ran after them." Ehrhardt was maneuvering to use his torpedoes when he abruptly spotted a German recognition signal and secured from firing position. He received a signal to come on board and encountered an enraged admiral who had been on a mission to attack Riga when he sensed his fleet being infiltrated by unknown ships.

"*Herr Kapitänleutnant,* how does it come that you never made any recognition signals?"

"*Herr Admiral*, I took your ships for Russians."

"*Donnerwetter*, I did the same for you. In the future let me know where you stand and what you're doing. On your account I had to break off my entire mission."

"*Herr Admiral*, I must ask you in the future to also give recognition signals," Ehrhardt said with his usual truculence.

Both officers spoke further and discovered that the admiralty staff had forgotten to inform either of them that the other was in the area.

"We agreed with one another and in conclusion we drank a glass of port wine to the health of the Admiralty Staff," Hermann Ehrhardt remembered with wry humor.

Other memories were not humorous. Losses to mines and groundings had a depressing effect on men who would have gladly fought visible enemies. Sailing in consort with the "*Augsburg*," Ehrhardt saw from his torpedo boat that the cruiser was off course and headed for a suspected minefield. He had a warning flashed to the "*Augsburg*" but the cruiser struck a mine before her crew could change course. Ehrhardt's friend, First Officer Gernot Goetting, was on board the "*Augsburg*" and escaped, perhaps due to Ehrhardt's warning, reinforcing "a friendship that one can call a true friendship." Goetting was destined to be an Ehrhardt supporter in his second military career as well—the one that landed Ehrhardt in prison.

Kapitänleutnant Ehrhardt and his half-flotilla were transferred to the coast of Flanders in the winter of 1915 to take part in coast defense—and to cover the U-boat bases against retaliatory attack. As the British blockade clapped down on Germany's military supplies and food, the Germans ramped up their U-boat campaign and moved to sinking ships in the "war zone." They also ventured into bombardments of the English coast by surface ships in a drastic protest against the blockade. On January 15, at the battle of the Dogger Bank, the British pursued one German squadron of four cruisers and sank the "*Blücher*," referred to as a "five-minute ship" by the Germans because that was how long they expected the "*Blücher*" to survive in a general surface action. The Germans also suffered damage to the battlecruiser *Seydlitz*, and lost almost 1,000 sailors but damaged the British flagship, HMS *Lion* with 15 dead. The Germans began to improve their gunnery and damage control and kept the big ships in port. But a few days later, the German Navy began Zeppelin attacks with giant dirigible airships converted from pre-war excursion liners. The British public labeled the German Zeppelin raiders "Baby Killers." The Naval Zeppelin Squadron commander, *Fregattenkapitän* Peter Strasser, came from Hanover—the ancestral home of the British monarchs and British territory before 1837. He was completely fluent in English. Strasser also wore his beard cut in the same manner as Hermann Ehrhardt—a fashion among younger Naval officers. Strasser bristled at the "baby-killer" insult. The Zeppelin crews explained their attacks by dropping scraped-down hambones with notes that read "from the starved-out German people." Ultimately the Zeppelin raids would kill 557 English civilians and

British servicemen and wound 1,358. The British starvation blockade would kill 772,736 German civilians, thousands of them after the Armistice. Most of the Germans were elderly people who literally starved to death or new-born babies and small children from poor families born with fatal birth defects due to fetal malnutrition.

German battlecruiser SMS Seydlitz

The second propaganda disaster befell German on May 7, 1915 when the U-20 sank the "*Lusitania*" within sight of Ireland and 1,201 passengers and crew were killed—including a number of celebrities and 128 Americans. The *Lusitania* was described as an innocent ocean liner by British propaganda. Her identical sister ship, the *Mauretania*, was regularly used as a British troopship. The *Lusitania* and *Mauretania* were both built as British auxiliary cruisers, designed with gun mountings for cannon. The guns were not mounted when the *Lusitania* sailed from New York. But at the time of the torpedoing she was carrying 1,248 cases of three-inch shrapnel shells, four shells to the case, weighing 58 tons and 4,927 boxes of .303 rifle and machine gun ammunition, at 1,000 rounds per box, consigned from Remington Small Arms to the Royal Arsenal at Woolwich. Five million rounds of machine gun ammunition and 5,000 rounds of exploding shrapnel shells could have killed a lot of Germans. This part of the cargo was legal contraband.

The *Lusitania* also carried food to Britain when the British blockaded food to Germany, where poor people were already feeling the pinch. The U-boat commander, *Kapitänleutnant* Walther Schwieger, fired from a submerged position because he was afraid of being rammed—British merchant captains had standing orders to ram U-boats that surfaced to order them to their lifeboats—but he hit the jackpot when the ship blew apart due to a mysterious secondary explosion and sank in 18 minutes. Schwieger himself was no great humanitarian—he had once fired at a British hospital ship, but missed. He had also once been damaged by ramming by a British ship and barely made it back to Germany. But he denied firing a second torpedo at the *Lusitania*. Extensive research by Marine Archaeologist Robert Ballard suggests that Schwieger's single torpedo polarized the coal dust between the Lusitania's double steel hulls and a spark touched off a secondary blast bigger than the torpedo, which capsized the ship before most of the lifeboats could be lowered. Keeping 5 million rounds of .303 ammunition and 5,000 shrapnel shells out of British hands was a tactical victory. But the deaths of the women and children and other civilians was a moral tragedy and a strategic propaganda catastrophe.

The Lusitania at the end of a record voyage

Kapitänleutnant Ehrhardt encountered a related case: The Germans had learned that "innocent" British ships would sometimes stop when a U-boat surfaced, then suddenly attempt to ram the U-boat. Some of these ships, the so-called Q-ships, looked as dingy as possible, hardly worth a torpedo, but secretly carried concealed cannon and machine guns. When the U-boats had been damaged, the British Q-ships left the crews in the water—or sometimes machine-gunned them. In an incident that started the coast of Flanders, a British merchant captain named Charles Fryatt had been executed for obeying the ramming order issued by Winston Churchill.

After a previous chase by a U-boat in which his ship escaped, Fryatt was ordered to stop by the U-33 on March 28, 1915. Captain Fryatt complied. But when the U-boat was fully surfaced, he ordered full steam ahead and attempted to ram it. The U-boat crash dived and escaped. When Fryatt returned to England, he was a celebrity and received a gold testimonial watch by the British Admiralty—which would have tried him in court had he not attempted the ramming. Captain Fryatt, who had six daughters and a son, was a celebrated hero in the newspapers. Meanwhile in October of 1915, the Germans found orders issued by Churchill which stated that U-boat sailors were to be treated as felons rather than prisoners of war: the U-boat men could be shot if shooting them was more convenient than taking them prisoner.

Captain Fryatt's ship *Brussels* was ambushed and captured by five German destroyers—Ehrhardt makes no mention of his role, if any, in the capture—and Fryatt was charged with piracy, convicted, and shot by a firing squad. His death warrant was signed by Ehrhardt's commanding officer, Admiral Ludwig von Schröder, and confirmed by the Kaiser.

"A man who sails on a disguised boat is a pirate and must know what that means," Schröder wrote. Ehrhardt backed Schröder. The *New York Times*, however, called the execution "a deliberate murder" and the *New York Herald* called it "the crowning Germany atrocity." After the first Zeppelin attacks on civilians and the sinking of the *Lusitania*, the officially neutral press had become increasingly anti-German.

Schröder's squadron continued raids against the east coast of England with *Kapitänleutnant* Ehrhardt as the leader of his own half flotilla.

"On a winter night in the year 1916 I struck with my half flotilla through the Straits of Dover, deep into the Channel," Ehrhardt wrote. "Such a breakthrough into the English security zone also brought us to full attention and nervous expectation. Everything was blacked out. Not the smallest light could show through a crack. In the high waters of the North Sea the gun and torpedo stations were covered with sailcloth that could be pulled away quickly. The guns and torpedo launchers were manned by one sailor each, with the watch crew in the forecastle near the bow of the ship. But they had to stand their watch in darkness, so their eyes would adjust to the night quickly when the alarm came. The boilers were under full pressure so we could shift to all 3,000 horsepower. The searchlights were secured because one ray of light could attract the enemy. The commander and both watch officers were on the bridge, with rapt attention by eye and ear....

"As soon as we reached our assigned station in the Channel we sighted the English transport steam '*Queen*.' I came alongside in the lead boat and called out to the English captain: '*Stop! All Hands in five minutes into the boat!*'

"The fellow scolded us and wanted to know our name and home port. But our call '*German destroyer!*' put an end to the conversation. I could sympathize with the paralysis that fell upon the English seamen. The fellows had to extinguish all lights. Then they went into the lifeboat with all possible speed. A torpedo shot sent the transport into the air. The fortress of Dover mobilized in nothing flat. Giant searchlights probed the sea—six, seven, eight, it was as bright as day. But we took off at extreme speed. Before the Tommies could direct their aging cannon at us, we were out of the lighted area with the speed of lightning.

"To our good fortune it was an extremely dark night. But it was clear we wouldn't get away without a fight, because the other side threw everything in our way that they could, whatever they had to hand. And right, we soon had an English destroyer flotilla closing on us from behind. They came within 50 meters—searchlights lit up—a short command of the torpedo officer, a light clatter, a torpedo left the tube. In a few seconds it ripped the English lead boat apart. The shock wave from the explosion was so strong that the deck crew was knocked sprawling and the heat from her boilers threw up a giant heat wave. I myself was thrown so hard against the railing of the command bridge that I got a concussion and broke an eardrum. A second destroyer was knocked out by our deck guns. The battle zone grew white because burning oil spread out over the sea. Individual black points were visible: the corpses of the English seamen, who had tried in vain to out-swim death, and had covered their faces with their arms [*when the burning oil caught up with them.*] The way was clear now and we pounded with all speed under full power to the east, to the protection of the coast batteries in Flanders."

Ehrhardt noted that his flotilla and others kept up nuisance raids on the English coast and also noted that the British fleet made no such attacks on the North German coast. The British, in fact, had constructed large landing barges during the first 18 months of the war for a projected invasion of the Baltic coast of Germany, but the invasion was never launched. The landing barges were used in the later stages of Winston Churchill's famous invasion of Turkey at Gallipoli in the Dardanelles, which also flopped.

"On one voyage I actually sailed up the Thames," Ehrhardt said. "I think the old Dutch Admiral [Michiel] de Ruyter was the last enemy that sailed through this water at least 250 years before. The former Hapag steamer '*Königin Luise*' had attempted to mine the Thames on August 5, 1914, and was sunk, but not without results: The mines she laid sent the English cruiser '*Amphion*' to the bottom."

Ehrhardt's own venture up the Thames was precarious but uneventful. The fog was so dense that neither side made contact. His half flotilla returned unscathed. "The English fleet didn't come out and the German fleet couldn't venture any further from their base than we did."

The Germans, Ehrhardt said, wanted to bring the British to battle in one great shoot-out that they hoped would break the blockade. On May 30 of 1916, Ehrhardt's half flotilla was part of a general consolidation. *Kapitänleutnant* Ehrhardt's IX Torpedo Boat Flotilla was grouped with the II and VI Flotillas under overall command of Admiral Franz von Hipper, whose flagship was the light cruiser "*Regensburg*." After a conference on the "*Regensburg*," which served as the mother ship for the torpedo boat flotillas, Ehrhardt sent a blinker signal to Karl Tillessen, another torpedo boat commander: "*Heute Himmelfahrt*"—"Today we go to Heaven."

"About 4 a.m. we were launched in battle formation toward the Skagerrak, the torpedo boats serving as submarine security. The bulk of the fleet would follow. The men had the feeling that there was something in the air, because there had been a lot of tension since the bombardments of Yarmouth and Lowestoft [*Dogger Bank*] and there had been increased traffic back and forth to the flagship. The departure of the whole fleet with the support of all available light craft gave this vague presumption an appearance of authority.

"The sea lay flat as a mirror. In order not to be destroyed by enemy submarines or the minefields, we moved through the darkness in a north-northwesterly direction along the Danish coast and prepared to show ourselves in the vicinity of the Norwegian coast. The battle watch was on duty and the free watch stood by on deck and enjoyed the beautiful weather. The men sang and played music, played cards, and talked and joked. The continuous advance, without encountering an actual opponent, had the men so astonished that only the increase in vigilance kept them from seeing this as just another day.

"About 4:30 p.m. Torpedo Boat 109, the ship on the extreme point, reported spotting some individual fighting ships in a westerly direction. The enemy must have seen us as well, because hardly had the light cruisers turned on him but that he turned and steamed on a northerly course. The cruisers took up the pursuit. About an hour later the armored cruisers sighted heavy enemy battle craft. Immediately we gave up the pursuit of the English light cruisers and turned toward the English battle cruisers, which is what they turned out to be."

"The enemy turned south-east. We took the same course in echelon and drew nearer. The battle stations were manned. The distance finders kept constant track of the distance. The gunnery officers had given the necessary orders. The battle cruisers were to open fire at 13 kilometers distance. The first muffled salvo rolled over the sea.

"The shells fell heavily on the battle zone. Our flotilla was on the lee side of the fire, our battle cruisers served as a cover for us."

HMS Indefatigable sinking

The British policy of keeping the bagged powder in the turrets and the magazine doors open to improve their speed of reloading came at a terrible cost. About 5 p.m. German time, the first German shell hit the British battle cruiser HMS *Indefatigable*. The battle cruiser blew up at the beginning of the action, 30 seconds after the first shell hit. Two British sailors survived out of 1,019 in her crew.

HMS *Lion*, the battle cruiser group's flagship, took a turret hit and the bagged powder inside the turret burst into flame. Closing the magazine doors barely prevented the sort magazine explosion that probably sank the *Indefatigable*. HMS *Lion* dropped out of the action.

HMS Lion

"The fight between the battle cruisers lasted about a half hour," Ehrhardt remembered. "One Englishman [*HMS Indefatigable*] was already sunk when a strong English squadron burst on us from a westerly direction at flank speed. These were the newest ships of the English fleet, which exceeded ours in speed and gun caliber. At 20 kilometers they opened a well-placed fire. The distance between the fighting battle cruisers had also diminished. The situation for our battle cruisers had become doubtful because of the attack by the newly arrived Englishmen. The torpedo boats would have to attack to relieve the battle cruisers. My flotilla, in echelon behind the battle cruisers, was the first one available. Covered by thick clouds of coal smoke and thick yellow powder smoke we struck under the cover of the heavy gunnery from the battle cruisers. But the English soon recognized the danger. They left the battle cruisers alone and fired on us with a steel hail of all sorts of shells. At the same time they sent all their available destroyers against us. We charged ahead at triple power. No sooner had my flotilla launched its torpedoes than my own boat was struck. In the blink of an eye the boat was immobilized and wrapped in thick clouds of steam. The other boats had to go ahead. A heavy fight with the English destroyers broke out a short distance away. All I could hear was the thunder and hissing of the battle. Suddenly there was an overpowering detonation. A monstrous smoke column stood over the place where the '*Queen Mary*' had been. A single good hit had neutralized this giant of the English fleet with her 29,000 tons and her armament of eight 38-centimeter guns. Two German battle cruisers, SMS *Seydlitz* and SMS *Derfflinger*, concentrated their fire on the battle cruiser HMS *Queen Mary*. A German shell struck the *Queen Mary's* center turret and she blew up and sank within two minutes."

Destruction of the HMS Queen Mary

"There seems to be something wrong with our bloody ships today," the British Admiral Sir David Beatty remarked.

Beatty was right: the battle cruisers had been designed to combine the hitting power of a battleship with the faster speed of a cruiser, but they were under-armored for general fleet actions. Leaving the magazine doors open in action turned them into floating death traps.

"It's not out of the question that one of the torpedoes we fired at the English fleet had this effect. Apparently, however, it was a hit from the heavy guns" Ehrhardt wrote. "My own boat was already engulfed in steam and smoke. There was a great danger of losing the boat to the enemy, since an English destroyer could pop up out of the battle haze in the blink of an eye and board us. I gave the order to sink the boat. We placed scuttling charges in three places and blew three holes in the hull. The men stood in their cork life vests on the forecastle, and we waited with a waving flag and three cheers for His Majesty for the sinking of the boat. Suddenly a cheer broke out from their midst. In the blink of an eye we saw the German battle cruisers were free. They had reversed course. South-easterly of them the lead ship of the German battleship fleet, S.M.S, '*König*' was in sight. Now the English turned and ran in a north-westerly direction. The barking of the torpedo boat guns had stopped. One of my other torpedo boats came alongside and picked me and my men up in the middle of enemy fire. I resumed command of the half flotilla. Two of the most modern English destroyers had been rendered unmaneuverable, two or three other had already sunk. Besides my boat, V 27, V 29 was also lost, but the crew had been rescued by the rest of the flotilla despite enemy fire. The purpose of the attack had been achieved. The battle cruisers had gotten some air. We got back into the sheltered position on the lee side of the enemy fire. The heavy guns took control of the fight again. In another blink of the eye the whole of the German fleet had been encountered. The German salvoes hammered on the iron of the ships, giant water columns rose around the fleeing Englishmen. A row of hits were observed. The visibility got worse. Ships' smoke and thick powder smoke clung to the water. Toward the north and the east there was nothing more to see. Gradually, the faster English ships got away from us. The pursuing cruisers swung in maneuver to follow a course to the northeast. The smaller cruisers were caught up in another fight. '*Wiesbaden*' took heavy hits and stayed unmaneuverable and lying exposed to enemy fire. Our torpedo boats attacked again. This time we succeeded in getting closer to the enemy battle line. Our torpedoes were fired under favorable conditions.

SMS Derfflinger

"The enemy turned immediately, covered by their thick smoke. We had struck at the vanguard of the bulk of the English fleet. The heaviest fire of the English battleships swelled up. Toward 9 p.m. it reached its greatest strength. The main power of the English fleet under Admiral [Sir John] Jellicoe joined in the battle. With their greater speed the English attempted 'crossing the T,' that is to say, they attempted to be the top stroke of the letter T, lying in front of the German fleet, so as to be able to concentrate their fire on one advancing ship after another. Admiral Scheer gave the order to the battle cruisers, in defiance of the well-recognized

English intention, to charge with full force against the enemy battle line. Salvo after salvo fell in the battle cruisers, which were damaged and impeded in their fighting power. The point ship, SMS 'Lützow' was heavily damaged, and Admiral Hipper had to be taken off by a torpedo boat, and later set his flag on the 'Moltke.' 'Seydlitz' and 'Derfflinger' were both hit and the torpedo boats were sent to relieve them. Together with the VI Flotilla we charged against the center of the enemy line. The English immediately turned their entire battery fire against us. The S 35 under *Kapitänleutnant* [Friedrich] Ihn took a direct hit from a 30.5 and sank immediately. The other boats kept a straight course and launched their torpedoes before they turned off. We covered the visibility of our own line from the sight of the enemy gunners by making a heavy cloud of smoke. The English battery fire was silenced. Apparently the enemy had turned away to cover themselves from our torpedo attack. At the same time the German line turned to starboard on a westerly course. The orders rang out simply, but years of practice maneuvers had gone into the ability to carry out these turns in unison and under control. The case was especially difficult because the flagship had not simply turned starboard but had actually reversed course. Nevertheless, this maneuver was carried out exactly. The fleet was now on a southwesterly course. The result of the maneuver was that the ships which had been in the rear of the column were now on the point position and the heavily damaged point ships followed in the rear.

Damage to the SMS Seydlitz

"Twilight arrived. We had not observed as much in the heat of battle as we did now that the guns had gone silent. Shortly before 10 p.m. the fleet went on the south course. The mission of the torpedo boats was to take over night security. The battle lines had separated from one another, but now we could expect to hear from the enemy again by night.

"We had to watch out from all sides, in order not to run into the whole body of the enemy. The English destroyers became very active. A whole flotilla was recognized and exterminated. The point ship 'Westfalen' alone knocked off six of them. In two minutes there was a burning oil slick on the sea that gigantic flames flared up from all through the night.

SMS Westfalen

"The recognition of the enemy ships and our own was here, there, and everywhere. One of my comrades, *Kapitänleutnant* [Otto] Ehrentraut, commander of S 32, saw a couple of torpedo boats and joined them until he suddenly realized —'Damn it, these are Englishmen!' He slipped off and turned away. The English saw their mark and sent a salvo after him. But he had already become invisible in the darkness.

"Searchlights hardly came into use during the night all through the action. Nobody wanted to attract the enemy's fire. The English apparently mistook our ships for their own. It was long after midnight when an older English battle cruiser tried to link up with the ships of our first squadron. She was recognized as the enemy and was shot to pieces from less than 1,500 meters. There was no return fire. In nothing flat she was glowing from end to end. Mighty detonations accompanied her going under a few minutes after the German fleet opened fire. (This may have been the British battle cruiser 'Invincible,' launched in 1907, which was struck in the turret, broke in half and sank, though the British record the sinking somewhat earlier.) Then the 'Nassau' accidentally rammed and sank an English destroyer.

Battlecruiser Lützow

"Ships lay around all over burning like torches. We lost the [SMS] '*Pommern*' due to an English torpedo boat attack. (The *Pommern*, launched in 1905, was an elderly ship with outmoded 11-inch guns) Our '*Elbing*' did an evasive turn to avoid an English torpedo boat attack and struck a German battle cruiser, and had to be scuttled when she lost power. But the heaviest loss was when the battle cruiser '*Lützow*' was no longer capable of moving. We took the crew off and gave the lordly ship, like a noble stag, two torpedoes as a final shot.

SMS Pommern

"On the return voyage the approach of 12 English ships of the line was announced. This was a new enemy, and our ships of the line gladly welcomed the chance for action. Everyone ran to his battle station. But they turned off at a distance of 60 sea miles."

Each side claimed victory in the battle the British called Jutland, after the Danish peninsula, and the Germans called Skagerrak, after the expanse of sea where the fighting took place. Among the dead was 35-year-old Gorch Fock, the maritime poet born as Johann Wilhelm Kinau in Hamburg. Gorch Fock wrote sea stories and poems in the vanishing North German language. He drowned in the sinking of the *SMS Wiesbaden*. Swedish fisherman found his body, along with that of many German and many English sailors and buried them all side by side.

"At the battle of Skagerrak the English ranged 1,184,450 tons against 639,200 German tons," Herman Ehrhardt wrote. The weight of the heaviest German shell of 30.5 centimeters was 390 kilograms, the size of the biggest English shells was 38 centimeters and the weight was 885 kilograms. The weight of the combined English broadside would have been four to one to the combined German broadside. The official figures are that the German side lost 60,300 tons and the English side lost 121,350 tons. But English prisoners reported to us that they lost 172,830 tons. Correspondingly the English prisoners told us that their losses were three or four times greater than ours. We also fished out a lot more English prisoners than the English did of our Germans....

Painted in 1917 by the German artist Willy Stower, this dramatic image shows a German destroyer torpedo boat in action at Jutland

"When my wife came home from visiting acquaintances, the children ran up and told her, 'Mommy, do you already know that father's boat has been sunk?' She lived through a hard night of uncertainty. Only my arrival home the next morning delivered her from the fearful thoughts that she had become a widow."

Ehrhardt said that the battle of Skagerrak was celebrated by the German people as a resounding victory, despite the fact that the English fleet had not been destroyed—and, as it developed, the blockade remained in force. The British saw the numbers a little differently: The Royal Navy had sailed to battle with 37 battleships or battle cruisers against 27 German battleships or battle cruisers. The weight of the British broadside was acknowledged to be two to one against the Germans, though not the four to one Ehrhardt calculated. The Royal Navy had lost 14 ships, including the three battle cruisers destroyed by turret explosions with virtually all hands, and had suffered a total of 6,784 casualties. The Germans had lost 11 ships, including one battleship, the elderly '*Pommern*' and one battle cruiser, the modern '*Lützow*,' whose crew was taken off. The other German losses were light cruisers like the *Wiesbaden* or torpedo boats. The Germans had suffered 3,058 casualties. The Germans had inflicted a tactical defeat on the British in terms of gunnery and damage control. This was no Battle of Trafalgar, in which the British had defeated the French and Spanish Navies in the same day in 1805, to rule the waves for another 111 years. But Royal Navy Captain Herbert Richmond summed up the message of the Battle of Jutland from the British standpoint: "It is absolutely necessary to look at the war as a whole; to avoid keeping our eyes only on the German Fleet. What we have to do is to starve and cripple Germany."

Chapter Nine

Hunger and Hatred

The triumphs of German gunnery over British damage control may have encouraged German naval officers, but the British losses at Jutland could be replaced. The German High Seas Fleet had failed to break the British blockade. By the middle of 1915 food shortages had already been palpable both in the quality of the food and in the amount of food available to those Germans who were unable to buy on the 'black market.' Potatoes replaced wheat and rye, Then turnips, pre-war stock feed, began to replace potatoes. Sausages were 2.5 percent fat with vegetable matter augmented by water for bulk filler. The official British figures eventually listed 772,736 German civilian deaths due to the British blockade—a figure comparable to the battle casualties of the entire British Army. The German U-boat attacks on merchant ships had seen largely suspended after the 'Lusitania' incident had threatened to bring the United States into the war, despite arguments from the German admirals that they could knock Britain out of the war with U-boats. The Germans fell back on subterfuge. Franz von Papen, a former cavalry officer and the German military attaché in the United States, continued to buy up advance orders of munitions destined from Britain to keep them out of British hands, as he had even before the Lusitania was attacked. He also fell back on sabotage and reportedly conspired with the mysterious Curt Thummel, a bi-lingual German agent, to blow up the

Black Tom ammunition storage depot opposite New York City on a New Jersey promontory in the Hudson River a month after the Battle of Jutland. Six months later, Thummel blew up another ammunition plant in Kingsland, New Jersey. American officials had looked the other way during the shipments of ammunition to the Allies. The fact that the two huge explosions were sabotage remains unproven.

The explosion at the Black Tom storage dump on the Hudson River opposite New York was heard as far away as Philadelphia and Baltimore.

The aftermath of the Lusitania, however, led to what started out as a constructive proposal. Rosika Schwimmer, a Hungarian-born suffragette and pacifist, and Louis Lochner, a German-American journalist and pacifist, convinced Henry Ford, the automobile manufacturer, to sponsor and lead a nautical expedition to organize peace negotiations without the approval of President Woodrow Wilson, whom Ford dismissed after an interview as "a small man." Ford was encouraged by Jane Addams, the settlement house humanitarian, by the devoutly Christian Secretary of State Williams Jennings Bryant, by Thomas Edison and by department store mogul John Wanamaker—but none of them sailed on the 'Oscar II' with Henry Ford, Rosika Schwimmer, and The Reverend Samuel Marquis, an Episcopalian clergyman and later a critical yet sympathetic biographer of Henry Ford. The ship's departure from New York received huge publicity with a crowd of about 15,000—and huge ridicule. Rosika Schwimmer, a driving force behind the expedition, was an avowed atheist and ardent feminist. Samuel Marquis was an Episcopalian minister. Louis Lochner was the son of a Lutheran pastor and an active churchman. James Addams was a lifelong Presbyterian. Many of the supporters left ashore hated war on Christian grounds. But they pursued a conservative social agenda which supported marriage and the family. Schwimmer, who advocated birth control and easier divorce, claimed she had all kinds of documents which showed that the warring powers were eager to negotiation with Henry Ford—and herself—but she refused to show them to anyone. She also barred reporters from the ship's radio shack. The flu broke out aboard, one man died, and Ford himself got sick. Influenced by Pastor Marquis, he quietly dropped out as the expedition's leader and returned to the United States on another ship. The net result of the expedition appears to have been to turn Henry Ford from a rustic technocrat without a great deal of malice to a bitter and outspoken anti-Semite. He blamed the radical Rosika Schwimmer for having befuddled him and ruining the chances for peace—which had been unlikely in any case at that particular time, since both sides still thought they could win. Schwimmer had been raised in a secular Jewish family before she became an outright atheist. Her pacifism was consistent and she was never a Bolshevik. But she had proved a divisive force by pursuing a liberal or radical agenda that had nothing to do with ending the war on basic Christian principles.

Jewish soldiers and a nurse serving in the German and Austro-Hungarian Armies celebrate Chanukah in occupied Russian territory in December 1916. (Contrary to Nazi propaganda, most German Jews supported the Kaiser's war effort and 14,000 were killed in action. Two of the men in the photo have visibly received the Iron Cross, second class.)

The message of the "Peace Ship" did not go entirely to waste. In December of 1916, as the British blockade began to take serious effect and neither side could break the deadlock on the Western Front, Kaiser Wilhelm made an overture to the

Vatican asking for advice in restoring peace to Europe. The other Central Powers—Austrian-Hungary in particular—were also interested. There was no immediate response and in February of 1917, the Kaiser allowed himself to be talking into re-introducing unrestricted submarine warfare around the British Isles. The submarine warfare was the catalyst for American involvement. The actual trigger was said to have been the 'Zimmermann Telegram' in which the German Foreign Minister, Arthur Zimmermann, offered Mexico an alliance with Germany *only* in case the Americans declared war on Germany—"Make war together. Make peace together." Zimmermann urged the prostrate Mexicans to convey the same offer to their friends, the Japanese. The Japanese had been proud British allies since 1902, still digesting Shandong in China and the Pacific islands confiscated from the Japanese defeat of Germany in 1914. The Zimmermann Telegram may have made an unpopular war more palatable to the Southwest and California, where the Americans had long experienced social resentment from prostrate Mexico after the annexation of California, Texas, New Mexico and Arizona. The telegram may also have made the war acceptable on the American Pacific Coast because U.S. restriction on "cheap Oriental labor" has stirred up Japanese hostility of behalf of all Asians. The Japanese had absolutely no interest in changing sides to join a predicable loser after a quarter-century of the Kaiser's rants against "The Yellow Peril."

The key event in New York and Washington is usually overlooked. Paul Warburg of the patriotic German Jewish banking house, which had helped New York City's Jacob Schiff and the Cassell family of London bankroll the Russo-Japanese War of 1904-05, was now a loyal American and a member of the Federal Reserve Board. Warburg had warned in November of 1916 that Wall Street was dangerously over-invested with the Allies: a German victory followed by an Allied default could have a catastrophic effect on Wall Street. The fiscal correction touched off a crash—$1,000,000,000 was wiped out within a week. America's timely declaration of war saved Britain, France, and some U.S. speculators from outright bankruptcy.

This combination of indignation and investment brought the United States into the war just as the first Russian Revolution broke out. Tsarist Russia, the largest of the Allied powers, gradually ceased to be an effective brake on the German military desire to seek a final victory on the Western Front.

Pope Benedict XV now began to work actively for a peace agreement where both sides, he hoped, would be disposed to a reasonable settlement. The Germans, with America in the war, had to know that no easy victory would be possible: Britain and France, with Russia now a largely useless ally, and Italy still a military liability, also had to known that crushing the Germans on their own territory in the manner of Napoleon a century before would also be unlikely. The contact with the Kaiser was the Papal Nuncio, Eugenio Pacelli, later Pope Pius XII. After some initial tension—the subject of Belgium came up—Pacelli believed that the Kaiser was entirely serious: The Kaiser was also impressed with Pacelli and called him aristocratic and dignified in manner. Wilhelm drew up a draft of a peace resolution to be read before the Reichstag, the lower and more important house of the German parliament. The German General Staff made several small amendments and then approved it. The Kaiser gave the peace proposal to Pacelli who brought it before the Pope.

On August 1, Pope Benedict XV made his peace proposal to end the war. "The supreme spiritual charge entrusted to Us by Christ, dictates to us." The one-page proposal is written in diplomatic language, but most experts have identified seven points: he most important were that the moral force of right should be substituted for the material force of arms; all sides should reduce their inventory of armaments; international arbitration should be established; war indemnities should be renounced; and there should be an examination of rival claims. The proposal was interpreted as suggesting negotiations for an independent state of Poland and for the return of Alsace-Lorraine to France.

Kaiser Wilhelm had worked on the draft of the proposal. The Kaiser had favored an independent Poland friendly to Germany even before the war began. The Kaiser agreed immediately: "The emperor sees in this most recent step of His Holiness a new proof of the noble and humanitarian sentiments, he entertains the lively hope that... success may come to the papal appeal."

The new Austrian Emperor Karl, who succeeded the aged Franz Joseph, was even more favorable, and the Bulgarians and Turks concurred.

The French and the Italians, however, bristled with hostility. The French, led by the avowed atheist and politically anti-clerical Georges Clémenceau, asked the British to discourage any further interference by the Vatican to intervene between the belligerents. President Woodrow Wilson's reported response was to simply ask: "What does he want to butt in for?" Unofficially, German Chancellor Bethmann-Hollweg has quietly hoped that President Wilson would help negotiate a peace agreement at the end of 1916 before the United States entered the war but Wilson, who had run on the slogan "He Kept Us Out Of War," had shown no interest.

U.S. Secretary of State Robert Lansing, speaking for President Wilson, summed up the official American response in August of 1917: "No part of this program can be carried out.... The object of this war is to deliver the free peoples of this world from the menace and the actual power of a vast military power controlled by an irresponsible government which having planned secretly to dominate the world, proceeded to carry the plan out. This power is not the German people. It is the master of the German people.... We cannot take the word of the present rulers of Germany as a guaranty of anything that is to endure, unless explicitly supported by such conclusive evidence of the will and purpose of the German people themselves."

Put bluntly, the Germans and the Austrians had offered to end the war even at the probable cost of Germany losing Alsace-Lorraine and Austria losing its share of Poland to create an independent Polish buffer state. The Austrians, in fact, had already offered to give Germany part of the Austrian province of Galicia, which was mostly Ukrainian, Polish and Orthodox Jewish, if the Germans gave Alsace-Lorraine back to France as their own concession to end the war. The somewhat desperate Austrians covertly told the Germans that they would give the Italians South Tyrol, even though the Italians, fighting uphill in the Alps, had yet to win a major battle and needed constant bail-outs from Britain and France. The British had actually waffled on the offer instead of rejecting it outright—probably because they were more concerned about a revitalized Russia minus the stultifying Romanov rule than they were about Queen Victoria's favorite grandson, for all his strident posturing and his flamboyant taste in uniforms. The French and the Italians—and the United States, which had yet to face the redoubtable German Army in any significant battles—had flatly turned down a chance to an equitable peace. "It was ignored by the French and the British, who did not feel it was generous enough to them."

After Jutland, Hermann Ehrhardt's torpedo boat flotilla was re-fitted and send to guard the advanced U-boat bases in Flanders at Ostend and Zeebrugge. "Ostend and Zeebrugge were the flash-point. Over and over again the English made air attacks on the homes of the sea-wasps. Even the bravest man might have had some reflections about being stationed in these harbors." In early 1917 he was promoted to *Korvettenkapitän*—the pay grade of a major in the Army. While he was waiting to go on leave with his family, he received an invitation from the Admiralty to do some liaison work with the Army. "I took it gladly, to familiarize myself with modern land warfare."

Korvettenkapitän Ehrhardt reported to Libau, the scene of his first action in the world war, and was delighted to meet an old comrade from South West Africa, now General Ludwig von Estorff. Like many Germans who saw the Russian Front toward the end of the war, Ehrhardt felt that his side was actually winning.

"Our navy had completely surpassed the Russians and the morale of the Russian land troops was shattered, our operations were a walk-over. Among the new formations of the Russian Revolutionary Army a Female Death Battalion came out against us. Perhaps they kept up their name, because at the first artillery hits that fell behind them, they struck out with deathly fear—at one another."

But shortly after the Russian forces of Alexander Kerensky lost the will to fight, the German Western Front under pressure from Britain, France, and the fresh United States troops also became to cave in. *Fregattenkapitän* Peter Strasser, commander of the German Zeppelins now rendered vulnerable by British tracers and exploding bullets, set off on a suicide mission with the last Zeppelin bomber attack rather than live on after sending so many of his men to their deaths. Strasser and his crew were killed to the last man.

"The confrontational and legendary torpedo boat battles in the North Sea went on. But there came the sad day, when the battles on the Western Front rolled backwards and at last we had to pull back our torpedo boats and U-boats from the

Flemish coast. My flotilla was assigned to taken them up and guide them through the minefields."

Both sides, he said, had laid down such extensive mine-fields that nobody on either side knew the exact locations of all of them. His torpedo boats were so light and the drafts were so shallow that it was hard to control them at low speeds in heavy seas, let along to clear a path through the mapped German minefields and the unmapped English minefields. But a feeling of intense comradeship for the U-boat men kept Ehrhardt's surface sailors at their stations through work that was miserably hard and extremely dangerous.

"One of my boats ran into a mine and was lost. In the middle of the next morning I lost a second boat that was torpedoed by an enemy submarine in the middle of a minefield. I went alongside to help with my lead boat. In the blink of an eye somebody shouted 'Torpedo wake' and I saw the white bubbling course of a torpedo headed for our stern. While we were gazing at this the British [*submarine*] L 10 suddenly surfaced 200 meters from us. She lay in such a fortunate position that we could put a dozen shells into her. She turned up straight as a candle and sank by the head. The crew of my sinking torpedo boat shouted "Hurrah!" with full spirit. They felt themselves avenged."

British L 10 Submarine

A month after *Korvettenkapitän* Ehrhardt's sailors cheered as the L 10 sank and the minefields were behind them, Hermann Ehrhardt was hit with an emotional shock worse than any mine or torpedo. The German Navy's battleship crews in Kiel and Wilhelmshaven had mutinied, the Kaiser had fled to the Netherlands, and Germany signed an armistice.

Chapter Ten
Whelping the Bloodhound

"The most dismal days of my soldier's existence dawned," Ehrhardt wrote of the Armistice and what followed. "The order to lay down our arms extended to the fleet. At the same time the right to revolution was acknowledged, no resistance to the movement was to be attempted, and the Soldier's Councils were acknowledged.

"For the younger officers there came a bitter realization that the older officers, men who had held up like iron in the actual battle, collapsed before the phantom of a revolution. Among the younger officers a proposal was made to the leaders of the fleet that the responsible posts would only be filled with officers who were loyal to the Kaiser to maintain our cause. The proposal was rejected."

Ehrhardt had been caught as flat-footed as most of his senior officers. The torpedo boat squadron and the U-boats they covered had been based in German-occupied Flanders, near the Netherlands, farm and dairy country with a nearby seacoast for fish. Food was relatively adequate and morale was relatively high. The fact that the battleship fleet, which had not put to sea since Jutland two years before, was the incubator of the mutiny that actually deposed the Kaiser was not just surprising but embarrassing to the entire German Navy.

What had actually happened was that on October 29 and October 30, 1918, the III Battle Squadron of the High Seas Fleet had been mustered to set out for a raid to support Ehrhardt's torpedo boat squadrons and the other destroyer groups during the evacuation operations in Flanders. The battleship sailors believed that their officers wanted to force a decisive battle with the Royal Navy and they refused to sail. There was initially no bloodshed because the naval commander in chief, Admiral Franz von Hipper, called off the operation. But the sailors, now officially mutineers, demanded an end of the war. Within the next few days, both German Army and Germany Navy units were wearing red cockades, in the Russian mode, and urging that the war be ended and that Kaiser abdicate. The Social Democrats, the largest German political party, with the support of the moderate Catholic Center and the small German Communist Party, demanded that the Kaiser and his eldest son both renounce the throne. The Kaiser, who was at headquarters behind the front, first tried to take shelter with his soldiers, most of whom wanted the war to end as soon as possible. Then the Kaiser fled to the Netherlands. The kings of Saxony and Bavaria also fled. The Reichstag members of the left and the center now constituted the German government. Ehrhardt was appalled that many of his seniors accepted the change from a personal monarchy to a seemingly chaotic group of revolutionary councils. Their baffled careerism filled him and many officers his own age or younger with thinly suppressed disgust.

"I was ordered before the chief of staff, who pulled a piece of paper out of his pocket and read out: 'The Kaiser is dismissed.' *Kapitän* Taegert stepped forward immediately and said 'I request a more respectful tone. For us the retired Kaiser remains His Majesty the Kaiser of Germany.'"

Hermann Ehrhardt, a romantic royalist and feudal loyalist, can only have know that the war as not going well. Ehrhardt had probably expected a negotiated peace between the various monarchs and the American president. He had not foreseen the outright downfall of the monarchy. The man who was destined to be his functional political adviser during the *Freikorps* days had had a ringside seat to the impending collapse. Leftists would later call him "The Bloodhound," not realizing that real bloodhounds are gentle creatures which sometimes nuzzle the felons they track down rather than biting them. The man with the nickname combined the functions of a bloodhound in seeking out prey and an attack-trained Doberman. His name was Waldemar Pabst.

Ernst Julius Waldemar Pabst was born on Christmas Eve of 1880—he remained a lifelong Christian, raised as a Lutheran but subsequently a convert to Catholicism after middle age at the behest of his second wife, who, given his reputation, was understandably worried about his salvation. His father, Dr. Arthur Pabst, Ph.D., was the assistant director of the Prussian Royal Museum of Arts and Crafts in Berlin, and came from a family where one ancestor had fought the Turks and helped lift the siege of Vienna in 1683. The family had Dutch as well as German provenance. One of Pabst's ancestors became famous for winning 100,000 *Gulden* in a lottery, but later went bankrupt. Another tried his luck in America, failed, returned to Germany all but broke, and was the victim of an unsolved murder and robbery.

Waldemar's grandfather, Julius Camillo Pabst, was somewhat more respectable. He served as a medical advisor and equipment supplier during the Franco-Prussian War of 1870-71. One of Waldemar Pabst's great-uncles, however, became notorious within the family when his widow went through his portfolio after his death and found that he had promised respectable marriages to a number of rich widows and spinsters if they sent him their savings to invest for their mutual future happiness in America—though he himself never left Germany. The family gave the money back at considerable hardship to their own finances.

Waldemar's father Arthur, the eldest son of Julius, was cut out for respectability. He studied hard, and became a museum official while still in his 20s. At 27 he married Margarete Lemonius, (originally Lehmann —which means 'concubine'), who came from an old Prussian family from Brandenburg, the home state of Berlin. Her own ancestors included foresters and minor bureaucrats and small but stable land-holders—and perhaps a distant royal by a natural connection. Perhaps due to his family's somewhat colorful background, Dr. Arthur Pabst was a strict father. But Waldemar found a supporter for his antics in none other than Crown Princess Victoria, the eldest daughter of Britain's Queen Victoria and wife of the German Crown Prince Friedrich. The

Waldemar Pabst

Crown Princess was a frequent visitor to the museum with the royal children. Victoria appeared to find the palpably bright and energetic little boy and his younger brother Heinz amusing. Pabst called her *"Tante Kronprinzessin"*—"Aunt Crown Princess." Like Hermann Ehrhardt, who owed the Kaiser his career as a naval cadet, the waiver of a court-martial for poaching, and possibly the command of his first torpedo boat, Waldemar Pabst remained a staunch monarchist for the rest of his life. He flinched at nothing to protect the dynasty, the nation, and the personal respectability that they represented to him. Pabst loved the Prussian royal family as people he had known personally, and almost as family. He also needed them for his own credibility.

Pabst's father took a job offer that moved the family from Berlin to Köln (Cologne in English), one of the oldest cities in Germany, originally "Colonia Agrippina" after the wife of Germanicus Caesar, who had stayed there during his failed attempt to conquer all of the Germanic tribes in New Testament times. The city had been a center of the early German Empire of the Middle Ages, had also been a center of medieval Catholicism and had had a respected Jewish population since late Roman times. The Lutheran Waldemar Pabst took some kidding from his new Catholic and Jewish neighbors about his last name—it means "Pope."

Young Waldemar attended public school for the first three years of his education, starting at seven, and made friends from many different social strata. At 10, he was admitted to the *Apostelgymnasium*, a prestigious secondary school that concentrated on classical languages. As his mentor's mother Queen Victoria might have said, he was not amused. By the second year, Pabst had made such slim progress in Latin and Greek that his father reluctantly decided to heed Waldemar's request and send him to cadet school. In 1894, at the age of 14, Waldemar Pabst entered the cadet school at Bensberg, near Köln, to begin his training to serve the second family he loved—the Hohenzollern family.

Pabst did better in the cadet school with its pragmatic training in how to be a future officer than he had at the complicated classical languages—also the bane of young Hermann Ehrhardt. But when he was 15 he received a shock: his father died after a short illness in 1895 and his mother almost immediately married Dr. Steiner, a Jewish physician and sanitary counselor of the city of Köln. Pabst was more shocked by the lack of any deep personal grief or prolonged formal mourning than he was by his step-father's Judaism, widely accepted in Prussia. After the serious nature of Dr. Arthur Pabst, the household of Dr. Steiner and his new wife was noted as rather boisterous and Pabst never felt at home there. He never bonded with his step-father, perhaps for Freudian reasons, but on his rare visits home he enjoyed playing impartially with his birth sister Charlotte and his half-Jewish half-sister Martha. Waldemar concentrated on his studies and did well. His younger brother Heinz was made of "softer wood." Heinz Pabst dropped out of the cadet corps after a single year, built something of a career in the theatre, and spent most of his adult life in Cincinnati, Ohio. Waldemar Pabst himself never felt comfortable in Cologne. Once, during the masquerades of the notorious *Karneval* before Lent, he had the bad judgment to let somebody, probably his younger brother Heinz, talk him into disguising himself as a Japanese *geisha*. He encountered an inebriated colonel who thought he was actually a woman and wanted to do something about it. Pabst was embarrassed but escaped unscathed.

Ernst Julius Waldemar Pabst, as earnest as his given first name indicated, also encountered some embarrassment from his grandfather and second namesake, the successful pharmacist Julius Camillo Pabst. Unhappy at home, perhaps for reasons Shakespeare's Hamlet would have understood, Waldemar spent vacations and holidays with his collateral relatives, mostly in Berlin. Once, in uniform as a newly minted lieutenant at a family reunion on Berlin, he was assigned to get the 85-year-old Julius to his train back to Halle, dropped him off at the railroad station, and later returned for his own train and found his grandfather arm in arm in arm with two fashionably dressed "ladies" who were obviously upscale streetwalkers.

"Grandfather, what are you doing here?" the young lieutenant asked in amazement.

"You damnable dork!" the old man shouted. "I can stay up all night and your 'aunties' will take me home and put me on the train tomorrow morning. Now don't call me Grandfather and get out of here!"

Julius slipped his grandson Waldemar 20 marks in hush money and returned to the fashionably dressed 'aunties.' Pabst later investigated the evening's adventure. One of the 'aunties' told him that the old man sat up drinking coffee with them at a well-known café and nothing else actually happened.

The adolescent Pabst formed a counter-reaction to his family's romantic or sexual flamboyance. When he joined the 23rd Infantry Regiment in Upper Silesia as a *Fähnrich*, ensign or junior lieutenant, four days before his 19th birthday in 1899, he made it known that he was looking for a future wife and not casual encounters of the type most younger officers enjoyed as often as possible.

Shortly, he was sent to further his military education in Metz, a largely French city annexed to the German Empire after the Franco-Prussian War in 1871. His fellow officers invited him to a social event where most of the guests turned out to be ladies of rather

Waldemar and Heinz Pabst as schoolboys

dubious reputation. They told him they would not let him out of the apartment until he had "coupled" with one. Proof of manhood was part of the initiation in some regiments. Pabst, none too sober but a man of his word, slipped into a room with a girl and promptly escaped out a window instead. The designated coupler covered up for him and told everybody that he had done his duty as a young officer. Other than his escape, Pabst enjoyed French society in Metz although his new French friends requested that he change into civilian clothing before he visited them at home. With some few exceptions, the people of Metz would have preferred that their city be returned to France.

Pabst had done well in his exams and his first years of service—German officers generally served with their home regiment for their entire career, or until they made colonel. Pabst presided over courts-martial, where he tended to sympathize with the defendants, and in mobilization planning, where he showed a natural gift for detail and determination. In 1907, he achieved a distinction that led to some personal pride and some peer envy. He was posted to the *Kriegsakademie* (War Academy) in Berlin, known as the pre-school of the German General Staff. The General Staff officers acted as liaison between the regimental officers and headquarters. They had enormous prestige and authority. Like German generals, staff officers of any rank wore red tabs on their collars and red stripes on the seams to their trousers. This led to a wry joke among the less distinguished regimental officers:

"Who was the first General Staff officer?"

"Joseph, because he wore a coat of many colors and thought himself better than his brothers."

Back with the 23rd Regiment in Upper Silesia, Pabst won the approval of his division commander, the incongruously named Prussian General Franz Martin Charles de Beaulieu when he organized a voluntary fund drive to buy the regiment a new flag for a regimental reunion because the old battle flag from the days of Frederick the Great was falling apart. Pabst worked on a subscription list so that the envied General Staff captains would contribute three Marks each, captains not of the General Staff staff two Marks, and lieutenants 30 Pfennigs. Prussian pay was notoriously low —cavalry officers had to buy their own horses or bring them from home. The French had a droll expression: *"I'm working for the King of Prussia."* This meant that they were not expecting to be paid for volunteer work.

Waldemar Pabst as a lieutenant in the 23rd Infantry based in Silesia, circa 1905

When the assassinations of Archduke Franz Ferdinand and his wife Countess Sophie Chotek touched off World War I, the 12th Infantry Division from Upper Silesia was rushed across Germany by railroad and took part in the non-violent incursion through Luxemburg and scattered woodland fighting in the Belgian Ardennes. The Schlieffen Plan—such as it was—aimed at knocking France out of the war before the Tsarist Russians had a chance to complete their mobilization and invade Prussia from the east.

The shock of the first clashes was enormous, especially for the Germans, who had seen very little colonial fighting and had not fought a European war since 1871. The machine gun and bolt-action magazine rifles led to slaughter even in battles like those in the Ardennes, where the artillery could not be deployed because of the heavy forests and broken ground. After one attack, the Bavarian Lieutenant Ernst Röhm—later to figure in the Hitler movement—shouted to his men to get to their feet. "I wanted to see how many I had left. The bugler who has remained by my side like a shadow, said sadly. 'Herr Leutnant, there is nobody there anymore.' And in truth nobody is standing on the whole front line. Only three men are unscathed, everybody else is dead or wounded." It was at about this time that Hitler, later Röhm's acolyte and finally his political murderer, decided he would rather be a dispatch runner than a front-line infantryman. The German Army reportedly lost 142,503 soldiers from August to December of 1914—six times as many as their losses in the entire Franco-Prussian War—and 5,842 of these were officers, whose average age was 23 years old.

Panic followed in the wake of slaughter. During the 23rd Infantry Regiment's first real battle, at Rossignol, the horses pulling the wheeled field kitchens with their soup boilers and smokestacks rushed forward, dragging the heavy wagons behind them as they charged out of control toward the enemy lines, as Pabst wrote "with the speed of an English runner and the cry '*The Frenchmen have broken through.*'" Pabst, the second General Staff captain, helped Captain Friedrich von Miaskowski, the senior General Staff officer, to ride down the escaping field kitchens on horseback and throw the brakes before they blundered into the enemy ranks. Captain Miaskowski and Captain Pabst also helped some Uhlans—Polish-style lancers in Prussian service—to find their way when they got lost during a reconnaissance mission. The next day, when a different regiment spotted the escaped field kitchens standing unattended, they opened fire on them. Captain Miaskowski and Captain Pabst disabused them before the "Goulash Cannons" were destroyed.

Pabst also performed a humane service: His complete fluency in French enabled him to interrogate French stragglers and prevent a number of them from being shot as snipers during the panic often described as the "Rape of Belgium," where there was little if any rape, but a lot of executions of accused sharpshooters caught out of uniform. Pabst actually saved a number of Frenchmen, soldier and civilian alike, from unwarranted executions by his command of their language.

Pabst and his regiment were moving to surround Verdun, two years before the mutually catastrophic battle in 1916, when the order came to withdraw some of the German troops from France and send them to stop the Russian invasion of eastern Germany. The Germans in Russia shortly won a series of astounding victories over the Russian Army but they lost momentum in the drive to capture Paris before the end of the year, as they almost had in 1870 and did in January of 1871. The war broke down into trench warfare and the grisly attrition of the Western Front.

Within the first six months of World War I, Pabst had won the Iron Cross, Second Class, then the Iron Cross, First Class. His skill as a staff officer led him to be transferred to several other regiments on an emergency basis. He served with the V Reserve Korps, with regiments made up largely of ethnic Poles from the vicinity of Posen—and excellent soldiers for Germany either due to martial pride or because odious Russia was a major opponent. Later he served with the *VI Armeekorps*—with large numbers of Social Democrat working-men called back from reservist status from what later became known as "Red Saxony." Captain Pabst found that virtually all of them accepted their role as German soldiers. "Here I met for the first time with soldiers who, in overwhelming numbers, were represented as Social Democrats, and opening acknowledged this during their military service. They were also *Landwehrleute* (reservists) out of the Saxon industrial area. These Social Democrats were, in their will to victory and their feeling of duty to the Fatherland, at least early in the year 1915, the full equal of the young, active soldiers of the Silesian regiments, whom I had been with earlier in the war." Social Democrats seldom bought into the mandatory atheism of Marxism and the blasphemous worship of Karl Marx, but they were otherwise a Marxist party—yet they wanted no part of the Russian system. Those Poles who aspired to independence also considered the absolutist and anti-Catholic Tsarist Russians, rather than the over-organized but tolerant Prussians or the lackluster Catholic Austrians, as their real enemies. Many Poles, especially those accepted into the elite Prussian Guard, were intensely proud of their uniforms and units.

Constant nervous tension and a lack of sleep caused by chronic rheumatism saddled Pabst with a nervous breakdown in April of 1916, during the Battle of Verdun. He was ordered to take a rest cure at Wiesbaden, a center for therapy. Here he met and ultimately proposed to Helma Corneli, a young woman of French ancestry whose family had relocated to the Rhineland during the bloody days of the French Revolution:

"This lady came from and old family of the French nobility, whose family left their homeland in 1790 due to their royalist convictions to find a new homeland in the Rhineland," Pabst wrote after her early death in 1945. "Through her life before its much-too-early end, she was my brave and loyal comrade, my unusually clever and selfless adviser, my better self. She surrounded me with an unprecedented love and concern. In all the difficult situations, and my life contained a great many of them, she remained loyal to me with self-sacrifice and fought for me, and through the special grace of God she was able to save my life more than once. If ever a woman gave a man every reward for his choice of a spouse, it was my wife."

French FT-17 Silesia tank and French-equipped Polish soldier in Poland. France supported secession to equalize populations between France and Germany

Befriended and consoled by his fiancée Helma Corneli, Pabst recovered and within two months he was back on the Western Front for the vicious end-game at Verdun. He later deplored the tendency to *Kadavergehorsam*, literally "corpse obedience," which supposedly stemmed from the time of Frederick the Great: Pabst dismissed this as a folk tale nd believed in flexibility and self-motivation rather than turning men into automatons. His staff work continued with the 7th Reserve-Division—more Social Democrats, but good soldiers all the same. They later fought admirably in the counter-attacks against the French offensive of 1917—the one that led the mass mutiny in the French Army. French soldiers in a number of units proclaimed that they would defend their own trenches but would not launch another offensive without "the tanks and the Americans."

In October of 1917, the Tsarist Russians were in much worse shape than the French—some of the Tsarist Russian reinforcements sent to France had taken a prominent part in the French Army Mutiny of 1917—and Pabst was ordered to report to the VIII Reserve Corps as the senior General Staff officer. He went gladly, and began to organize the so-called *Friedensturm*, the "Peace Offensive" to split the French and British Armies before the Americans and the new French tanks became a real threat. Pabst had his doubts: the Germany troops transferred from the Russian Front "were no longer first-class (they had absorbed too much of the Bolshevik poison there) and their training had not prepared them for the tactics necessary to deal with the French and the English. Above all, even at the beginning of 1918 the overwhelming material superiority of our opponents had become obvious."

The tactic that the Germans relied on heavily was the *Stoßtruppen*, or "Shock Troopers." Captain Willy Rohr of the Guards Rifle Battalion originated special units in 1915 which were intended to slip into the enemy trenches more by stealth than mass assault and destroy resistance by combat at point-blank range. As organized in 1917 for the *Friedensturm* of early 1918, *Stoßtruppen* had to be full-grown men but under 27 years of age, above average in intelligence and strength, and capable in hand-to-hand combat. They wore steel helmets, sometimes in bold, angular camouflage patterns, and also wore comfortable shoes with puttees rather than jackboots, with knee pads on their trousers for crawling or creeping over wreckage. The Shock Troop units were backed by special light machine guns with bipods and sometimes by special light artillery.

Shock troops for the Kapp Putsch (wearing gaiters instead of boots)

A typical assault Shock Troop would begin with a short barrage—or none at all if night or fog provided cover. The *Stoßtruppen* would slip into the enemy trench at a weak point. The first man would sometimes carry the steel shield from an old-fashioned machine gun in his left hand and an entrenching tool, a short sharpened spade wielded like a sword or a battle ax, in his right hand. The target was the enemy's throat or the side of the neck for a quick and bloody kill to demoralize the other opponents. The second man would carry a Bergmann submachine gun, newly issued in 1918, or a Mauser C96 or Luger 08 automatic pistol to cover the man with the shield and the spade. The third man carried two full bags of stick hand grenades and threw them over the heads of his two lead comrades into the further part of the trench, where they blew up and killed, wounded, or stunned the enemies behind the contact point. German hand grenades could be used either for concussion or for both concussion and fragmentation by slipping a scored steel sleeve over the explosive charge in the explosives can atop the stick. When the Germans later encountered the U.S. Marines at Belleau Wood, they described the tough and experienced Marines, often veterans of clashes in China or Central America, as "a shock unit"—their highest designation.

The *Stoßtruppen* assaults during the *Friedensturm,* officially called Operation Michael, were devastating and almost split the French from the British. But the attacks were costly in using up the lives of the best fighting men the German Army had left.

Pabst was shortly assigned to transform the Guards Cavalry Division, the elite unit of the whole Imperial German Army now released from mounted service in Russia, into an infantry division for use on the Western Front. In a short-term stroke of genius that turned out to be a long-term blunder of epic proportions, the German General Staff had sealed Nikolai Lenin and his wife, Leon Trotsky, Karl Radek and some of their key followers to the Leningrad Station in a sealed box-car with ample money and told them to take Russia out of the war. Lenin speeded up the impending collapse of Alexander Kerensky's Russian Revolutionary Army, to the delight of the Germans. The Germans politicized their victory by declaring the independence of Estonia, Latvia, the Ukraine, and the formerly Russian part of Poland—leaving immense good will behind them as they turned virtually all their forces against Britain and France, and assuring themselves of an ample food supply after the next autumn harvest in eastern Europe.

The commander of the Guards Cavalry Division was General Heinrich von Hofmann, born in 1863, a cavalry general whose outlook had not changed since the afterglow of the Franco-Prussian War of 1870-71, when the sometimes armored horse cavalry had been badly used and outmoded but nobody wanted to admit this. General von Hofmann had taken part in the German part of the expedition to China in 1900 during the Boxer Rebellion. The first General Staff Officer was Captain Waldemar Pabst. *Hauptmann* Pabst described Hofmann as "a brave and chivalrous cavalry officer of the best Prussian school" and it was obvious that most of the officers and enlisted men were also of top quality, and made up what they lacked in infantry training by never having known a serious defeat in three years' fighting in Russia.

"It would have been impossible to reorganize a cavalry division into an infantry division and to train them to such military achievements as they piled up on the Western Front had it not been for the desire all the officers and enlisted men had to give the best they had for the defense of the homeland," Pabst wrote. "The non-commissioned officers in particular were fanatical in their zeal for the energetic training, even though the combat on the Western Front was entirely new to them."

The Guards Cavalry Division (Infantry) with Waldemar Pabst as the chief General Staff officer became operational on the Western Front on May 31 and supplied a number of *Stoßtruppen* units. The unit broke through several French positions during the *Friedensturm*. But the *Stoßtruppen* initiative, despite great success and the near-rout of several allied units, actually helped destroy the German Army as a whole, Concentrating the toughest and most aggressive soldiers in "shock" units left the rest of the German Army weakened and demoralized. As the Germans themselves admitted, the breakthroughs by the shock troops brought the second-string regular infantry—many of them older men or under-fed boys—into contact with Allied ration dumps where the hungry Germans looted food and then looted liquor. The offensive broke down. August 8 was what Quartermaster General Erich von Ludendorff called "the black day of the German Army."

"Disastrous news from the Somme," Admiral Georg Alexander von Müller wrote in his war diary of life at the Kaiser's court. "The French, British and Canadians have broken through our front to a depth of 12 km. The Kaiser was in very low spirits this evening. He said 'It's very strange that our men cannot get used to tanks....'"

The Empress suffered a heart attack and the Kaiser, who sincerely loved her and depended on her emotionally, was too distracted by her medical problems—she eventually recovered—to keep a close track of events. But the situation continued to worsen. The Guards Cavalry Division (Infantry) distinguished themselves as one of the most reliable units the Imperial Germans had left. On August 19, the German Army report said "...Set in a difficult position right on the Front and during 12 days of exceptionally heavy fighting, they fought with courage and daring and made repeated counter-attacks. They were surrounded on three sides and hit by constant artillery and flanking machine gun fire. They knocked out a number of tanks and broke through the enemy lines repeatedly to bring in many prisoners. Ordered back by superior officers, they established a new position and held it despite bloody losses and a greatly reduced battle strength. Heavy attacks were repulsed and lost positions were recaptured. These achievements were accomplished by the iron will and sharp leadership of the company officers and down to the youngest replacements among the enlisted men, including some who had barely been under fire before." General Hofmann received the oak leaves for his Pour Le Mérite, first awarded in 1917, and Pabst received the Knight's Cross of the Royal Prussian Order of Hohenzollern with Crowns and Swords and the Royal Saxon House Order with Swords. But the stubborn battle by the Guards Cavalry Division (Infantry) at this point was almost a last ditch stand against hopeless odds.

On September 4, Albert Ballin, the Jewish shipping magnate and the Kaiser's close friend, was able to slip through the protective screen of courtiers and explain the situation based on his contacts in shipping, banking and diplomatic around the world. Ballin urged that the Germans maintain military resistance but at the same time offer to evacuate the occupied countries, join the League of Nations then being formed, and help the Western nations defeat the growing power of the Bolsheviks in Russia. Ballin believed that the Labor parties in the Anglo-Saxon countries and in France actually had some sympathy for Germany, or at least for German workers. He believed that the British did not want to acknowledge too great a debt to the Americans for helping to win the war against Germany, and that Woodrow Wilson was a pure idealist but might become embittered due to heavy U.S. casualties and U-Boat sinkings. Ballin urged the negotiated end of the war with all possible speed but he also opposed an outright surrender. Walther Rathenau, the enigmatic German Jewish multi-millionaire and economic mainstay of Germany's economic war effort, had long advocated essentially the same policy—prepare to negotiate but avoid an outright surrender at all costs.

Waldemar Pabst as a General Staff Captain during World War I

Erich Ludendorff, the eccentric Prussian general widely seen as the brains behind the honorable but stolid Paul von Hindenburg, agreed with Ballin and Rathenau. But the speed of the Allied advances and the morale of the German troops in most regiments unnerved him. The Germans believed that they could build a line of defense behind the Meuse that could hold out indefinitely if they had the support of the German people. The hungry German people, already hit by an influenza epidemic as well as chronic food shortages, reacted to the news of negotiations with the Allies by urging that peace be concluded as quickly as possible. Ludendorff, already under the care of a psychiatrist, essentially had a nervous breakdown at the end of October and resigned as quartermaster general. When his resignation was shown on the screen of a movie theater, many soldiers in the audience cheered. Ludendorff shortly fled Germany disguised with a false beard and glasses and took shelter in Sweden.

The Kaiser was also experiencing some mental health problems: a few days after Ludendorff resigned, he confidentially told Admiral Müller and the rest of his staff that he was negotiating with the Japanese—a staunch British ally whom Wilhelm had been insulting with his "Yellow Peril" rhetoric for the past 25 years—to take trains across Siberia into Germany and attack the United States troops in France. The Japanese attack obviously never materialized. The Japanese Prime Minister Terauchi Masatake had already been approached by the British and the Americans and asked to attack the Bolsheviks in Siberia, as all three nations shortly did. Japan in 1918 was somewhat pro-American and extremely pro-British, and wanted no part of helping Germany. Members of the German cabinet began to urge Wilhelm to nominate a successor—not the lecherous Crown Prince Wilhelm, who left his dainty wife weeping at home during his numerous conquests, and who was even more hated that the Kaiser himself, but perhaps a younger son or a grandson so that the Hohenzollern dynasty could at least be saved: Wilhelm II and Crown Prince Willy were clearly beyond all dynastic hope. At the same time, the German battleship fleet, suspecting they were about to be sent on a death ride against the Royal Navy after two years of awful food, mutinied in Kiel. Some workers and some soldiers on leave and many deserters joined the Red sailors. By November 1, newspapers were predicting the Kaiser's abdication. On November 6 the new government of Prince Max von Baden, the Kaiser's cousin, sent Matthias Erzberger, a leader of the Catholic Center Party to try to negotiate for the best peace terms he could get with the Allies in the Forest of Compiègne. Erzberger was a poor choice—except perhaps as a future scapegoat. He had supported the war in 1914, but by 1917 he had become an open advocate of peace at almost any price. Sending him as a delegate seriously undermined the hope that the German people

Young Walther Rathenau in a Guards Cuirassiers uniform, A Jewish multi-millionaire, he was a strong German patriot who took the post of foreign minister expecting to be murdered. He was right.

might be able to fight a strictly defensive war as suggested by Albert Ballin, by Walther Rathenau, and by Erich von Ludendorff before his nervous breakdown.

The Germans had hoped for something like Wilson's 14 Points but the terms backed by France were much tougher. On November 8, the Social Democrats, the largest party in the Reichstag in Berlin, said that they would leave the government if the Kaiser did not abdicate by noon and take the Crown Price with him. Ship-to-ship fighting was expected in Kiel.

That night, Captain Waldemar Pabst was sent to headquarters for a desperate meeting with 39 officers from different German divisions at or near the Front. He described the results.

"After a pessimistic briefing from Hindenburg over the situation of the last few days, the General Quartermaster, Colonel Heye, asked the assembled gentlemen: '1. How do the troops stand for the Kaiser? Would it be possible that the Kaiser at the head of his troops could reconquer the homeland?' One officer said yes, 15 said it was doubtful, and 23 answered no. The second question was 'How do the troops stand with regard to Bolshevism? Would they take up weapons against Bolshevism in their own homeland?' Eight said the troops would not fight, 12 wanted more time to see what happened, and 19 called the situation doubtful." The old generals and colonels were furious but the situation spoke for itself. The next day the Kaiser abdicated and showed up in the Netherlands asking for a good strong cup of real English tea.

The briefing about Allied intentions and Germany's options was Ballin's last meeting with the Kaiser. On November 9, when the Kaiser fell from power and a hostile armistice was obvious, Albert Ballin was roughed up by young Marxist thugs. He reportedly committed suicide with an overdose of barbiturates. Walther Rathenau, however, remained active in trying to hold together whatever was left of the Germany he had grown up and prospered in. And so, in a very different way, did Captain Waldemar Pabst.

Chapter Eleven

Red Flags Over Germany

With the Kaiser departed for the Netherlands and Ludendorff hiding in Sweden, the Social Democrats, the majority party in the Reichstag, quickly reached an agreement with the German Army, still commanded by the oaken Paul von Hindenburg, an officer veteran of the Franco-Prussian War of 1870-1870 and a man who boasted that he had never read a book except for military history. Friedrich Ebert, the new chancellor, was a working-class liberal who had lost two sons in the war. He was not a fanatical Marxist. "I hate revolution like sin," he famously said. Ebert told the German officer corps that they could retain their ranks and authority as long as they supported the government. But officers who returned to Germany alone or in small groups found themselves reviled by leftists and plagued by delinquents who snuck up behind them and ripped off their epaulets.

A month after the Armistice of November 11, 1918, Ebert would offer the troops a sort of victory parade in Berlin, even though there had been no victory. He told the officers and soldiers "You have not been beaten on the battlefield"—a half-truth often attributed to the monarchists. While German territory had been spared an actual Allied invasion, the Kaiser fled only after most of his troops refused to fight for him and some naval units staged armed mutinies.

But before that parade, *Korvettenkapitän* Hermann Ehrhardt received the telephone call that brought the reality of defeat home to him as the last battles never had.

"Lowering the flag without a fight seemed to be not only a crime against the flag, but a negation of one's natural manhood," Ehrhardt wrote. "But the changes in the fleet were such that people who dreamed of a final battle on Monday took one look at the drunken Red horde on Tuesday and were convinced that the reality was different than the dream.

"Then suddenly the demand came from the Entente: the German Fleet will be handed over. This was a hard blow for the ship's crews, who had believed the phrases of the Red Brothers and believed that the world revolution was on the march... It evoked defiance in at least some of the sailors, for whom the ship was like a mother."

Captain Ehrhardt was ordered to take the 17th Torpedo Boat Flotilla to England. He said that the demand that his undefeated ships be handed over caused the sailors themselves to protest. But had had his orders from Gustav Noske, the new Social Democratic Defense Minister, to hand the ships over or the enemy would occupy Hamburg, Kiel, and Wilhelmshaven and the peace treaty that followed the November 11 armistice would be worse than otherwise.

"At that point I was not yet a political officer," Ehrhardt said. "If Noske grabbed me by my sword-knot, an order was an order and had to be carried out. The Imperial officer's education and the Prussian obedience had settled deep into my bones."

The men were depressed, Ehrhardt said, because they began to understand that this was a voyage into imprisonment. The weather was grey and dismal. When the English escorts joined the German ships, an English blinker signal warned: "Any German ship that shows a red flag will immediately be sent to the bottom." Ehrhardt noted that the English were practical. The revolution in Germany that ended the war without a final show-down with the German fleet was a good thing for the Royal Navy. But the English wanted no part of Communist propaganda or of Communism itself.

The High Seas Fleet clustered in Scapa Flow in the Orkneys, north of Scotland. The Germans handed over 74 surface warships and 174 U-Boats during the last days of November 1918.

"The bay of Scapa Flow took us in," Ehrhardt wrote. "The coast was comfortless, the land rocky. We took our orders from the English fleet that controlled maritime traffic. Foodstuffs were scarce and rationed, no alcohol, no tobacco, the men slept a lot and the ships were neglected."

The end of the war had ended the U-Boat blockade of Britain, but the Royal Navy continued the food blockade of Germany, and ultimately an estimated 100,000 German civilians or wounded soldiers would die of malnutrition between the Armistice and the Treaty of Versailles. The food for the German fleet came from Germany, where food shortages were rampant. The sailors caught seagulls and fish to supplement their sparse rations. Most newspapers that arrived from home were two weeks old.

"What was happening in Germany?" Ehrhardt asked. "The question tortured me. It made me sick, but I saw nothing I could get a hold on. We couldn't get newspapers. We were condemned to stultification."

General Staff Captain Waldemar Pabst received his initiation to the new German government as the Guards Cavalry Division (Infantry) was marching back from the Western Front to the German border through neutral Luxemburg. He received signed orders that the regiments should be allowed, even encouraged, to form "soldier's councils" which could out-vote the regimental officers. Pabst called together a council of senior non-commissioned officers, some of them 20-year men, whose main question was whether their long service would still count toward pensions and civil-service jobs with the post office or local police force, as it had in the Kaiser's government. The senior non-coms also agreed with Pabst that maintaining law and order in Germany was more important than soldiers from peasant or skilled-trades backgrounds playing at national politics.

When Headquarters asked Pabst to confer with the soldier's councils in his regiments, he replied: "*WE* have no soldier's councils. I took the order to form soldier's councils and threw it in the wastebasket."

Shortly, the Guards Cavalry Division (Infantry) was ordered to take the trains from the Rhineland near the French border to Berlin. The orders struck Pabst as strange and suspicious. The regiments were to unload at Berlin-Zehlendorf and at Wannsee, with the Pioneer Company (*combat engineers*) in the first element, then the combat companies, then the machine gun squadron, and finally the division staff. It sounded to Pabst as if the authorities wanted to separate the division's staff from the combat units in case of trouble.

"Listen, you've sent me a transport order that I can't use to travel to Berlin," Pabst said on the telephone. "What shall we do if, as we have heard, that there's an actual revolution going on in Berlin... Whereupon the officer who gave the transport order—he was a lieutenant colonel—tried to shit on me. But I said, shouldn't we just have a nice friendly conversation? This is *NOT* a transport order to Berlin."

Pabst had known Berlin quite well, both as a youngster and as a young officer. "The first unloading station in Berlin is the Wildhof Station, and after that Gentin. I'm going to make sure that there isn't some propaganda stunt at the stations (*mentioned in the first order*) where guys can come to the stations by trolley or bus to make fun of the Division. If you don't send me to the stations that can't be used for propaganda opportunities—we're not going." The lieutenant colonel scolded Captain Pabst for his rudeness but assigned the Guards Cavalry Division to Wildhof and Gentin.

Captain Pabst rearranged the order of the transport. The first coaches carried the Division staff. Pabst also rounded up some of the Guards Cuirassiers, the heaviest of heavy cavalry, muscular six-footers who corresponded to the grenadier companies in old-fashioned infantry regiments. Each man had been issued a rubber truncheon, a sort of over-sized blackjack, or some other bludgeon that could be used to knock people silly without deliberately killing them, along with his service weapons. The Guards Cuirassiers served as the bodyguards for the Division staff.

On November 26, a grey, foggy morning with scattered rain, Pabst and General Hofmann got off the train and encountered pretty much what Pabst had anticipated.

"We go off. And as we got off I saw a horde standing at the station, a horde in grey coats, red armbands, red cockades, rifles dangling with their muzzles in the mud, and a bunch of civilians. A civilian from among them stepped forward and said to my commanding officer, General von Hoffman, a splendid man who was sick at heart and held himself back from everything: 'You, come on over here!'

"His Excellency was naturally unaccustomed to this sort of friendly greeting, whether in war or in peace, and he looked to me for help. I said '*Excellenz*, stay where you are.'

"He stood where he was. The gentleman in civilian clothes, with two other civilians, and the threat of their comical-looking soldiers, repeated: 'I said you should come on over here!'

"Then I said, '*Excellenz*, may I speak to this man in Berlin-ish?' He gave his agreement.

"I said 'Who actually are you?'

"'I am People's Commissioner Barth.'

"'Ah so,' said I, '...and what is a People's Commissioner?'

"'You don't know that?'

"'Where should I have known that from? I never heard the silly word before.'

"'I am their representative.'

"'You, and my representative? Have you ever been a soldier?'

"'That has nothing to do with you,'

"'It damn well has plenty to do with *me* if you were ever a soldier. In any case, I don't recognize you as my representative. What do you actually want here?'

"'We want to greet the Division.'

"'Ah so,' said I. 'And who has authorized this?'

"'I myself as People's Commissioner!'

"'That's beyond my conception. What sort of guys are these guys here?'

"'These are the representatives of the Soldier's Council of Greater Berlin.'

"'Ah,' said I, 'Very sympathetic-looking guys you've brought me meet me here—who's that one there?'

"'That's the representative of the Deserters Council of Greater Berlin,'

"That split my paper collar. I said: 'If you swine, who presumed to come here and stand around the station when a respectable combat division arrives, haven't disappeared from the station within five minutes, you'll be knocked silly and taken into custody. What happens next is for you to find out!'"

Pabst noted that it took a lot less than five minutes for the delegates and Soldiers' and Deserters' delegates to leave the station platform.

Pabst and the other officers found that they had been quartered west of Berlin in Nikolassee. Pabst had just become functional commander of the Guards Cavalry Division (Infantry) due to the shattered physical and emotional health of General von Hoffman—whom Pabst continued to treat with great respect and almost filial affection. Captain Pabst's first official act was to have the divisional bivouac surrounded with barbed wire entanglements—against his own official government in Greater Berlin. Pabst, a General Staff captain, was now the functional commander of the whole Guards Cavalry Division (Infantry) due to a vacuum of authority elsewhere. He was what the poet Friedrich Schiller a century before had called *"ein mutiger springer"*—a bold usurper.

"Hardly were we in Nikolassee when People's Commissioner Barth appeared again, seated in an expensive automobile that he had probably stolen from somewhere," Pabst wrote. "Now he told me that he had heard we had built our entanglements because we were hostile toward Berlin."

"Toward Berlin?" Pabst said. "No, against your brothers, because we don't trust what they may have in store for us."

Barth replied that it was not possible to maintain a state within a state.

"The state within a state must remain, until we know where we stand with you," Pabst replied.

Barth repeated that Pabst had to obey his orders from the high command. Pabst said he would consult with headquarters. He fixed Barth with a glare of utter contempt. Barth left.

The next day Pabst received an order to report to the headquarters of the high command, now located in Cassel, and to report to General Heinrich Schëuch. Pabst put on his full dress uniform with his spiked helmet, both Iron Crosses and the Hohenzollern and Saxon house orders. He kept the endangered Imperial epaulets in place. He took along his quartermaster officer, his ordinance officer, two staff officers and some of the husky cuirassiers with the rubber truncheons.

"We drove through the suburbs. Nobody actually bothered us… Then we came to the *Leipzigstraße*. There stood a double guard, red armbands, red cockades, rifles upside down. We got out. The guard post failed to salute us. Instead of having them shot like dogs I had two old non-coms beat them into submission with the rubber truncheons and they very quickly gave us a proper salute. We told them they should keep their snouts shut about it or we'd taken them away with us. Then we went in and made out report to the Minister of Defense."

Pabst and his officers encountered two scared-looking bureaucrats in civilian clothing—General Schëuch and Paul Göhre, a Social Democrat politician.

"How did you get through from Berlin with epaulets on your uniforms?" General Schëuch asked incredulously.

"Of course we came through with epaulets on our uniforms," Pabst retorted. "I'd like to have seen anyone try to stop us."

General Schëuch asked Captain Pabst for his 'Ordre de Bataille' the French expression for a table of organization. General Schëuch asked a few more questions and then asked about the mood of the troops.

"We're waiting for the moment when we can break loose," Pabst said. "A pigsty like we found when we got back to Berlin can never happen again."

Pabst told Schëuch that the revolution might have started with good economic intentions but it had broken down into a mess. "There's no more discipline.... We've seen that unhealthy elements in society have put on military uniforms to exercise their demand for mass socialism, as they imagine it to be."

"Had we been here with the Guards Cavalry Division and three or four other division in or near Berlin, with the same spirit that we have now, there would have been no revolution. No command from higher authority could have held us back from using our weapons on these mutineers if we had a clear order. When I look at the troops of these revolutionary soldiers, I don't have the impression that that these men would have been ready to die on November 9.... I ask—for what ideal would the revolutionaries have been ready to die? I have yet to discover any revolutionary concept of the world among them. They were driven only by war-weariness and hunger, and they were very different from the French revolutionaries of 1789.... Short and sweet—we're waiting for the moment when we can break loose against them."

General Schëuch looked frightened. He gestured to Pabst not to speak further and motioned him to continue their conversation behind closed doors.

"We can't have that sort of conversation in my office right now, the gentleman in my office is the representative of the Soldiers Council of the War Ministry, and believe you me, what you've said already is enough to raise a fearsome rumpus," General Schëuch said.

"Excellenz, it's all sow sausage to me whether there's a rumpus or not," Pabst said bluntly. "We're really here, I stand at any time at your disposal, and I have set the guards on duty at our position."

"You've done that, you're set your guards on duty?" General Schëuch asked. He seemed surprised that there were still loyal troops within Greater Berlin.

"Yes, as long as we're here, it's all secure," Pabst said.

"Make sure you get out of here safely," General Schëuch said.

"Does the Representative of the Soldiers Council also get to see the 'Ordre de Bataille' and does he also know about the accommodation of the Division?" Pabst asked.

"Yes, he'll soon request them."

"Had I known that, I wouldn't have given them to you," Pabst said. "But in other respects, it's all sausage anyway."

When Pabst and his staff officers and bodyguards got back to the Guards Cavalry Division (Infantry) in Nikolassee, he organized what he called a "press section" to indoctrinate the enlisted men about what he saw coming. The Guards Cavalry Division officers were not Marxist material. German cavalry officers had to buy or bring their own horses, and most of the Guards officers were of the nobility, the aristocracy, or the moneyed upper middle classes with university training, professional skills, and anticipated inheritances. These men, usually brave and often snobbish, were reliable monarchists or supporters of a constitutional government. The enlisted men, Pabst understood, needed a different perspective. He worked out a regular speech based on his own respect for the fine fighting qualities of the French soldiers he had observed and the increasing concept that men who had served at the Front, whatever their nationalities, had more in common with one another than they did with slackers or trouble-makers who stayed home from the war. The official speech:

"You've served in the field for more than four years against the English and the French, and you've rejoiced when you put a bullet through the skull of Corporal Jean-Jacques Meunier of the First French Infantry Regiment. This man hadn't done anything to you, this man was defending his country the same as you were defending yours. Now under different circumstances you may have to shoot people out of hand who speak your own language, just as you had to shoot the brave defenders of their own country, because these are the people who are dragging our country deeper and deeper into Bolshevism."

Pabst said that the enlisted men grasped that simple explanation and that, starting even before the end of 1918, the soldiers of the Guards Cavalry Division were entirely ready to shoot other men in *Feldgrau*, as long as the targeted enemies were wearing red cockades.

Chapter Twelve

Murder of the Mad Lovers

Pabst didn't have long to wait. In late December of 1919, the German government of President Friedrich Ebert—six Social Democrats, six leftist Independent Social Democrats, and 12 soldiers who were mostly Social Democrats—attempted to subdue 3,000 Red sailors from Kiel who had come to Berlin and had demanded, and not received, a bonus so they could celebrate Christmas, a rather odd interpretation of Marxist atheism. The army, under hazy authority, turned out after the Red sailors occupied the Chancellery. The Red sailors then occupied the empty palace of the former Kaiser. The army attacked the palace, blew some holes in the wall with artillery, and killed or wounded about 60 Red sailors. Crowds turned out in support of the sailors and the army officers in charge reportedly had a narrow escape from lynching.

On December 27, 1918, the USDP wing of the German Social Democratic Party was headed by members known as the Spartacus League, an illegal group dedicated to violent overthrow of the government. The Spartacists took their name from the rebel gladiator of Roman times, described by Plutarch and extolled and misunderstood by Karl Marx. With some other left-wing members of the USPD, they rolled out the German Communist Party. The primary leader was Karl Liebknecht, a second-generation Socialist who had migrated to Berlin from Saxony, where the Wettin Dynasty had dragged its feet in promoting the sort of economic reforms that had long existed in Prussia. Liebknecht was not Jewish, though his name, his pince-nez spectacles and his curly black hair convinced some people otherwise. He was of mixed Germanic and Slavic ancestry. He had won war-time fame by being the only *Reichstag* member to vote against war credits in 1914, and later spent years in prison as a war-time pacifist.

Most of the other Spartacist leaders were apostate Jewish Marxists who had grown up inside the Russian Empire, like Rosa Luxemburg and Eugen and Rosa Leviné, or inside the Austro-Hungarian Empire like Karl Radek—born Karol Sobelsohn in Lemberg in Austrian Galicia, later known as Lwów. These people spent a great deal of time applauding one another's genius. Many of them had grown up overprotected by doting parents in countries where violent anti-Semitism was only too common: as a result, they seem to have turned into the sort of infantile egotists who could never bear to hear praise or adulation for anyone but themselves. Outside opinions—even by the voters—were disregarded as irrelevant. A number of ethnic Germans or ethnic Poles also joined up, most of them people who had never been able to function economically even in the most socialistic monarchy in Europe. By 1911 German legislation begun under Otto von Bismarck—accident insurance in 1881, sickness insurance in 1883, and old age and disability insurance in 1887, later expanded under Kaiser Wilhelm— had set the world's highest standard for worker protection, far beyond anything

Rosa Luxemburg

in Britain, France, Italy, or the United States, not to mention Russia. Protestant, Catholic, and Jewish charities took up most of the slack. Nobody starved in the Kaiser's Germany until the British Starvation Blockade and the black market for food reinforced the difference between the rich and the poor: the rich could pay the absurdly inflated prices for meat and fruit and real wheat or rye bread: the poor could not pay and ate rationed turnip-and-sawdust bread and sausages made of water and fat. The British continued the blockade to make sure that the resentful Germans signed the Treaty of Versailles, which was much tougher than anything the Germans had expected. This tactic inadvertently increased hostility between the urban poor and unskilled workers of Germany and those Germans with family farms or house gardens, investment properties, and savings accounts. Neither the

British nor the Americans wanted to see Germany taken over by Bolsheviks. They simply wanted Germany to be neutralized militarily and perhaps industrially and to pay the full cost of a war that even the Socialists and Spartacists admitted was mutual fault: Austria-Hungary and Serbia were seen as the primary culprits and Britain, France, Tsarist Russia and Germany all about equally to blame in the second rank of culprits.

Mainstream German Social Democrats like Friedrich Ebert were appalled by the defection of part of the USPD and the Spartacus League—now the German Communist Party—from the mainstream Social Democrats. The Ebert faction was also appalled to see that the Reds had targeted their own mainstream Social Democrats—essentially a Marxist working-man's party which advocated free speech and tolerated religion and small-time private property—as right-wing oppressors.

"The shameless doings of Karl Liebknecht and Rosa Luxemburg besmirch the revolution and endanger all its achievements," the Social Democrat Ebert wrote. "The masses cannot afford to wait a minute longer and quietly look on while these brutes and their hangers-on cripple the activity of the republican authorities, incite the people deeper and deeper into a civil war, and strangle the rights of free speech with their dirty hands. With lies, slander and violence they want to tear down everything that dares to stand in their way. With an insolence exceeding all bounds they act as though they were masters of Berlin."

The new German Communist Party was largely composed of people who were not born as German citizens. Rosa Luxemburg, the brightest and least violent of their leaders, was born in Russian Poland, the fifth child of a middling-successful Jewish family who lived in Warsaw. Brilliant but tiny and crippled, with curvature of the spine and one leg shorter than the other, she graduated at the top of her class in high school and joined a left-wing workers' group. She fled Russian Poland to avoid detention after the group organized a strike that left four of its leaders executed by Tsarist authorities. Rosa Luxemburg completed her education in Zurich in Switzerland with a doctorate in law. When she returned to Poland, she alienated the Polish independence movement by advocating the brotherhood of a World Socialist Revolution, which would have entangled the suspicious Poles with Germany, a country many Poles disliked, and Russia, a county most Poles detested. August Bebel, a German-born ethnic Pole and foe of Kaiser Wilhelm, and the leader of the Social Democrats in the Kaiser's time until his death by a stroke in 1913, had not appreciated their views, and favored parliamentary means toward social justice. To Babel, even the Kaiser was far better than the Tsar. The vast majority of German-born Jews agreed with him.

Rosa Luxemburg became a German citizen through an arranged marriage of convenience within Socialist circles to Gustav Lübeck, a fellow Socialist. She moved to Berlin but maintained constant contact with Leo Jogiches, her rapturous long-term lover, though her correspondence suggests she had serious affairs with four other men as well, and perhaps even with her lawful husband. Just what any of them did is anybody's guess because Rosa Luxemburg was so fastidious that she wet her postage stamps with damp paper instead of licking them with her tongue.

Luxemburg had her roots in the rabbinate, as did Rosa Leviné, born Rosa Broida, the twelfth daughter of an impoverished rabbi in Białystok. Rosa Broida, a beautiful girl who became an attractive woman, dropped out of formal Judaism in favor of revolution. After a forcibly chaste upbringing, Rose Broida achieved mandatory sexual liberation at 19 as a revolutionary: "I met [Moissaye] Olgin in Vilno. His beautiful mistress had left him and I was to take her place. I did not love him but he introduced me to a new world, to the socialist elite of the town, which I had always admired from afar. He brought into my life a sense of security and stability." Not to mention sexual initiation and status climbing. Bluntly, a poor country girl who, as a rabbi's daughter, had been expected to remain chaste before her marriage, if possible to another rabbi, let party organizers pass her around like a door prize to energetic young Marxists with leadership potential. Moissaye Olgin, born the Ukraine, was a specialist at translating revolutionary documents from Russian, German and English into Yiddish, which was a Germanic language written with Hebrew letters.

In Heidelberg, while studying at the most famous university in Germany after having fled Austria for revolutionary activities, she met Eugen Leviné-Nissen, the son of a rich Russian Jewish family who had emigrated to Germany in 1905 to escape violent persecution both as a Jew and as a leftist. "For me it was love at first sight," Rosa said. They married in 1915—"we have reached a degree of happiness which will never be surpassed," Rosa wrote. They had a child together whom they both doted on and Leviné was shortly drafted into the Kaiser's army as an interpreter for the German military police guarding the half-starved Tsarist Russian prisoners held behind barbed wire. His primary job was to help investigate food thefts, brawls, and homosexual rape. Eugen Leviné never got near the Western Front, but two years of this contact with criminality behind barbed wire sent him home as a German Army veteran with an honorable discharge minus any medals, a nervous breakdown, and a deep-dyed disbelief in the ability of the "masses" to run their own lives. Leviné's one refuge from ugly reality was his Rosa, the wife he utterly adored —but his Rosa, as the daughter of a rabbi, had transgressed beyond return to Orthodox Judaism, and neither of them wanted to become a Christian. They saw Judaism as repressive and judgmental, and having growing up around sometimes violent Russian persecution of Jews, they saw Christianity as the hostile religion of snobbish, spoiled aristocrats, fanatic monks who looked like Rasputin, and drunken, dangerous peasants. Their new absolute faith was the World Revolution.

On January 1, President Ebert appointed the tough-minded Social Democrat Gustav Noske as war minister. Noske had actually once been a workingman—a basket weaver in a baby carriage factory—before he discovered politics, and he had supported the war because he believed that Germany was being punished for her progressive economy. On January 4, the Social Democratic President Ebert dismissed Emil Eichhorn, Berlin's chief of police and a member of the USPD.

"A member of the Independent Socialist Party [*USDP*] and a close friend of the late August Bebel, he enjoyed great popularity among revolutionary workers of all shades for his personal integrity and genuine devotion to the working class," Rosa Leviné wrote. "His position was regarded as a bulwark against counter-revolutionary conspiracy and was a thorn in the flesh of reactionary forces."

Rosa Broida Levine and Eugen Leviné with their son, 1916

Emil Eichhorn, a Saxon émigré like Karl Liebknecht, had been appointed police chief only after he himself had led the occupation of police headquarters on 9 November, 1918, during the flight of the Kaiser and the SPD take-over. He had disarmed the soldiers on guard and released 600 leftist political prisoners. A glassmaker by trade, Eichhorn had no prior law enforcement experience. But Police Chief Eichhorn refused to vacate police headquarters and declined to accept the decree of the Social Democratic government. President Ebert ordered General Arnold Lequis, a veteran of China, Southwest Africa, and the Western Front, to recapture the police station. But leaflets were distributed calling for a mass demonstration of armed workers to show their support: a huge demonstration of 200,000 workers (leftist sources say 700,000) marched through Berlin. Some of the workers showed up and rallied around Eichhorn at Police Headquarters. When some trainees from the Potsdam Regiment of the German Army were sent to evict him, they failed and were hooted out of the neighborhood.

The Spartacists—now the German Communist Party—saw their opportunity. With the financial encouragement and weapons from the new Soviet Russian embassy, they formed a secret "insurrection committee." Karl Liebknecht and Wilhelm Pieck voted for an immediate revolution. Rosa Luxemburg, who doted on spring flowers and birds and had some serious reservations about Lenin's ruthlessness, voted against the German revolution—partly out of natural kindness, but mostly because she doubted the Spartacists would be able to defeat what was left of the German Army. Luxemburg's ever-reliable lover, Leo Jogiches, and Karl Radek, who actually understood something about Germany outside Berlin, also voted in opposition. They lost the vote. On the morning of January 5 the Spartacists occupied what they thought were strategic sites around the city: none of the Spartacists appeared to have had any military experience whatsoever. Heinrich Noske called out the Germany Army and the new *Freikorps* units to restore order.

Karl Liebknecht

When the Armistice ended the war on November 11, 1918, some German units virtually melted away. The married men, the kids, the land-owning farmers, the licensed professionals and the reluctant conscripts left for home, family, and jobs. Many if not most of the men who stayed in uniform were there for the food, the warm beds, and the clothing issues. They were what Americans call "Army bums." Anti-Communist officers, on their own initiative, began to sort out the shirkers and to form units of soldiers who were ready to fight the leftists and—if necessary—the Russian Bolsheviks. Waldemar Pabst had essentially turned the Guards Cavalry Division, minus those men who quietly slipped away, into a sort of pioneer *Freikorps*, though Captain Pabst and the division remained part of the regular army, now known as the *Reichswehr*, or "realm's defense." His speeches and indoctrination on the need to shoot German traitors rather than French patriots proved to be prescient.

The first government assault took place on January 12. War Minister Noske ordered the Potsdam Regiment to re-capture the Socialist *Vorwärts* newspaper office from the Reds. The regiment's commander, Major Friedrich von Stephani, reportedly slipped into the *Vorwärts* building and discovered low morale and an obvious lack of military skill. His regiment—about 300 soldiers recruited from Guards regiments and sergeants on training assignments in Berlin—used field artillery to besiege the newspaper building. When a Red machine gun fired from inside the building, the artillery quickly knocked it out. The shock troops stormed the *Vorwärts* building room by room, virtually without opposition, as the terrified defenders threw down their rifles and put up their hands. Some Reds were no doubt shot at point-blank range or slashed down with sharpened entrenching tools and knives by keyed-up shock troops during the heart-thumping assault. War Minister Noske authorized a communique intended to inspire terror which reported all 300 Red defenders had been executed right outside the newspaper officer. The Reds picked right up on it to inspire frantic resistance. The figure inspired frantic terror instead, just as Noske had hoped. But the legend of 300 dead Reds shot on the spot began a long series of tales in which atrocities by the Right—the right wing of the Social Democrats, in this case—were grotesquely inflated: the legend that the Right routinely murdered Red prisoners became a staple of Red propaganda.

Writing in the surviving Leftist newspaper, *Rote Fahne*—"Red Banner"—Rosa Luxemburg shrieked her defiance: "Order rules in Berlin! You stupid hangmen! Tomorrow the Revolution will drag you out in chains and muffle your screams at your own executions with kettle drums that proclaim: 'I was, I am, I will be.'"

The Social Democrats targeted January 15 for the conquest of Red Berlin, such as it was, by the *Reichswehr* national army and the *Freikorps* groups. The invasion was led by two captured British tanks taken by the Kaiser's army in Flanders, a Mark IV with two six-pounder guns in side sponsors, and a high-speed Whippet tank powered by two London bus engines and topped by a steel siege tower bristling with machine guns instead of a revolving turret. Pabst and his division arrived on January 13 as two other *Freikorps* units and some half-trained regulars finally disarmed the dismayed Red sailors of the *III Volksmarinedivision*, with eight military dead among the regulars and unknown losses among the Red sailors, who were not well-treated by the dead men's friends. The political leaders of the revolution put in a disappearance rather that fighting to the death. After the military flop, the Reds needed martyrs to inspire future uprisings. They claimed more than 1,000 heroic fighters, possibly 1,200 including woman machine gunners, murdered by brutal reactionaries. The official figures from hospitals and cemeteries suggest about 156 people killed on both sides, though the Reds got the worst of it by far. A few Reds who had been seen firing on regular troops were marched into back alleys and shot in the head. When the word got out, resistance vaporized. The Whippet tank was parked outside the Hotel Eden on the corner of Budapest Street and the Kurfürstendamm just in case.

Pabst and many other Germans who read French found their own revolutionaries something of a disgrace compared to the Paris Commune of 1871. The Germans on both sides took comfort from the fact that most of the Red leaders had not been born in Prussia. The Reds from Saxony and Russia had misunderstood Prussia and Prussianized Germany completely: two hundred years of religious tolerance under the Prussian kings, trained in a Calvinist outlook by French Huguenot tutors, and five decades of complete civil rights and vast financial opportunities under the German Empire had fostered a Prussian Jewish subculture that was overwhelming loyal and in many cases extremely patriotic. The new German Republic's constitution had, In fact, been cobbled together from elements of the old German Empire's constitution by Hugo Preuß, a Jewish liberal—he added voting rights for women—and a lawyer and teacher of law. The name "Preuß" means "*Prussian*" and Hugo Preuß more than lived up to it. Hugo Preuß, born in 1860 to a commercial family in Berlin, was raised as an observant Jew. When his father Louis died, his uncle Leopold married the brother's widow and raised Hugo as his own son, as required by a literal interpretation of The Old Testament. Hugo Preuß studied law at the University of Berlin and at Heidelberg and graduated with a doctor of law degree from Göttingen in 1883.

Hugo Preuß had four sons: Dr. Ernst Gustav Preuß served at the front, won the Iron Cross, First Class, and was promoted to serve as an officer; Dr. Kurt Preuß served at the front and won the Iron Cross Second Class; Gerhard Preuß served at the front and was severely wounded and crippled for life. He later died of complications of his wounds and influenza in 1921. Only the fourth son, Hans Helmuth Preuß, missed front-line service because he was under-age. Patriotic Jewish liberals like the Preuß family wanted Germany to survive as a slightly more liberal republic, not as a Communist satellite of hated Russia. The Preuß family was far more typical of the German Jewish population than the would-be revolutionaries who declared war on the Socialist Democrats they had once supported as soon as they saw an opening. Some German Jews fought in the *Freikorps* or alongside the *Freikorps* among the *Reichswehr* troops against the communists. Eric Warburg, all of 18 years old and the scion of an eminent Jewish international banking house, was still in military training with the Potsdam Regiment when his unit was sent in to attack the Red sailors at the Palace. The company lost eight men. "God grant that we may yet have the monarchical form of government, which is the only correct one for Germany," he later wrote. While some patriotic German Jews fought the communists, others rendered medical aid to the nationalists or used their financial influence to help Germany: Eric's father Max argued for more lenient peace terms after the Armistice as an official representative of the German Republic to France.

Still others simply kept out of the line of fire—but these too sometimes proved useful.

Captain Waldemar Pabst set up division headquarters at two large rooms of the Hotel Eden, the small hall and the small salon. He had already decided what would have to be done with Karl Liebknecht and Rosa Luxemburg.

"I had heard both of them speak," he wrote many years afterward. "Twice I had slipped into their assemblies in civilian clothes and I can only say that Adolf Hitler couldn't speak nearly as well as Rosa Luxemburg could." Pabst remembered that a regimental commander from a Westphalian Cuirassier regiment, a count, a devout Catholic and a political reactionary, had come to headquarters and described to Pabst how hearing Karl Liebknecht speak at a rally had changed his whole view of the future—from now on things would all have to be different. As soon as the beguiled count had left his office, Staff Captain Pabst telephoned divisional head-quarters and, using the power he had cheerfully usurped from Major General Heinrich von Hofmann starting at the railroad station, arranged to have the count relieved and sent to the *Ersatz-Eskadron* [cavalry replacement unit] and replaced by the senior *Rittmeister* [cavalry major] in the regiment. When the colonel reappeared, Pabst explained to him that he had brought this on himself—and the colonel agreed with him.

"It would have been a form of *Harakiri* to have allowed the leaders of the opposition to speak out and to intoxicate any part of our men. In any case I didn't have the sort of esteem for Liebknecht's oratory or his personal courage to see him as a Messiah, but rather saw him as an unthinking and unscrupulous politician.... They must go! When we get our hands on them, we have to act without flinching and take the law into our own hands."

Pabst was not alone. Leaflets published in Berlin since December 1918 had urged that the Spartacist leaders be killed. The rumor went out that Liebknecht and Luxemburg had escaped from Berlin, but Pabst didn't believe it. He tried to trace them through the telephone and the mail. An official order went out from the Social Democratic government under President Friedrich Ebert charging Liebknecht and Luxemburg with high treason. Philipp Scheidemann, a Social Democratic official who had been accused of triggering the premature German surrender, and Georg Sklarz, whom Pabst bluntly described as a crooked military contractor and war profiteer, put up "head money" of 50,000 marks for their arrest.

Pabst was delighted when, on night of January 15, Karl Liebknecht was ushered into his office, soon followed by Rosa Luxemburg and Wilhelm Pieck. All three of them had been staying in a dwelling owned by *Frau* Marcussohn in Berlin —Wilmersdorf and had been captured. Pabst never quite gave the deed a name but it appears that *Frau* Marcussohn, whose name suggests she was Jewish, had turned them in for the Sklar's and Schiedemann's reward—and perhaps because, like a great many German Jews even in leftist Berlin, she refused to hand the city over to

Russian-based revolutionaries. A third motive may have been simple fear of incrimination.

"First Herr Liebknecht was put forward," Pabst wrote many years afterward. "As soon as this happened I telephoned to senior authority, Herr Noske, and General of Infantry Baron [*Walther*] von Lüttwitz. Noske ordered that both prisoners be transported to the holding prison at Moabit: it was absolutely necessary to take precautions that the spiritual leaders of the revolution not be freed on the way to Moabit by their retainers, or that they should succeed in an attempt at escape."

Pabst had asked Noske in an early conversation if he had formal approval to kill Liebknecht and Luxemburg if they fell into his hands.

"That's not my concern,' Noske said gruffly. "The [Social Democratic] Party could break up because the masses would never understand the need for that sort of measure. The General should take care of that, they're his prisoners!"

Pabst now had to make sure he had the right people in custody. Rosa Luxemburg, with calm courage, admitted who she was with precise phrases and, perhaps recognizing that Pabst was highly intelligent, actually tried to explain to Pabst the logical inevitability of Communism. Liebknecht tried to lie his way out and told Pabst that he was actually Herr Marcussohn. Liebknecht's identity was only confirmed because Liebknecht had his initials stitched into his underwear as laundry marks.

Wilhelm Pieck, the third party under arrest, was so terrified that he was shaking when he was placed before Pabst, because he had already seen Luxemburg and Liebknecht interrogated and knew they were headed for Moabit Prison. Pieck, an ethnic German, was a minor figure in the uprising and Pabst contemptuously promised Pieck his life if he produced a list of all the important German and Russian leaders, notably Karl Radek, and told Pabst where to find them. Pieck scribbled a frantic betrayal of his comrades and Pabst handed him a safe-conduct and let him out the side door. Pieck later lied and denied his betrayal of his comrades: he said he had escaped from the Hotel Eden, which was guarded by the Whippet tank and swarming with Rightist troops at the time. Pieck became the puppet ruler of East Germany under Russian auspices after World War II ended and remained in power until 1960.

Pabst now had Rosa Luxemburg and Karl Liebknecht to deal with. In documents assembled by Rüdiger Konrad and an a 1962 interview with *Der Spiegel*, the (then-West) German news magazine, Pabst portrayed himself as struggling, almost agonizing, over the question. "What should happen now to the guiltiest instigators?" Pabst wrote afterwards. "I spent heavy, lonely hours in the night of January 15 to 16 going over it thoroughly in my mind and finally decided to take the law into my own hands. The decision was not made at all lightly.... I went back in my office and considered with a certain quiet calm, how to render the execution of these two assassins and traitors comprehensible...."

Karl Liebknecht during the January Uprising in 1919—only a few days later he and Rosa Luxemburg were murdered.

Then he asked for volunteers for what would come next. He got plenty of them. Waiting outside the plush Hotel Eden near the car was Hussar Otto Runge, a 43-year-old long-time regular soldier with sad eyes and a mustache that dropped like the wings of a vulture, seconded from the horse cavalry to railroad watch duties. Runge's Mauser was unloaded and without a bayonet. But as Leibknecht walked pass him to the closed car, Runge seized the muzzle and swung the butt of the rifle down on Liebknecht's head like an executioner's axe and fractured his skull. Liebknecht was bundled into a closed car. Several other angry officers punched him in the face to share the guilt or credit for his death. When they felt his pulse and knew that he was still alive, they shot him in the head. His body was dumped near the

Tiergarten without identification. An hour later, the officers reported that their car had suffered a break-down and when they tried to walk Liebknecht to the prison but he broke away and fled. After shouting three times for him to stop, they shot him. That, at least, was what whoever read the report was asked to accept.

Rosa Luxemburg was led out just before midnight. As she left the revolving doors, Runge once again swung his Mauser like an axe and inflicted a massive skull fracture. A naval officer jumped into the car and finished her off with a pistol shot to the head. Her hands and feet were trussed up and ballasted and her corpse was thrown into the Landwehr Canal. The report this time was that some unknown person from the crowd had shot her.

When Pabst awakened the nominal commander of the Guards-Cavalry-Division, General von Hoffman, Pabst reports that the general said; "My dear friend, I thank for you for taking the decision out of my hands, I would never have been prepared for it."

"I know that, Your Excellency, and because I knew it I took the entire responsibility of having both of them shot on myself."

"Of course, I myself will assume all responsibility for what happened that night," General Hofmann replied. "That is my duty as a comrade and of my place as the commander of the division. But let me tell you as an actual friend: You will have to carry this burden for your whole life."

Chapter Thirteen
Despair and Defiance

Through the month of November 1918, Hermann Ehrhardt was stuck at Scapa Flow in Orkneys north of Scotland watching the discipline of the fleet fall apart as the ships themselves lapsed into disrepair. Most of the sailors were simply apathetic, though active Reds among them engaged in trampling around atop the officers' sleeping quarters. Despair and a bad diet brought Ehrhardt near the edge of depression.

"Then came the order: the greatest part of the crew and the officers should return to the homeland, and only a weak watch crew had to remain," Ehrhardt said. "Without being obvious about it, we officers wanted to leave only the most loyal and best petty officers with the watch crews. Only a few senior officers remained behind. I sailed back to the homeland."

With the order to return to Germany in early December, Ehrhardt said he felt stung by the contempt of the British officers because the High Seas Fleet had ended the war without a final showdown. Trained as he was in a tradition that honored the Royal Navy, to the point that cadets were given their training orders in English, Ehrhardt felt this disgrace deeply. His anger and depression deepened by the day.

"At last the transport steamer came to take us back to the homeland. Every officer had to carry his luggage by himself, and we had to use a rope ladder instead of the side port."

Leutnant zur See Franz Maria Liedig, born in 1900 and all of 18 years old, noted dryly that the British sailors had a lot of trouble getting the German sailors onto the chartered steamer from pinnaces that the British themselves could barely handle properly. Once aboard, some of the German enlisted sailors, Liedig said, fantasized about a world revolution but most of them were more concerned about how much they could get paid in the merchant service. Some of the senior officers actually had cabins, while the junior officers and the sailors were quartered in the freight holds of the chartered rust-bucket steamer. Ehrhardt and several other command-rank officers sulked in their grubby cabins, ominously unseen and obviously in deep despair.

"On the second day, nobody knew where we actually were," 18-year-old *Leutnant* Liedig said. "We knew there were minefields in the vicinity and many of the officers seemed to think we would be better off running into a mine and going to the bottom.... Our transport was accustomed to be ballasted with many thousand tons of freight. Without this burden it rode high in the water. The Plimsoll line was at least a man's height above the surface of the sea.... When we saw the English minefield markets we dropped the anchor and the old South American freighter swung on the anchor chain and toward the minefield like a horse headed for oats. In absolute silence, we headed for the minefield."

The Red sailors incongruously chose this moment to declare a strike: "We won't go any further unless we get a raise of five marks!"

"This was the moment when the previously unseen *Korvettenkapitän* Ehrhardt finally decided things had become unbearably dumb. Suddenly he stood on deck, legs set apart, and without even considering that officers of higher rank were aboard he took control and had his own flotilla crew piped. Ehrhardt's destroyermen stood up and assembled. In five minutes they had the anchor hauled back in, the machinery working, and the bridge crew replaced. The confusion was ended and the steamer was back on course for Wilhelmshaven and around the minefields.

Having collapsed in the crunch, the Red Sailors' Council vanished. The hired crew of the steamer went back to their duties under the strict supervision of Ehrhardt's torpedo boat flotilla sailors.

"It was *Korvettenkapitän* Ehrhardt that got us out of that mess," Liedig said. "His bearing won my whole heart.... We younger officers were all inspired by the new 'commander.' Fearlessly he gave us his opinion of the revolution and his opposition to it. He had no civilian concerns, he thought so much like many of us younger officers that we considered him one of us."

Liedig returned to his former base in Kiel and Ehrhardt returned to Wilhelmshaven, where his father-in-law had built him and his wife a house. He found that 25,000 battleship sailors and unattached soldiers had virtually taken over the city, and the civic authorities were all but powerless. His wife, to his intense relief, had taken his step-daughter and their two little boys home to her mother in Hamburg, a city-state since the Middle Ages which had some local government still in operation. Ehrhardt also found, to his intense indignation, that the Reds had looted his wine cellar.

"I didn't let this plundering slide," Ehrhardt said. "I put on my full uniform and went to see the brothers in the Red headquarters. The Police Presidium, where the Soldiers Council resided, showed the typical Revolutionary picture. Screaming drunks lay around in the corners. Others were drunk with delusions of their own importance, growled and fidgeted, each one putting on a stage show for the others. My accusation was handled by a petty officer with a red scarf wrapped around his neck. He said he expected that my wine had been needed to help the sick. I declared to him to his face: that is not true. I set down in writing what had been looted."

Ehrhardt began to weld his 300 young officers into a *Stoßtrupp* unit for any eventuality. The gentlemen, he says, met in the meadows outside the city, passing through the city streets in civilian clothes, usually around 1 a.m. to avoid being noticed. The handful of anti-Communist shock troops from the Army or from the German marines, called *Seesoldaten* (sea soldiers) instructed them in the use of infantry weapons and in street fighting tactics.

"On January 27, the Kaiser's birthday, every sparrow on the roof knew that the Communists had taken the power unto themselves and set up a Soviet of Wilhelmshaven. Now the previous power brokers, the Socialist non-commissioned officers, who were in league with the professional soldiers, became concerned. They came to ask *us* what should happen. The clueless question came to me. I told them the best thing would be to hit the Communists before they established themselves, and to smash them this very night. I told them I had 300 disciplined men I trusted to bring it off, and I would undertake the action with them.

"Among the deck officers there was one who always had the biggest mouth. He asked: 'But who will take the responsibility if blood flows?'

"I answered him, 'I will, if I lead the thing.' He nodded quietly."

Ehrhardt briefed his 300 naval shock troops and they surrounded the Thousand Man Barracks in Wilhelmshaven where the official new ruler of Wilhelmshaven, a grade school teacher named Karl Joerns, presided over the Soviet. About 300 *Reichswehr* soldiers joined Ehrhardt's sailors.

"A wild mob of humanity besieged the barracks," Ehrhardt said. "And when the Communists fired first, a regular fire-fight was soon in progress. The Socialist dock workers went wild—before this they didn't know whether the Communists were their friends or their competition. Naval guns were even fired, but the whole night was a joke. Rifles and machine guns cracked from all sides. The Communists lost seven dead and we lost three, I think from our own gunfire.

"Toward morning, the Communists gave themselves up.

"The captured Red soldiers were badly mishandled until I personally intervened. They beat the teacher Joerns almost to death. I rescued him out of a sense of duty. The professional soldiers were very grateful that I had taken command and responsibility and had put the fear so quickly behind them. The next morning, their top man made us a speech: the issue is now resolved. We thank the gentlemen officers; but now we have the matter in hand, and we can take care of it ourselves.

"Then you can kiss my ass," Ehrhardt replied.

Naval Engineer August Maertens of the Ehrhardt Brigade upper left, with Naval uniform and bow tie joined the Bremen civil defense group in the re-capture of Bremen from the Bremen Soviet Republic. The battle cost the lives of 29 civilians.

The counter-revolution in Wilhelmshaven triggered another Nationalist revolt against the Marxists in Bremen, an older Hanseatic city rule by a local Senate. The majority parties were the Social Democrats with 42 percent of the vote, and the Democrats with 33.5 percent. The Independent Social Democrats—essentially Communist—had 18.2 percent but tried to seize power by controlling the street and taking middle-class hostages. The anti-Communist troops and citizens, having heard about the walk-over at Wilhelmshaven, overthrew the local Reds, deposed them and closed the Communist newspaper. The death tally was 24 Nationalists and 28 Reds killed, with 18 civilian men, five women and six children killed, and many wounded. A number of Communists were arrested and jailed. Other North German cities, especially those with medieval traditions of self-rule, also began to push out the Communists. Ehrhardt and his men at Wilhelmshaven had started a trend to armed resistance that began to sweep across northern Germany.

In nearby Bremen, a former medieval city-state and a major German seaport, Communists led by Johann Knief, a schoolteacher, and Heinrich Vogeler, a celebrated avante-garde artist, formed a Soviet Council supported by armed workers, most of them under-employed munitions workers. On February 4, Noske ordered regional Freikorps units to disarm them. The brief battle killed 28 Communists, 24 anti-Communist soldiers and 29 civilians including five women and six children.

Ehrhardt noted, however, that the effort of the Army enlisted men based in Wilhelmshaven fell apart within five days because nobody would stand guard duty, and the professional soldiers asked him back "but it broke up due to the laziness and lack of discipline of the men.... Nobody wanted to stand guard duty, if they were ordered to do so, they either left the area or sat down in the nearest tavern." Most of them, he said, seemed to be members of the Independent Socialist Party, which was farther left than the Social Democrats and tended to bond with the Communists.

"Then [*War Minister Gustav*] Noske issued his call for the *Freikorps*," Ehrhardt said. "I said to myself: that's the field for you. Get a respectable troop on their feet, to defend the Fatherland internally and in the East. Then you can go to Berlin and help fight Bolshevism. I understood the enormity of the Communist danger, because I had seen how the Revolution was fostered by making use of worthless and criminal elements.

"I spoke with *Kapitänleutnant [Karl]* Tillessen many times about this theme.

"Tillessen said: 'As an officer I can never serve with this band of oath-breakers.'

"I responded: 'Tillissen, if you'd lived in a house a long time and the landlord let it catch fire through sloppy maintenance, would you shrug while his cakes burned and leave your own luggage behind in the fire?'

"I informed Tillessen: 'Above all we must have order again. When Germany first begins to think for itself, it will have enough of the blessings of this so-called Revolution. But first we have to grab hold, join together, and save the country.'"

Karl Tillessen was won over and joined up, and so did most of the other younger naval officers and some of the petty officers and sailors.

"I renounced the alarm station of regular soldiers to let them squat in their own dreck and put out a call to recruit the Ehrhardt Brigade," Ehrhardt wrote. "The 300 young officers all joined the shock troop company. I also signed up about 300 of the regular soldiers. These were in part old soldiers from the front, in any case energetic and brave men, who, as opposed to the majority of the non-commissioned officers, weren't ready to go home and sit around in the nice warm kitchen with Mother.

"Immediately after the founding of the brigade we heard from an agitator; the recruiting of this reactionary troop could not be tolerated. I countered the apprehension of this man with a measured response, that my unit would carry the Imperial Battle Flag that had waved over my boat at the Battle of Skagerrak.

"My answer was to fortify the Sea Battalion Barracks, and to secure the entrance with armed guards. Within eight days soldiers from the army brought the strength of my troop to a thousand men."

One of them was Rudolf Mann, an *Oberleutnant* (first lieutenant) in the German Army with battle and front-line supply experience on the Western Front. Mann had grown up in Berlin, born into the gentry though not to the aristocracy, and spoke fluent French and some English. He had seen a fellow officer sentenced to prison by the Reds for refusing to cut up an Imperial flag to make red cockades. He had also seen—as a participant—the fighting at the Thousand Man Barracks and the excellent showing made by Ehrhardt's naval officers and petty officers as compared to most of the regular army men. Mann was not independently wealthy and was concerned about drawing pay, but the defiant Imperial flag over Ehrhardt's headquarters drew him with nostalgic fascination. The first thing Mann spotted was that the officers were still wearing their forbidden epaulets. Meanwhile, soldiers in *Feldgrau* and sturdy young men still in civilian clothes were actually cleaning the corridors after several months of filthy living.

"Up in a small sergeant's cubicle the commander of the Brigade sat on a stool writing signatures. I went inside and said my little speech, as if a stranger journeyman in a skilled trade was introducing himself to the master.

"Field experience?"

"*Jawohl!*"

"I can use you, I can tell that right away, as long as your papers get here fast."

"He kept signing his name while he was eating and I saw this and knew that this young, stiff-backed officer with the hard face and the hard manner was the right man for hard times like these."

Mann had long been devoted to the analytical power of handwriting analysis and studied Ehrhardt's penmanship on the documents as Ehrhardt kept signing them all through his lunch. "The signature tells the expert a great deal," Mann said. "The wide, distinct capital letter shows security and the essence of self-confidence, and the quick writing with the quickly crossed Ts and *umlauts* showed an active nature. The rounded descenders beneath the letters showed goodness of heart. Open script, open character. I saw that from the first glance, and it delighted me."

Ehrhardt also knew some psychology. He ordered Mann to write out an oath of allegiance to the Social Democrats and watched Mann's face. When he saw the artfully concealed distaste, Mann was immediately appointed as regimental adjutant.

Mann set to work organizing the paper work and the supplies—he ultimately served as the Ehrhardt Brigade's quartermaster as well. *Oberleutnant* Mann rounded up a couple of enlisted men with legible handwriting, had a few iron bedsteads moved around and found an artistic soldier to paint an emblem for the door:

"Business Office of the Regiment." Soldiers followed the old Imperial flag to the office and those who passed muster signed on.

"Orders came from Berlin recognizing the foundation of the Brigade as soldiers of the government," Ehrhardt said. "I arranged the understanding as to pay, uniforms, and food. The organization was complete."

Chapter Fourteen
Counter-Attack

In April of 1919, Hermann Ehrhardt's newly organized Second Marine Brigade received its marching orders from General Ludwig Maercker, father of the *Freikorps* movement, approved by War Minister Gustav Noske.

"We were sent to the railroad station at Charlotteburg and were intensely curious," Rudolph Mann remembered. "The written orders that were sent to us said:

"1. The Brigade will roll out to an undertaking within the borders of Germany....

"2. [The second item listed the numbers of the train cars and the railroad officials.]

"3. Horses on the excursion will stand with harness or saddles.

"4. Exit ramps or other means of de-training will be loaded aboard."

Mann understood that the horses had to be kept saddled or harnessed because the enemy might be near, but he said bluntly that nobody appeared to have the materials to build exit ramps even in Berlin, where the situation was still turbulent. He had no idea where the train was taking the brand-new Ehrhardt Brigade, except that it seemed to be headed east in the general direction of Berlin. In the evening, the train had reached Stendhal, and the Brigade's ordinance officer, whose family lived there, invited the other ranking officers to a splendid quick supper he had quickly arranged in a private room at the railroad station.

"I'll never know whether it was any good or not, because a courier officer with a black dispatch pouch showed up, and we had to get up immediately and climb into the train, which was already moving slowly away from the station and leave the meal uneaten

"As we shortly reached our destination, sheer joy took our minds off our hunger."

Braunschweig—called Brunswick in English and in the North German *platt* spoken in the old Saxon country—was one of the oldest German cities without the Roman imprint of the oldest German cities—the ancient Roman cities along the Rhine like Cologne. Founded in the 900s, Brunswick was for many years the capital of the old Saxon dynasty which had founded the medieval German Empire after Charlemagne's French-German empire broke up in the 800s. Henry the Fowler was their founder in the days of the wars with the Slavic tribes and the Hungarians. Henry the Lion, their most famous subsequent leader, was driven into exile in a power struggle with Friedrich Barbarossa during the Crusading era and fled to England, where his daughter, "the Empress Matilda," married King Stephen. Brunswick then became a key trading city of the Hanseatic League in the later Middle Ages and, like Hannover, the ancestral seat of the British kings, developed a strong affinity with England. Two Dukes of Brunswick had been commanders of the Prussian Army: the first was killed fighting Napoleon in 1806. The Brunswickers formed a regiment of avengers, volunteers in their famous black uniforms with the skull-and-crossbones emblem and the motto *Sieg oder Tod* (Victory or Death). The British who fought side by side with them in Spain said that the whole unit looked like an endless hearse when it was on the march. The second Duke, son of the first duke and an ally of the British, was killed at Waterloo in 1815, as noted by George Gordon, Lord Byron. The Duchy of Brunswick, a small independent state, was located inside Hannover, the third largest German state, and had a long and proud history of independence.

On November 8, 1918, when the Kaiser fled Germany, a Social Workers Council demanded that the last Duke of Brunswick, Ernest Augustus—nominal heir to the non-existent throne of Hannover—give up his throne. The Independent Social Democrats were virtual Communists, unlike the mainstream Social Democrats, who tolerated religion and some private property. The Independent Social Democrats were too extreme for citizens accustomed to self-government and proud of their city's heritage. A free election on December 22 of 1918 brought in a coalition of Social Democratic splinter groups along with some Democrats, a smaller moderate party of educators and intellectuals. The moderate-liberal coalition in Brunswick restored some semblance of representative government and civility. The Reds struck back with a take-over by the Spartacists, unabashed Bolsheviks who took their orders from Lenin and Trotsky in Moscow.

When Mann and his fellow officers heard they were headed for Brunswick, their trigger fingers started to itch. "The Red Sailors of Kiel had set aside the rightful rulers and drove around in automobiles in the Russian manner, threatening the country people with being burned out of their houses if they didn't hand over food. All of Germany would be looking at us, because all sorts of stuff had been written about what had become of this peaceful and hard-working dukedom during the past few weeks. We were delighted that the administration in Berlin had given us the honor of reducing this Soviet Republic to nothing.

"In reality it was not a Soviet Republic but a Selfish Republic. A reign of terror would be too strong a word for a few hundred dim-witted sailors, who lived in lordly delight, as representatives of the Workers Council, Sepp Orter and [Ernest Augustus] Merges."

Orter and Merges, the figureheads of what was a desperate Bolshevik seizure of power, were political oddities. Josef 'Sepp' Orter hailed from Saarbrücken, in southwestern Germany, nowhere near Brunswick, and had been an anarchist rather than a Marxist until opportunity knocked with the flight of the Kaiser. Ernest August Mergers, a local tailor, somehow managed to get fat even while the British starvation blockade kept the poor near starvation while middle-class people were vegetarians, not by choice, and poor people ate sausages with dog and rat meat in them. The Minister of Culture, Minna Faßhauer, was an abortionist posing as a midwife—abortion had been illegal in the Kaiser's Germany—and Minna had a certain appetite for life. She surrounded herself with good-looking young men in need of job appointments. When a strapping young reserve officer who had just passed his accounting examinations knocked on her office door and asked for an appointment as an assessor, she immediately promoted him based on his good looks: "Assessor? No, administrative counselor, that's you, that's you!" When the Spartacists took over, the young officer defected to the Guards Cavalry Division instead. He briefed Waldemar Pabst on the new administration in the proud old city of Brunswick.

Newspapers carried a story that someone had posted a placard on the pedestal of the Duke of Brunswick's statue.

> *"Lieber Herzog, steige nieder*
> *Und regiere du uns wieder;*
> *Laß in diesen schweren Zeiten*
> *Lieber Schneider Merges reiten!*

Roughly:

Dear old Duke, please climb on down
Rule again in our old town.
Say in this, our time of woe
Tailor Merges has to go!

Rudolf Mann quoted the poem but noted that all the oppressed Bruns-wickers had done before the *Freikorps* units arrived was to write poetry, unlike the the people of Bremen and their home-ground veterans who had taken matters into their own hands and routed the Reds.

"A courier told us that the forward section of the Reds had set up a defense position in Helmstedt. His pouch contained an exact account of the information from the spies: The strength of the Red army was 15,000 men, with 156 cannon de-ployed—defense trenches, machine gun nests covered with barbed wire—that really looked impressive.

"I myself took the thing lightly. I had experience in Flanders and knew that the coolie is no great fighter on land. His heraldic beast is the hare and not the tiger."

Rudolf Mann was now a staff officer of sorts and stayed up most of the night going over the maps and writing out the orders. The Ehrhardt Brigade was assigned to attack from the northeast—Maercker, the commander in chief had used railroads from all directions to pull off a classic Prussian double envelopment, modeled on Hannibal's victory over the Romans at Cannae. The ultimate target was the city of Brunswick's infantry barracks, where the Reds had their headquarters.

"So one sat there hungry and thirsty by flickering candlelight and wrote at-tack orders based on the most recent intelligence," Mann remembered. "Just like a war—by German regulations everything had to be written down. How much nicer it had been to drink cognac and sleep, to sleep and in the morning let loose the thunderbolts like old Blücher.

"But General Maercker was the commander in chief, and he loved clean work, quite exact and from all sides—that's how he had already taken many cities without a musket being fired—a noose, that one pulled tight."

When the train pulled in at Schandelah, the soldiers were enchanted to see cavalry with lances in the moonlight moving like ghosts, with only the snorting of the horses and the jingle of the swords and the spurs to show they were real. This was the *Freikorps Lützow*, a horse cavalry unit named after a famous unit of the Na-poleonic Wars, an anachronistic dream of past romantic glory.

"Our unloading in Schandelah was a masterpiece in itself," Mann noted wryly, jolted back to reality. He saw once at the station that there was no way to get the baggage wagons carrying food and ammunition off the train because the exit ramps had not been built.

"*Wat sin moet, dat moet sin*," Mann said in *platt*. (What must be, that must be—*que sera, sera*, as it were.) Mann shouted to the *Pionierkompanie*, the combat en-gineers: "Build ramps." The pioneers ripped rails and ties out of the vacant track "as if they were digging peat" and used them to bridge the space between the train cars and the ground. They covered the rails with large slabs of the railroad station itself. "Order carried out, ramps finished."

The baggage wagon came down from the first flatcar, and the all-important field kitchen, a cook-stove with vats and a pantry on a four-wheeled wagon, came off the second flatcar without incident. Some of the other wagons broke their brake housings in being unloaded but they all kept their wheels. The staff officer supervising the operation keep looking at his brand-new wristwatch as the pioneers cut down trees and used the logs to bridge a shallow stream. A wheel cracked and a nervous horse fell in the water and was rescued as quietly as possible. Mann scolded the driver—he loved horses—and ventured into the outskirts of town where a large farm-holder and his wife met him at the front door of their country estate. "They dragged me into the house as if I were a long-lost son," Mann said. He finally got a quick and hearty breakfast which he ate on the run.

"The Schandelah farmers had to sing their sad song about the food raids by the Brunswick sailors," Mann said "But who can come up against peasant shrewdness and win? They stuck all the best food up inside the chimneys. They had three dozen decorated Easter eggs along with ham and local sausages and they gave them to our soldiers with a free hand, and a lot of other food besides. The farmers harnessed up their best teams and used their wagons to carry our packs into town. Every landowner found it an honor to help us.... These people won me over to love them."

The *Freikorps* had other help. A dis-employed railroad worker with a bad reputation led one Red military column against the *Freikorps*. The assistant station master, a former soldier, wanted no part of fighting German national troops and he pulled the plug on the electrical system before he left the central railroad station though the back door. The Red leader had some sort of grudge against the senior station master, a white-haired old man—"You traitor, you cut the wires, down with you!"—and beat the old man half to death before he ran for it himself.

"He opened holes in his scalp, stepped on his face and broke some of his ribs," Mann reported after his unit arrived and gave the poor old man some medical help. "I note this to support my position that not every Communist is a friend of humanity and a pure idealist."

One hour later, the Red ex-railroad worker was found hiding and dragged through the streets with his hands cuffed to the stirrups of a horse. He got four weeks in prison. The enormous casualties of World War I had conditioned most of Germany against arbitrary capital punishment.

"We marched up and down the hills. Orter and Merges escaped by airplane. There was no armed resistance. The pretty girls popped up like asparagus and waved to us."

The Guards Cavalry Division had been detailed to help encircle Brunswick but more trouble in Berlin kept them from marching. Waldemar Pabst, by now becoming the division's tactical commander as General Hoffman slid into premature old age, noted that the Ehrhardt Brigade had done an excellent job with the help of local units. Pabst drew his own message:

"This showed that the Communist party-goers were in no way ready to die for the world revolution," Pabst wrote. "They threw their weapons in the river or the fishponds and greeted our troops as if they themselves were harmless citizens and workers. The previous administrative chiefs, Tailor Merges and Sepp Orter, hopped on an airplane as quickly as possible at the beginning of our undertaking and disappeared. This as-good-as battle-free 'conquest' of the land of Brunswick convinced me that the standpoint that I had rolled out before War Minister Noske and President Ebert was the correct one. The cowardice and the indecisiveness of the so-called citizens was responsible for the ease with which some 'dictators' of the Soviet councils could live comfortably when nobody dared to oppose them."

Ehrhardt himself saw the Battle of Brunswick as a walk-over and reported not losing a single man. He too favored a tougher stance with the Reds.

The only resistance Mann recalled was when some of the wives, mothers and girlfriends of the Reds turned out to shriek curses, shake their fists and spit at the soldiers in uniform—female workers out of the jam factory, wearing wooden shoes, he said, like a scene out of Émile Zola. A cavalry unit made a mock charge at them and the women fled screaming in all directions, with no bloodshed except for bumps and scrapes when some of them lost their footing. The soldiers laughed uproariously.

For Rudolph Mann, victory felt good after the defeat on the Western Front and the Kaiser's abrupt departure.

"Our entry march into Brunswick was like the first ray of sunlight after a long, rainy night," Mann remembered. "The family that had adopted me couldn't do enough to show their love of me," Mann remembered. "I celebrated my birthday in a *Weinstube* and received a vintage bottle of extremely old red wine. I don't know how much it would have cost because the landlord wouldn't take money for it."

Hermann Ehrhardt, older and more analytical, was less impressed with the results.

"Maercker was highly praised for his bloodless operation," Ehrhardt wrote in 1924. "The effect of the bloodlessness showed up later. The administration put out an arrest order for Orter and Merges but nobody carried it out. Tailor Merges and Sepp Orter came back later and took up their political roles again. An Independent Socialist administration once again pressed the citizenry to the wall. Finally a representative of this administration was put under lock and key for criminal activities.

"Maercker proved later that he was no man of action. *Oberst* [Colonel] Bauer tried to move him to oust the national assembly in Weimar and send them all home. The counter-attack against the Left would have succeeded much more quickly. But Maercker shied away from a confrontation, he feared that such an act could provoke a contrary effect."

The Second Marine Brigade soon restored order in a number of different central German cities. Ehrhardt said that his main concern was that people would feed the sea cadets and ensigns and young Army lieutenants too màny chocolates or invite them to upper-middle-class ballroom dances where irregularities might occur. His careful recruiting had weeded out "the worst bums, the psychopaths, the wanderers arrested for vagrancy, and the people who actually belonged in jail."

"I gladly let the youngsters enjoy the dances with the Saalfelder ladies and the delicious gourmet chocolates, especially the sea cadets and the ensigns, but that was not the point of our troop. As a regular *Landsknecht* leader I spied around to find something to do in the *Reich* that was honorable."

Chapter Fifteen
Blood Week in Berlin

Four days after the murders of Karl Liebknecht and Rosa Luxemburg, the Germans held their first election since the end of the war on January 19, 1919. The German three-tier voting system that was the bane of liberals under the Kaiser because it gave an advantage to large land-owners had been eliminated. For the first time in history, German women of 20 and over were allowed to vote along with German men of the same age: the prior voting age had been 28 years of age for men only, probably due to a suspicion that the youngest men might vote military conscription out of existence.

The Reds in Berlin were still screaming about the Liebknecht-Luxemburg murders—which had been covertly approved by the Social Democrats Friedrich Ebert and Gustav Noske—and partisan newspapers were filled with inflated Red casualty figures when the German voters cast their ballots on January 19. The Social Democrats, accused of supporting the Liebknecht-Luxemburg murders, received 11.5 million votes and 163 seats in the Reichstag. The Democrats, a moderate-liberal party popular with lawyers, professors and teachers, received 5.5 million votes and 75 seats. The *Zentrum* (Center) Catholic Party of Prussia and Hanover, Otto von Bismarck's former critics and sometime nemesis, received 73 delegates. The Bavarian People's Party, also largely working-class and politically dedicated to Catholic interests, received 18 delegates. All four of these parties supported the

Republic, all four either tolerated or sustained religion, and all four were anti-Bolshevik. None of them wanted any Soviet presence in German politics.

On the Right, the Nationalists—which is to say, the tacit monarchists—received 3 million votes and 44 seats, the conservative German People's Party received 1.6 million votes and 11 seats, and smaller conservative parties received a scattering of votes and 7 more seats.

Gustav Noske, Weimar's War Minister in January 1919

The USDP—the Independent Socialists who had largely supported the Red uprising of the Spartacists—polled 2.3 million votes and received 22 seats. The Reds were officially outnumbered almost 3 to 1 by people who wanted the Kaiser back, and outnumbered about 8 to 1 by the moderate Socialists and moderate Catholics who supported the Republic and opposed both the Kaiser and Red revolution. The top four liberal or Catholic parties who rejected Communism and the three conservative parties who hated Communism had a combined edge of better than 10 to 1 against the Communists and their supporters. Even Count Harry Kessler, the homosexual future poet laureate of Berlin Decadence known as "the Red Count," wrote off Karl Liebknecht and Rosa Luxemburg off as a couple of dangerous trouble-makers who had gotten what they deserved. "Liebknecht and Rosa Luxemburg found a tragic and fantasized end. They had so many dead on their consciences through the civil war they started that their violent end seemed as a mandatory and necessary consequence of their deeds." Kessler was very much left-of-center politically and socially, yet a great many other liberals who didn't take their orders from Moscow privately agreed with him. Based on majority rule and the opinion of most intellectuals, liberals included, the Red Revolution appeared to be dead with Liebknecht and Luxemburg.

Captured Reds in Berlin, January 1919

An opportunity of sorts to re-kindle the public turmoil came up as the Social Democrats began an official investigation of the murder of Rosa Luxemburg. As General von Hofmann had predicted to Waldemar Pabst, killing a woman, any woman, was an extremely unpopular act in Germany. Even anti-radical Germans were upset. The soldiers of the Guards Cavalry Division, on the other hand, volunteered to resist any attempt to arrest Waldemar Pabst at gunpoint and cheered him whenever he strode past military formations on parade. But Captain Pabst insisted that everything proceed according to law—at least in public. One of the investigating judges was Naval Captain Wilhelm Canaris, an anti-communist who, like Pabst himself and like Hermann Ehrhardt, later became an anti-Nazi conspirator and would ultimately be hanged in a Nazi concentration camp at Flossenbürg after the July 20 bomb plot against Hitler. The gallery was packed until it buckled with soldiers. Pabst and Canaris pulled strings to deflect any suspicion from the staff of the Guards Cavalry Division. Luxemburg's corpse was still at the bottom of the Landwehr Canal, but the most consistent and devoted of her lovers, Leo Jogiches, showed up in Berlin and announced that he intended to avenge her.

Leo Jogiches, born to a rich Jewish family in Vilna, Lithuania, then part of the Russian Empire, seldom spoke about his past but he had been imprisoned as a young man in Tsarist Russia for trying to organize the Polish struggle against Russia—though most Poles made it plain that they did not share his and Rosa Luxemburg's Marxist view of the future. After a spell in prison, Jogiches was dispatched against his will to serve in the Tsarist Army—a very dubious way of reforming anti-Tsarist revolutionaries. He shortly made good his escape to Switzerland, where he first met Rosa Luxemburg, three years his junior, in 1890. Jogiches and Luxemburg became life-long friends and on-and-off lovers until she vanished. She wrote to him in 1900: "How I need you! We need each other! Truly no other couple has such a mission in this life, as we do, mutually to make a human being out of the other. I feel this with every step I take, and so I feel the pain of our separation more than ever." Leo Jogiches' reciprocal passion for Rosa Luxemburg and for the Revolution must have been genuine because he came out of hiding after her disappearance and Liebknecht's known murder to assume leadership of the KDP and made himself an obvious target for more of the same.

On February 6, the new National Assembly met for the first time in Weimar and confirmed Friedrich Ebert as president of Germany. Romantics said that Weimar had been chosen as the site of new capital because Johannes Wolfgang von Goethe, author of "Faust," known as "the last Renaissance Man" and friend of Friedrich Schiller, Ludwig van Beethoven, Johannes Wolfgang Amadeus Mozart, and the young Felix Mendelssohn, had been the municipal administrator of Weimar in the later 18th Century. Weimar was to German intellectual and artistic culture what colonial Boston or colonial Philadelphia had been to Americans and what recent Berlin had been to Prussian militarism. Pragmatists noted that Berlin was located on a flat plain near shallow rivers—virtually indefensible and much too close to Russia and Poland. The new democratic German government was informally dubbed "The Weimar Republic."

On February 12, back in Berlin, Leo Jogiches had an article published in "*Die Rote Fahne,*" the Communist newspaper, describing the murders of Rosa Luxemburg and Karl Liebknecht. Jogiches, who had spent years as a political prisoner and spoke Polish as well as Russian and German, had a knack for reporting. The Officers of the Guards Cavalry Division noted with some concern that he had the facts exactly right, though names were missing. Leftist newspapers began to protest that the Liebknecht-Luxemburg investigation was rigged. Pabst received a warning that the transport commander were about to be called into court, but he carried out his plans to get married anyway.

On February 28, Captain Waldemar Pabst and Helma Corneli, whose ancestors had come to Germany to escape religious persecution during the French Revolution, rode to Saint Hedwig's Catholic Church in Berlin in an open two-horse carriage. Announcements in the Leftist press had invited the public to view the procession with headlines like "*The Bloodhound Marries*" and "*Marriage of the Mass Murderer.*" The Red papers pointed out that Gustav Noske, the other "Bloodhound," had congratulated the bride and groom. The groom showed up in full dress uniform—"I went back to my chambers in Paderborn dug up my old cavalry uniform from the 8th Hussars." The church was surrounded by sawhorses strung with barbed wire. Pabst noted that the spectators who saw him and recognized him were tactful and considerate. The honeymoon lasted only 24 hours because Pabst had to get back to court to defend his comrades.

On March 3, *Rote Fahne,* the Communist newspaper, proclaimed "The Battle of the Marne of the Red Revolution."

Government forces had been tackling some of the Communist strongholds operating under cover in working-class neighbors in Berlin. The January election results where the Communists lost about 10 to 1 and the obviously rigged Liebknecht-Luxemburg trial in Berlin had convinced the Reds that it was now or never. The Red Sailors of the *Volksmarinedivision* had announced that they too supported the Weimar government, which only made matters worse. The Reds declared a general strike by workers and soldiers—with sporadic results. The day before in the nearby city of Halle, on March 2, *Oberstleutnant* [Lieutenant Colonel] Robert von Klüber, a holder of the *Pour Le Mérite,* the Kaiser's highest decoration for valor, had been caught in civilian clothes, savagely beaten, shot, and thrown into a stream and forced to tread water until he sank and drowned. The next day in Berlin, Red attacks on 32 police stations, killed five policemen and captured 1,100 rifles and a number of machine guns. Most of the Red Sailors of the *Volksmarinedivision* changed sides immediately and joined the uprising. The Ebert administration—themselves mostly Social Democrats or Democrats—declared Berlin in a state of siege. On March 4 they called in *Freikorps* units to help the Guards Cavalry Division put down the uprising "ruthlessly." About 15,000 *Freikorps* men, most of them from scratch units around Berlin, joined 16,000 soldiers of the Guards Cavalry Division against about 15,000 Red Sailors and Berlin military reserve units. Germany's conscription system had trained a huge number of reservists and many of them turned out in uniform to support the revolution, wearing red arm-bands if they chose to do so and backed by a far greater number of Leftists who sometimes fought out of uniform, spied on government units or acted as couriers. Some of the Reds use foreign-made dumdum bullets which expanded when they hit human bodies. The government troops fought back with captured British tanks, heavy 150-millimeter howitzers that could knock down a whole house, and *Minenwerfer,* short-barreled big-bore mortars designed for trench warfare that could produce death by concussion. The Reds sometimes shot government troops in the back or from sniper positions in apartments. They also sometimes slipped off their red arm-bands and put on white arm-bands.

Both sides shot prisoners on capture. On March 7, a government soldier who had become separated from his command was beaten half to death by one Spartacist mob: they used the soldier's own steel helmet as a bludgeon. The brutalized soldier was rescued by two members of the mob who were horrified and had a change of heart. He was carried, semi-conscious, to a hospital by the two humanitarians. The same enraged mob showed up at the hospital after their two decent friends left. The mob dragged the badly injured soldier out and stood him against a wall, where they had to fire at him three times before he was actually killed. Some of the female spectators, who had originally supported the revolt, or at least their husbands, collapsed in convulsions at the sight.

Two days later, on March 9, six government soldiers were shot and then mutilated by the Reds "so their own comrades won't be able to recognize them."

Noske, outraged by "the bestial way in which individual soldiers were slaughtered," issued a new order on March 9: "The increasing cruelty and bestiality of the Spartacists fighting against us obliges me to order: any person who is encountered with weapons in his hand fighting against us is to be shot immediately." Pabst, the hard-core monarchist and former playmate of the royal children, noted dryly that the Social Democrat Gustav Noske had done what the Kaiser's administrators should have done on November 9 in 1918.

Captain Pabst expanded the order as he passed it on to the Guards Cavalry Division soldiers: "Guiding principal: anyone who faces us with weapons, or who plunders, goes up against the wall. Every troop leader is responsible to see that this happens. Furthermore, in any house from which our troops are fired upon, the occupants are to be thrown out on the street whether they claim to be innocent or not, and the houses are to be searched in their absence, and suspicious persons, when actual weapons are to be found, are to be shot."

Pabst also took full advantage of the shoot-on-capture order. Leo Jogiches, the mad lover who tried to avenge Rosa Luxemburg with a detailed description of the Liebknecht-Luxemburg murders, had been arrested at his apartment in Berlin-Neukölln when the March Revolt broke out on the very plausible suspicion that he was one of the instigators. Jogiches's death on March 10 was reported as "shot while trying to escape." In fact, Jogiches was shot in the back of the head by a senior prison warden named Ernst Tamschick in Moabit Prison on March 10, the day after Noske's official order. The Communist agent Karl Radek had also been picked up but Radek used his Russian passport to claim immunity and was kept alive, perhaps for future reference.

On March 11, *Freikorps Oberleutnant* Otto Marloh ordered his men to shoot a whole detachment of 29 surrendered Red Sailors whom he felt had changed sides once too often.

"This order proved so effective, however harsh it might sound, that it brought about the final end of spilling German blood," Pabst wrote. "Our people had the means, at long last, to protect themselves from assassination. On the other side, fear and terror ruled, so much so that the bloody specter of back-shooting vanished in proportion to the fear. The battle now became more man to man and face to face. It must certainly be conceded that the troops may have taken a reckless advantage of this order. But we soon arrived at the actual intention, which was the avoidance of useless bloodshed."

The Battle of Blood Week in Berlin officially ended on March 13, with about 1,500 dead, including at least some women and children killed by stray bullets or concussion from the *Minenwerfer*. The Social Democrat Noske, whom Pabst dryly noted had never himself been a soldier, received a lot of criticism for his order to shoot armed rebels on capture, but Pabst—who was not fond of Noske—said that this tough stance was responsible for saving lives on both sides. Noske himself was defiant of his critics in the USDP, the political incubator of the Spartacists.

"I couldn't endure the bestial slaughter of individual soldiers any longer," Noske said. "I had to stop the bestiality through the threat of extreme terror.... I don't propose to engage in legal hair-splitting. When thousands of men raised their weapons against the government in the streets of Berlin, when plunder and murder were celebrated as orgies, then a condition rose up on the far side of justice, and the necessity of the state bid me to restore peace and order as hurried as possible.... I am answerable for what happened before the state, the country and the people. I do not shy away from the judgment of the nation."

Armed Spartacists take over parts of Berlin, January 1919

Noske then issued another order: householders had to turn in their weapons and were subject to search. Berlin might be pacified but other parts of Germany were still unruly. On April 12, Spartacists in quasi-independent Saxony, the former home of the Liebknecht brood, kidnapped Defense Minister Gustav Neuring, a

Social Democrat, out of the Saxon War Ministry building. Neuring, a left-wing liberal and a recent supporter of the Worker and Soldiers Council that had driven the Kaiser into exile few months before, had reportedly threatened to reduce the pensions of the war wounded. A mob of 500 to 600 former soldiers, some of them wounded or recently recovered, turned up at his office in Dresden. The crowd refused to disperse. Neuring ordered the police to throw concussion hand grenades into the crowd. The police refused to throw hand grenades at German veterans. The angry demonstrators set up machine guns and opened fire on the War Ministry building. After a few hours of shattered glass and ricochets, the mob stormed the building, rampaged up to the second floor and seized Neuring, who had gone into hiding. Saxon War Minister Neuring was beaten up and thrown into the Elbe River. He was 40 years old and a good swimmer and could tread water, even with his clothes on, so he was shot in cold blood before an audience of several thousand people when he tried to swim back to the bank. He vanished into the Elbe. The former bailiwick of the fugitive Wettin Dynasty, once the most conservative kingdom in Germany, was now known as Red Saxony.

On April 13, 1919, Hermann Ehrhardt and elements of the Second Naval Brigade arrived in Berlin from Braunschweig, much too late for Ehrhardt's own satisfaction. Captain Ehrhardt had been trying to get authorized to support the Guards Cavalry Division in Berlin since the fighting broke out, but his first two regiments were deployed elsewhere and the third regiment—his reputation was drawing large numbers of recruits—was still in training. The third regiment was rushed to Berlin while it was still being trained in street-fighting tactics and the use of hand grenades, and assigned to help patrol some of the upper-class neighborhoods. Ehrhardt noted caustically that the upper-class people, many of them *nouveau riche* rather than nobles or aristocrats, seemed to think themselves too good to offer shelter to the soldiers who were sent to protect them, even in the recent emergency: "The attitude of these elegant people showed me quite clearly how sick the attitudes and thinking in Germany had become."

The elegant people, however, made no bones about pointing out arms cache left behind by retreating Reds: Within a matter of days, The Second Marine Brigade had collected 2,060 rifles, 72 machine guns, 70 pistols and 1,800 swords and sabers, many of them perhaps family heirlooms from the previous occupants to the upper-class villas.

Now for the first time, Waldemar Pabst, the staff officer, made the acquaintance of Hermann Ehrhardt, the naval commander with some overtones of being a privateer, if not a pirate. Their mutual admiration would soon be fostered by the demands of the day. Eugen Leviné-Nissen, last of the mad lovers, was not to be found in Berlin either among the living or the dead. When he turned up again, Ehrhardt and Pabst would not be long in confronting him.

Chapter Sixteen
The Soviet Republic(s) of Bavaria

Bavaria, the second largest state of the German Empire, has a history of what is called "particularism"—the idea that the Bavarians are more German than the Prussians or Hanoverians or Saxons if and when they chose to be Germans at all.

The Bavarian Wittelsbach Dynasty was the oldest ruling house in the German states. The Wittelsbach rulers, however, had been mere tribal dukes until Napoleon Bonaparte promoted the dukes of Bavaria and of Saxony to the rank of king because he wanted each of his brothers and sisters to marry a prince or princess.

Before and after the promotion, the oldest dynasty in Germany had a certain reputation for eccentricity. One of the Wittelsbachs extended a confused and confusing German civil war during the Middle Ages when he murdered the recent winner—his own ally—for no reason that made any sense to anyone. He was shortly killed by the murdered man's infuriated bodyguard. But the dynasty stayed in power. Ludwig I, in early Victorian times, had shocked the world with his sexual appetites, which were abnormally normal. Ludwig's affair with Lola Montez, a 'Spanish' dancer who was actually Anglo-Irish, caused riots when he spent huge amounts of money on her jewelry. Ludwig II, the deceased sex maniac's son, was tall, handsome, black-bearded and very probably homosexual. But Ludwig II became a celebrated patron of Richard Wagner. People believed Ludwig II was insane when he built *Neuschwanstein*, a fantasy castle inspired by Wagner's opera *"Lohengrin,"* which features a prince turned into an enchanted swan. Ludwig's fairy tale castle in Bavaria in turn inspired the fairy tale castle at Disneyland in California and Cinderella's Castle at Disney World in Florida. Neuschwanstein castle was completely indefensible the day it was built—siege guns had come a long way since the 11th Century—but in the long run it turned out to be one of Bavaria's leading tourist attractions. Ludwig built more castles, but he brought Bavaria into what turned out to be the Seven Weeks War on the side of Austria-Hungary and Hanover against Prussia. The Hanoverians won one battle at Langensalza in 1866, but were deprived of their king. The brave, blind English-educated George V, The last Hanoverian king, who had been baptized by Jane Austin's brother, bluntly refused to accept Prussian dominance and withdrew into life-long exile. The Austrians and the Bavarians lost every battle they fought against Prussia. Hanover also gave up after the single victory at Langensalza. Ludwig II also knuckled under and, after Bavarian, Hanoverian and Prussian troops had served together in the victorious Franco-Prussian War of 1870 – 1871, he accepted King Wilhelm von Hohenzollern of Prussia as Kaiser Wilhelm I of Germany, with the understanding that Bavaria would continue to be semi-independent. Otto von Bismarck helped this merger to happen by looting the Hanoverian royal treasury and paying off some of Ludwig II's formidable debts. The rest of the Hanoverian Dynasty's *Welf* Fund, also known as the "reptile fund," was set aside to bribe journalists. Bismarck usually got a pretty good press except from Hanoverian dissidents—most of them Catholic gentry in a mostly Lutheran and Calvinist country. The last white survivor of Custer's Last Stand who fought with the seven companies that defended Reno Hill and lived to be almost 100 was Charles Windolph—a winner of the Congressional Medal of Honor and a fugitive Hanoverian who preferred America, Indians and all, to Prussia.

Ludwig II of Bavaria took the partial dissolution of his royal powers rather badly, but he was sustained by his overwhelming love of Wagner's operas and his castle-building program. Prussia also allowed Bavaria to maintain its own separate army in peacetime and its own postal and customs service. When King Ludwig, the Swan King, still single, reportedly shifted his propositions from stable boys to leery soldiers, he was taken away from his role as monarch, and died in 1886 in what some have called a very suspicious suicide by drowning. The Swan King's half-sister Elizabeth, one of the most beautiful royal princesses in European history, had married and produced a son after her wedding to Kaiser Franz Josef of Austria Hungary. The son, Rudolf von Habsburg, also had the Wittelsbach reputation. Rudolf and one of his mistresses, the teenaged Mary Vetsera, died in another suspicious suicide in Rudolf's hunting lodge at Mayerling in 1889—a topic of several romantic movies in French and English. Elizabeth, known to her friends as "Sisi," became increasing unhappy with her stodgy husband, slept very little, carried a leather parasol to protect her face from spectators and the sun and travelled as much as possible.

She was brutally and senselessly murdered by an Italian anarchist who stabbed her to death with a sharpened awl.

Tiny, inoffensive, sexually normal with a wife and four daughters, Ludwig III was the reasonably popular King of Bavaria when war broke out in 1914. The world war became increasing unpopular in Bavaria, perhaps more than anywhere else in Germany. As the casualties piled up and defeat became probable, Prussia and the Kaiser, who was also King of Prussia, also became increasingly unpopular.

The failure of the *Friedensturm*, the German peace offensive in the spring of 1918, and the increasing number of fresh, courageous American troops and improved French FT-17 Renault tanks on the Western Front led the Bavarian dislike of Prussia and of the war to reach threatening proportions.

Kurt Eisner

"The Army was very near the end of its tether," Admiral Alexander von Müller, the Kaiser's military secretary, wrote in his diary on September 29, 1918 after a conference with staff officers. "Bavarian and even Saxon troops have surrendered." On November 1 the admiral wrote: "The papers are full of the Kaiser's abdication. I hear from private sources that the King of Bavaria has been won over to the idea that His Majesty should do so."

Bavaria, the kingdom that attracted eccentrics as a magnet attracts iron filings, had attracted Kurt Eisner, a bearded, bespectacled man who had once been the drama critic of the Munich Post. Eisner was a Marxist but wore the trademark beat-up floppy black hat of an anarchist. He was a failed husband, the failed father or five children, and a failed journalist who came from Berlin after his divorce and his dismissal from his job in Berlin. The fact that he was both a Jew and a Prussian would seem to have been two strikes against him in Bavaria, where neither Prussians nor Jews were especially popular.

Kurt Eisner had supported the war until 1915, when he became a pacifist. In 1918 he had been jailed at Stadelheim Prison in Munich for supporting a strike at an ammunition plant. But the day after the King of Bavaria had reportedly suggested that Kaiser Wilhelm might want to abdicate, Kurt Eisner also made an abdication speech.

"There will be no *Reichstag* elections as scheduled," Eisner said on November 2. "The revolution will come first."

Eisner and his fellow speakers demanded bread and peace—the same slogans Lenin had used in Russia after the German General Staff slipped him into the country—and the abdication not just of the Kaiser but of all the German monarchs. The crowd he spoke before contained large numbers of military deserters or wounded men sent home to recover who had elected not to return of their units. In a worst-case scenario, these men would once have stood at least some chance of being shot for cowardice in the face of the enemy. Social Democratic workers from Berlin, long radicalized by years of agitation, were also in the crowd. King Ludwig III of Bavaria, 73 years of age, was taking his afternoon walk in street clothes when he saw how excited the crowds were. He hurried home and the queen, the four Wittelsbach princesses, and two loyal courtiers fled in an automobile with jewels and, in Ludwig III's case, a box of his favorite cigars. The car went into a ditch in the mountain fog and had to be hauled out by farm horses. The royal family headed for Hungary.

"If it were someone else who was taking over, I wouldn't say anything," Ludwig III murmured. "But as it is, nothing good can come of it. I am afraid there will be civil war."

In Berlin, once the Kaiser had escaped to the Netherlands, his cousin Prince Max of Baden—a South German *Swobe* and a Catholic though still a Hohenzollern—took over as regent until he could hand the regime over to Friedrich Ebert, leader of the Social Democrats.

Eisner and his supporters gathered at a beer hall where Eisner was elected the First Chairman of the Workers and Soldiers Council. The crowd marched to the Bavarian Diet and proclaimed that the Bavarian monarchy—actually reasonably popular with most Bavarians outside the factory-worker-and-deserter clique—had been abolished and that Bavaria was now a republic with Kurt Eisner as its Minister President. Red flags went up all around Munich. Eisner brought some of his coffee-house cronies into Bavarian politics. Erich Mühsam, son of a Jewish pharmacist from Berlin, and Gustav Landauer, son of a Jewish shoe shop owner, who had studied classics at Heidelberg, renounced God, and became a nominal anarchist who generally voted with the communists and part of the new Bavarian cabinet, though neither was a Bavarian. Landauer was another drastically failed husband—his wife died of pneumonia during the Starvation Blockade because he lacked the resilience to provide her with adequate food or medicine. Both men had done time in prison for opposing the war and supporting the ammunition strike. They had been released about the same time that Eisner had. They had no experience in administration or elected politics. This soon became obvious.

The Eisner regime in Munich lasted 100 days. No important legislation was adopted, but the worker-soldier councils heard many speeches and passed many resolutions. High schools were told to adopt Student Soviets and—in one of the few changes enacted by the Kurt Eisner regime—the Catholic Church's supervision of the schools was rescinded. Rightists and political Catholics were not slow to point out the background of the Eisner faction to explain this change of policy: in the German Empire, schools were traditionally supervised by Catholics, Lutherans, Calvinists or Jews, depending on the beliefs of the students and their families. And Bavaria was the most Catholic part of Germany except for Silesia, which was predominantly Polish.

The attack on Christianity did nothing to improve an economy already influenced by the collapse of the booming ammunition industry with the end of the war, or to improve the nutrition of people with no money to buy food on the black market. The British Starvation Blockade was still in force, reportedly to make sure the beaten Germans signed the impending peace treaty. The Red Guards were kept in line by flattery and access to women from the street. But most Bavarians came to detest Kurt Eisner and his anarchist clique. Voters were threatened by Red Sailors riding around in trucks and warning them to vote for Eisner or else. Erich Mühsam enjoyed waving a pistol he probably didn't know how to load. The election itself, however, was not violent and when the ballots were counted, the Bavarian People's Party—political Catholics and economic moderates—received 66 seats in the Bavarian Diet. Kurt Eisner's clique received 3 seats. Eisner's followers garnered 2.5 percent of the total votes.

Eisner refused to re-convene the Bavarian Diet. Right-wing officers demanded that he step down or be taken out. Support for extreme Leftists, weak all over Germany, was almost non-existent in semi-autonomous Bavaria outside Munich with its colony of unemployed ammunition workers and military deserters. Finally, under pressure from all sides, Eisner agreed to reconvene the Bavarian Diet on February 21.

Kurt Eisner set out for the Diet about 10 a.m. on February 21 accompanied by his secretary and a Red Sailor as a bodyguard. Eisner probably never saw Count Anton Arco auf Valley, a handsome 22-year-old Austrian-born aristocrat with an excellent war record in the Bavarian Army—and a monarchist with a deep-dyed hatred of Leftists. Count Arco Valley was also one-quarter Jewish—he had a rich and elegant Jewish grandmother, not uncommon among financially challenged aristocrats all over Europe. But he had encountered prejudice from the Thule Society, a group of occultist anti-Semites who dreamed of revitalizing the Germans by eliminating any Jewish influence and shunning marriage to people with any Jewish ancestry.

"Eisner is a Bolshevik, a Jew; he isn't German, he doesn't feel German, he subverts all patriotic thoughts and feelings. He is a traitor to this land," Arco Valley reportedly said. Count Anton Arco Valley was related to Lord Acton, the famous British exponent of democracy who said: "Power tends to corrupt, and absolute power tends to corrupt absolutely." Acton, incidentally, had dropped the Semitic Jews and Arabs from the list of important cultures in preference to the Aryan

Persians as predecessors of the Greeks, the Romans, and the Germanic peoples. What happened on the streets of Munich was triggered partly by monarchist politics and also partly by anti-Semitism, which was percolating in Bavarian politics due the Prussian-born Eisner Regime's attack on the Catholic schools.

Count Anton Arco auf Valley spotted Eisner walking toward the Diet, slid out of a doorway, and pulled a pistol. He shot Eisner in the back of the head, then dropped his aim and shot Eisner in the back. The second bullet punctured a lung. Eisner was dead or dying when he hit the ground. The Red Sailor bodyguard swung his rifle and fired at Arco Valley, knocked him down with the first shot, then shot four more times. Arco Valley was kicked by the 40 or so Reds who had been waiting the cheer for Eisner. The Count, amazingly, pulled through after medical help and stood trial. Eisner died on the street.

Murder on the street had turned Eisner the bungler into Eisner the martyr, at least in the opinion of some hysterical women who wept and dipped their scarves in Eisner's blood. Rumors also started that the Minister of the Interior, Erhard Auer, had engineered Eisner's murder: Auer was a Social Democrat, not a Rightist, but he had done considerably better in the January 12 election than Eisner had. The Diet convened an hour after the murder and as Auer was offering a favorable eulogy for Eisner, a saloon waiter and part-time butcher named Alois Lindner stalking in wearing a grey overcoat and carrying a U.S-made Browning Automatic Rifle, a

Count Anton Arco auf Valley, assassin of Kurt Eisner

light machine gun issued to U.S. Marines in the last days of the war. Lindner shot Auer twice with the BAR and Auer hit the floor, seriously wounded. Lindner then snapped off two more shots at the Catholics in the Bavarian People's Party's delegate benches. Major Paul Ritter von Jahreis, a titled aristocrat and adviser to the Military Ministry, courageously tried to block Lindner. The sometime butcher shot Major Jahreis to death at point-blank range. Then Linder walked out with no further attempts at an arrest. Another volley from somewhere inside the Diet killed the Bavarian People's Party's leading delegate, Heinrich Osel. Two other elected delegates, Albert Roßhaupter and Johannes Timm, suffered nervous breakdowns and had to be pried out weeping from under their benches. Bavaria now had no government.

Bells tolled for Kurt Eisner, the atheist who had tried to undermine the Catholic schools and set up Student Soviets. The bell-ringers sometimes had to be convinced at gun-point to toll them. Heinrich Mann, Thomas Mann's brother, offered a eulogy and called Eisner "the first truly spiritual man ever to head a German state." Otto von Bismarck, after a spectacularly unruly youth, had become a born-again Christian with Jewish friends, and Kaiser Wilhelm, who was also a devout Christian with Jewish friends, could read some Hebrew and enjoyed Biblical archaeology. But Heinrich Mann spoke fiction as well as writing it—or at least he knew who was holding the gun. Bavarians who had not appreciated Eisner found a way to turn his sidewalk shrines into a laughingstock: they poured the urine from bitches in heat onto the pavement and male dogs scent-marked the tempting sites with their own male urine.

A Worker-Soldier-Peasant Council took shaky control of the government, arrested right-wing hostages, and tried to run the city and the country without success. On March 11, the Bavarian Diet reconvened and elected Johannes Hoffmann, a mainstream Social Democrat as the new Minister President. Hoffmann, a former schoolteacher, was a liberal but not a Red. He tried to form a coalition government with the USPD and keep the outright revolutionaries out of power.

Four days after Hoffmann was lawfully elected, Béla Kun, son of a Jewish father and a Magyar mother, led a mostly Jewish contingent of extreme Leftists in a revolt that took over Budapest in Hungary. The Reds in Munich saw this as evidence that the World Revolution was expanding—never mind the national elections in Germany that gave the Reds fewer seats than the monarchists and far fewer than the mainstream Socialists. When the Bavarian politicians assembled in the vacant bedroom of the exiled Queen of Bavaria to form a left-liberal government, the communists, who had been invited, did not attend. Gustav Landauer, the theologian of atheism, advocated something like the Soviet system. Erich Mühsam proposed himself as People's Deputy of Foreign Affairs. But the post eventually went to Dr. Franz Lipp, who actually owned a respectable grey frock coat and put it about that he had been a diplomat for the Kaiser and was a close friend of the Pope. He later turned out to be a former inmate of several insane asylums. But he got the portfolio. Military Affairs were handed over to a former waiter—he knew how to take orders —but Ernst Toller, a dramatist who had fought on the Western Front, later took over the military duties. The financial system was handed to a team including a free-enterprise advocate and a Marxist, who were supposed to cooperate. Gustav Landauer himself was named People's Deputy for Education over the objections of some of the peasant delegates, who shrewdly wondered aloud how a Jewish atheist would interact with Bavaria's Catholic school bureaucracy.

Abruptly, history arrived in the person of Eugen Leviné-Nissen, the last Mad Lover from Berlin, who, having escaped the downfall of Rosa Luxemburg and Leo Jogiches, showed up in Munich. Leviné was accompanied by two more Russian Jewish Bolsheviks: Max Levien, like Eugen Leviné, was a fugitive after the Russian

Revolution of 1905 and a veteran for four years in the Kaiser's Army: he wore spurs and carried a riding whip. Tovia Axelrod, a long-term Russian revolutionary, had done time in Tsarist work camps in Siberia. These men took their orders from Lenin, not from Landauer. Their military expert was Rudolf Egelhofer. A lapsed Catholic from a poverty-ridden Bavarian family—his father was a sign-painter—Egelhofer was a former sailor and aviation mechanic in the Kaiser's navy with a five-year record including infantry combat, insubordination and desertion. At 23, Egelhofer had been one of the original Red Sailors from the battleship fleet who began the mutiny at Wilhelmshaven and Kiel in November 1918 which Hermann Ehrhardt had helped to crush in January of 1919. Now Egelhofer was back home in Munich threatening anyone who disagreed with the revolution with a clenched fist.

Max Levien

The knowledge that Reds from Berlin had arrived impelled Johannes Hoffmann, the lawfully elected President Minister of Bavaria, to get out of town with his handful of loyal troops. Hoffmann set up a legal government in the smaller city of Bamberg and put out a call for more Bavarian troops to take back the capital. For the next six days, the Soviet Republic of Bavaria operated as a mixed anarchist-and-communist regime. Dr. Franz Lipp declared war on Switzerland and Württemberg when they refused to loan Bavaria 60 locomotives. Ernst Toller, a 26-year-old Prussian Jewish war veteran invalided home with a nervous breakdown, who was serving as chairman of the central council, shortly told Dr. Lipp that Lipp would have to resign. Anybody who had served in the German Army understood that Munich was no match for Switzerland. Meanwhile, Dr. Gustav Landauer announced that anyone over the age of 18 could attend the University whether he was qualified or not. Elementary students should begin memorizing Walt Whitman, his own favorite poet—when translated from English to German by Dr. Gustav Landauer.

On Palm Sunday, April 13, The Republican Guards, troops loyal to the Hoffmann regime in Bamberg, arrested Erich Mühsam and some other Soviet Republic of Bavaria officials and took them under guard to the main railroad station. The government dithered but the hard-core Reds, led by the three Russians and Egelhofer, attacked and routed the loyal Bavarian troops with machine gun fire and mortars. The Bavarian loyalists crowded onto a train and made good their escape. The Russian Reds, having shown fight while the German-born Jewish anarchists tripped over their own rhetoric, declared the Second Soviet Republic of Bavaria—this time under absolute Communist control with tentacles firmly entwined with Moscow's. The Reds first called a general strike—more to show off than anything else—and middle-class children learned to do without milk. Red soldiers began looting middle-class homes and stealing food, liquor, and valuables at gunpoint. An army of about 20,000 workers and soldiers—unemployed munitions plant employees and deserters were heavily represented—was given a raise in pay. Privates received 25 marks a day—five times the pay of Nationalist troops—and non-commissioned officers received 130 marks a day, while Red officers received 200 marks a day. Since most German officers were the sons of property owners who had graduated from cadet school or reserve officers with *Gymnasium* or University educations, attracting them to the Reds involved a substantial bribe—which actually worked in a few cases. The Red soldiers were told that they could now elect and depose their own officers, as the Russian Army had after the first Russian Revolution during the war, which led Alexander Kerensky's Russian Army to collapse.

Hotels were turned into barracks where liquor flowed freely and women off the street were made more than welcome. The Reds tipped the table too far when they armed Russian military prisoners captured by the Kaiser's victorious army and added them to their Red Army. Prussian discipline had been able to make good use of Polish and Russian hardiness—the orderly who had helped the African porter save Hermann Ehrhardt from typhus during the Southwest African campaign was an ethnic Russian—but an army without discipline was more than the half-starved foreign-born prisoners could handle. The Bavarian civilians, especially the women, came to dread the sight of the Russians.

Rudolf Egelhofer, the Red Sailor with the clenched fist; executed by Bavarian authorities

Hoping that the Bavarians of Munich had had enough of anarchic foreign rule, the Social Democratic Minister President Johannes Hoffmann launched another invasion by his loyal Bavarian troops on April 20. Ernst Toller, who had been arrested internally by the Reds, was set free and told he was to command the Red Army of Bavaria—under the supervision of Rudolf Egelhofer, the former Red Sailor and now the war minister of the Second Soviet Republic of Bavaria. Toller was Jewish, like the three Russians, but he had been born in Posen, in the eastern part of Prussia. Toller made a better poster boy for the revolution than the Russian-born Leviné, Levien, and Axelrod. Set free from prison, but not from his dreams of a glorious future like the visionaries Rosa Luxemburg and Leo Jogiches, Toller borrowed a horse from a cavalryman who decided to maintain his neutrality and rode to the Red Army besieging a market town named Dachau. The Bavarian troops were taken in flank by some Leftist workers from a nearby munitions plant and most of the Bavarian loyalists either surrendered or ran for it. Bavarian troops had helped clinch the Prussian victory at Sedan in September 1870 that had helped established the German Empire. Some of them crossed the Meuse on ponton bridges and others swam to block a French escape. Ominously, the Bavarians shot a number of French civilians who had picked up rifle, and possible some who hadn't. The Bavarians' valor was remembered along with Prussian General Friedrich von Bredow's Death Ride two weeks before at Mars-la-Tour was a high point of German heroism in the Franco-Prussian War of 1870. But the war-time *Stoßtruppen* on the Western Front of 1918 must have syphoned off their best men, and other Bavarian soldiers undoubtedly went back to their families at the time of the Armistice. The Bavarian leftover troops of 1919 fought badly against the Reds. The Red Army captured five officers and 36 enlisted men. Red Sailor Egelhofer told Toller to shoot the officers but Toller had the decency to refuse. The Bavarian national troops shortly lost another engagement near Freising. Ernst Toller, fresh out of Red custody, was hailed as "The Conqueror of Dachau." Toller marched through Munich with a procession that included about 100 bedraggled Bavarian military prisoners and a few captured cannon.

"Egelhofer, the commander of the [*Red*] Army, spoke from an open window," the journalist Oskar Maria Graf wrote. "Decisive and unaffected in his sailor's uniform, he stood there, sometimes brandishing a clenched fist. Whoever heard him had to believe him."

Max Levien proclaimed the new internationalism when he spoke about the future of Germany with an opinion that was not especially German.

"It is necessary that Germany be humiliated, that the colonial troops of France and England march through the Brandenburg Gate," Levien said.

The Bavarian peasants, staunchly Catholic, staunchly 'particularist' in seeing Bavaria as a separate country, were unimpressed with Prussian or Russian Jews or home-grown Red hooligans trying to run Bavaria as a Soviet Republic. The peasants also did what they could to restrict the flow of food to the city, to the point where parents had to present a doctor's certificate to obtain milk for small children.

"What does it matter if for a few weeks less milk reaches Munich?" Eugen Leviné reportedly said. "Most of it goes to the children of the bourgeoisie anyway. We are not interested in keeping them alive. No harm if they die—they'll only grow into enemies of the proletariat."

And in Brunswick, Captain Hermann Ehrhardt picked up his telephone....

Chapter Seventeen

The Munich Massacres

Hermann Ehrhardt disliked Gustav Noske.

"In Berlin I learned to know Noske," Captain Ehrhardt wrote in 'Adventure and Fate.' "His whole type repelled me. His forehead was low and half overgrown with hair that he tried to keep cut short, like a criminal. He kept his shoulders held high on his long disproportionate body. This *Reichswehr* minister made a soldierly impression on everyone else. He acted jovial, as if we had sat ourselves down to drinks many times together. I have to acknowledge he had his act down pat. I never would have guessed he was ever a basket weaver.

"The administration kept trying to develop republican and socialist troops but the effort shattered, because half-baked officers turned into pancakes."

Inspecting the Marine Brigade, Noske stood before one strapping young *Freikorps* volunteer and asked him what he had done before he enlisted.

"I was a basket weaver!" The soldier said briskly.

"Keep up the good work," Noske said. "You can never tell what honest work will do for you. Just look at me!" Ehrhardt knew the young soldier was actually an upper-class university student but he managed to keep a straight face.

Noske seems to have understood that Ehrhardt disliked him, so the Second Marine Brigade was sent to protect a number of smaller cities in central Germany. A large number of the troops were sent back to Braunschweig, where they received a welcome fit for long-lost sons or nephews. Ehrhardt's one fear was that his men would eat and drink and dance too much and forget their drill and training. The Second Marine Brigade continued to grow—by the early spring of 1919 Captain Ehrhardt had about 3,000 men under arms, some of them Army or Navy cadets who were too young to have seen fighting in the war. Some of them were also working-class Social Democrats who realized that their own democratic revolution had been hijacked by foreign-born communist revolutionaries.

"In Munich the Red rule stood forth in full flower," Ehrhardt wrote. "This Soviet Republic seemed to be of a different type than the one in Braunschweig. I made an application to Berlin and asked for permission to take my troops, who lacked anything else do to, to Munich. The application succeeded and the Brigade rolled toward Munich at a strength of 3,000 men."

Ehrhardt's application was the last word on the subject. As soon as the Russian-born Bolsheviks had seized power from the bungling German-born anarchists, Bavarian President Minister Johannes Hoffmann was on the telephone to Berlin asking for national troops to save Bavaria from the revolution he himself had initially supported. Hoffmann's fellow Social Democrats in Berlin told him that if Bavaria wanted government troops from Prussia and the other German states, Bavaria would have to give up any pretense of 'particularism' and accept status as part of the new German Republic. Dr. Hugo Preuß, the German Jewish patriot whose three sons had served on the Western Front—two of them decorated for valor and one of them now at home dying of wounds—was in the process of cobbling together the Weimar Republic's constitution. Nobody among the Social Democrats including Ebert and Noske, among the Democrats like Dr. Preuß, or among the other liberals or the Nationalists wanted any more semi-autonomous German states within Germany—especially when the states couldn't defend themselves from Russian-born trouble-makers. Hoffmann refused to give up Bavaria's special rights. President Ebert, who had lost two sons of his own at the Front, declined to send troops.

After the Bavarian defeat at Dachau, Hoffmann went to Weimar and asked again for government troops—this time in desperation. He was bluntly told that Bavaria would have to waive all special rights and become an integral part of Germany. General Ernst von Oven, *Pour le Mérite*, son of an old military family, a 60-year-old Prussian general and former General Staff officer who had served on the Russian Front, at Verdun against the French, on the Somme against the British, and in the Argonne Forest against the Americans, would command the army of *Reichswehr* and *Freikorps* troops and hand Munich back to Hoffmann—with no questions asked about how it was to be done. President Minister Hoffmann glumly agreed.

Captain Ehrhardt's Second Marine Brigade left for Munich on the night of April 29-30 at express train speed and rolled through Thuringia and over the border into Bavaria.

"The country people greeted the soldiers who had come to restore order," Ehrhardt remembered. "In the cities we sensed a Red mood and a Particularist distaste for us: *'What do you sow-Prussians want here?'* It's a shame I'm just a South German. We lost some time because 30 to 40 kilometers from the city of Munich some rails had been ripped up and a wrecked locomotive lay as a barrier on the roadbed.

"A fully camouflaged armored train labeled Number 25 came up the track and passed us and this was a proof to me that heavy action was expected. His Excellency von Oven was the over-all commander. Our brigade was part of the group commander by Colonel Detjen. All the dispatches sounded serious. There were rumors of a Red army of 60,000 men. But I didn't believe it at all, because I remembered the exaggeration before the undertaking at Braunschweig."

As General von Oven's armored train and the side-tracked Second Marine Brigade train rumbled on toward Munich, Adolf Hitler made his first political speech.

Hitler, An Austrian national, had served for four years on the Western Front with the List Regiment, Infantry Regiment Number 16 of the Royal Bavarian Army. Wounded once by shell fragments, holder of the Iron Cross Second Class and ultimately of the Iron Cross First Class, Hitler had lost his eyesight near the end of the war either due to an enemy gas attack or a bout of hysterical blindness. Despite his obvious intelligence he never reached any rank beyond that of *Gefreiter,* or lance-corporal, Private First Class in U.S. Army terminology. He was now stationed in Munich with the Second Infantry Regiment. The Bavarian regular soldiers were called together in response to a frantic appeal from Red Sailor Rudolf Egelhofer to join the defense of Munich against the "Prussian White Guard." Due to the leftovers of the anarchistic Gustav Landauer administration, the soldiers actually got to vote on whether or not they wanted to fight.

Gefreiter Hitler climbed on a chair, displaying his Iron Cross First Class, worn large on the chest, and spoke out: "Those who say we should remain neutral are right. After all, we're no pack of Revolutionary Guards for a lot of fugitive Jews." Hitler had gauged the temper of his audience perfectly. The soldiers voted to stay out of the fighting on either side.

Red Sailor Egelhofer had ordered on April 14 that the populace of Munich hand over their guns on penalty of death. Then the Reds had taken hostages. A raid of the Thule Society, an occultist group with anti-Semitic and anti-communist overtones, captured seven Thule Society members: Prince Gustav Thurn und Taxis, a scion of one of the richest families in South Germany; Countess Hella von Westarp, an attractive 32-year-old proto-feminist and distant relative of General Ernst von Oven on her mother's side; Baron Friedrich Wilhelm von Seydlitz, whose ancestor had been a cavalry hero in the army of Frederick the Great; Baron Franz von Teuchert; and three educated professional citizens of Munich, Walter Neuhaus, Walter Deike, and Anton Daumenlang. The other three hostages in the first batch were Professor Ernst Berger, an elderly and respected Jewish academic who specialized in criticism of the fine arts and who had reportedly ripped down a Red poster in disdain, and two hussar officers, Fritz Linnenbrügger and Walter Hindorf, captured in a previous battle. Another 12 Rightist hostages were taken a day or two later.

On April 29, the Reds reported a letter a *Freikorps* volunteer had supposedly written to his kid sister: he claimed that 11 Red nurses had been executed at Gruenwald because they were caught carrying pistols. This may have been a bratty adolescent attempt to frighten little girls, but the report had serious overtones of the *Hunnenbriefe* of 1900, where unsigned letters to the newspapers bragged of German atrocities committed against Chinese women and children at a place where no German troops had actually been engaged. Attempts to substantiate the murders of the nurses after the fact came up blank—strongly suggesting that the murders never happened.

Rudolf Egelhofer clenched his famous fist and ordered that the first 10 hostages at the Luitpold Gymnasium be shot to retaliate for the (probably fictitious) murder of the 11 nurses. The obvious motive was to instill fear and also pretty obviously to express resentment: the seven Thule Society members, however quirky or demented their beliefs—the Thule Society claimed that the Revolution had been touched off by inferior races to destroy the Germanic Aryans—came from families with titles and money. Professor Berger was a respectable Jewish scholar who had won affection through his genuine ability and wholesome personality. The two hussars were officers in an army where the soldiers didn't expect to vote.

"The bestial murder of hostages was cried out inside the city and outside," the *Münchner Neueste Nachrichten* reported on May 3. "The tidings permeated in the early afternoon and shattered the peace of the city. Revulsion and indescribable excitement.... The murders took place in the garden of the Luitpold Gymnasium. The hostages were placed with their backs against the rear inner wall. The order to fire came from a man of the Red Army named Seidl. The soldiers of the *Leib-Regiment* [Royal Guard Regiment] refused to shoot, so Russians were induced the carry out the murders. Some of the hostages were finished off with the blows of rifle butts and the thrusts of bayonets. One of the murder victims was a lady. The corpses were robbed and mutilated in such a way that three of them have not yet been identified by name. Two corpses had the top halves of their heads missing. The corpses were brought to the coroner's office...."

The somewhat exaggerated degree of the mutilations was propaganda of the Right, just as the 11 murdered nurses who seem never to have existed were propaganda of the Left. Occultist sources also claimed that Eugen Leviné and two comrades had raped Hella von Westarp in her cell any number of times, but this too appears to have been propaganda fiction. Countess von Westarp was not kept in a cell but in a cellar, and given Leviné's revulsion to bad sex habits, his repressed nature and his near-slavish devotion to his wife, choreographed gang rapes seemed somewhat out of character. The actual instigator of the hostage murders was not Leviné, who appears to have already gone into hiding after telling everybody else to fight to the death, but Egelhofer, the lapsed Catholic, who hung around just long enough to order the murder of the hostages.

"I'm not going to do a rat run," Egelhofer anecdotally said. Then he did a rat run. He took shelter, perhaps not by mutual consent, with a female surgeon, Dr. Hildegard Menzi, who lived on the fashionable Maximilianstraße. The actual fighting began after the Red leaders had taken shelter and left their enlisted men and junior officers to fend for themselves.

The Second Marine Brigade arrived on the outskirts of Munich on the night of April 30-May 1 and got their artillery and trucks unloaded from the transport train as quickly as possible. Their achievements in Wilhelmshaven and Braunschweig had enabled them to acquire quick-firing naval guns, a battery of 77-millimeter field cannon that could put a shell through a window and a battery of 10.5-centimeter (105-mm) field howitzers that could knock down a wall.

"Message after message came out of the city, deputations of citizens, groups of young former officers with official papers," Ehrhardt wrote. "All of them asked for our help, and fast. The Communists had murdered hostages! The number of the murder victims turned out to be strongly exaggerated, but at the same time we heard how bitter the citizens were over the cruelty of the Reds and that they wanted

Countess Hella von Westarp

to get rid of them but were unable to do it on their own. I sent an officer to them but he didn't get through.

"Right then and there I decide to handle my troops on my own, slip out of my own assigned sector and break into the city. This involved some risk, because the Reds were expecting our attack on May 2, but this attack gave us the advantage of surprise."

Rudolf Mann was the officer who didn't get back and he certainly got a surprise. "I was sent out to the telephone at the railroad station, and I tried with some success to stay calm while I called a lot of wrong numbers to report to head-quarters," Mann wrote. "I found, when I got back, my horse and my orderly were standing at the garden hedge. My unit was gone, vanished without a trace."

"On the whole, everything went well. The troops who broke through to the Leopoldstraße and broadened out onto the Ludwigstraße, were received with joy as liberators. But the small detachment that broke through to the main railroad station took one on the nose. Our field telegraph truck was led into an ambush and the crew was murdered."

Mann described how the telephone truck was waved into an ambush by soldiers wearing blue armbands—supposedly Bavarian loyalists—and then took fire from the Reds in the Pioneer Barracks. Two of the wounded Ehrhardt Brigade volunteers were dragged into the barracks and beaten with rifle butts and a hammer "to make them confess." One of them made it to the hospital alive after the Red position was overrun. The other died under the beating.

Ehrhardt himself, who led from the front as he had in the torpedo squadron, shortly jumped into the *Kraftwagen*—a slab-sided armored truck with four-wheel drive, metal-spoke tires that shrugged off punctures, a largely bullet-proof cab and sides and a top turret that could carry three to five machine guns. By sheer coincidence, the armored truck was named the "Ehrhardt," after the manufacturer.

"I myself rode in the armored truck, failed to keep track of the right streets, and in the vicinity of St. Ursula's Church I drove right into the enemy camp. They stared at me with big eyes and one man gestured for me to put on the power and run for it. A growling crowd came out of the houses, but my driver shifted into reverse and we backed around a corner and got away. A few bullets whistled after us but we were never hit." Ehrhardt never found out if the supposed Red who gestured for him to escape was a secret friend of the *Reichswehr* or simply didn't want to risk a shoot-out with an armored truck.

Most of the Bavarians got in the way only by asking for weapons or truck rides so they could join in the fight against the Reds. But after the telephone truck ambush, the Prussian regulars and the Marine Brigade didn't trust them.

"They were mostly kids who looked like students but the instructions were not to tolerate any armed civilians," Mann wrote. "They all were considered capable of shooting us from the side or handing the weapons over to enemy.... It's for sure that a cross-cut of the Munich soldiers wouldn't have turned up any idea about Communist thought or for the Soviet program that the Independent Social Democrats wanted, and that they weren't active in promoting it like the actual Revolutionaries. They were a moody mass who changed with the wind, and what they liked best was to have a good feed and lots of new clothes without having to actually doing anything. People who came out of Prussia to hinder the hand-outs were their enemy"

Mann's company advanced into Schwabing, the famous Bohemian quarter of Munich, and set up house-keeping: barbed wire entanglements and machine guns in case of a counter-attack. He got a look at some of the abandoned Red positions: "I must say that the previous occupants, the Red Guard Battalion, seem to have been rather clean. The members seem to have spent most of their nights at home with their families, because their sleeping areas were not often to be seen. Everything was quite orderly, with signs on the door and on the bulletin board, notes that said 'Long Live Our Gruppenführer,' bowls of bean soup and fragments of ration bread on the tables. Guns and weapons belts lay all over the place, cartridges and hand grenades in small heaps. They had also looted old blunderbusses that shot soft led musket balls and antique dueling sabers. As a sign of their nastiness, a packet of dumdum bullets with split tips were passed from hand to hand."

The first day had not gone well for the Reds. On May 1, Rudolf Egelhofer was hauled out of hiding. Dr. Hildegard Menzi was apparently arrested but not executed, which suggests, given the mood of the times, that she may have turned him in. The Bavarian authorities took the Red Sailor with the clenched first to the Residenz, where they spent the next two days beating him to almost death with their own clenched fists. The Bavarians finished him off with a bullet to the head. In death, Egelhofer became an emblem to both sides: The future West Germans of the German Federal Republic portrayed the late Egelhofer as a scowling, bug-eyed monster and the future East Germans of the German Democratic Republic portrayed him as a starkly handsome hero and named a torpedo boat after him.

Toward nightfall on May 1, The Ehrhardt Brigade sent out three assault groups to locate the Reds, who had broken off contact. The Noske order to shoot any Reds captured under arms appeared to have made an impression on them.

The first group, commanded by Lieutenant Berlin and guided by a "White Guardist" who took them through the side streets, slipped past the Red defenders, captured the Protestant Church, and reinforced the "White Guards" who in this case happened to be dependable. *Leutnant zur See* Berlin and his soldiers went back to their sleeping quarters at the Academy of Portraiture. *Leutnant* Berlin, who was Jewish, made sure that his men respected the paintings and bedded down only in unobtrusive corners in case any spectators showed up at the museum.

Karl Tillessen, the young officer Ehrhardt had to convince to fight for the Social Democrats, followed Berlin's patrol out and found that the Reds had fled the vicinity of the Protestant Church. Some "White Guards" who knew the streets joined them. The burning remnants of a newspaper kiosk illuminated their faces: "Were they friendly to us?" Tillessen asked himself. "In any case they had the physiognomy of criminal types, and in the fever of fighting, if the Red Guards seemed to be winning, they might well change sides."

As Tillessen's 12-man patrol came around the side of the Protestant Church, they took fire from the Café Orient across the street. His patrol couldn't advance without losses. But a section under Corporal Wiegner ran around the far side of the church. Then their best man—"our matador"—threw long-distance stick hand grenades into the café garden. Two huge explosions detonated and the whole patrol rushed the Café Orient and captured it.

"The nest was empty. During the intensive search of the building we only found a waitress who had been terrified at the beginning of the fight and ran into the room 'where even the Kaiser has to go on foot,' and was slightly wounded in the arm. The Spartacists themselves had run away at the sight of the first hand grenade being hauled back for the throw."

Tillessen's detachment was returning when some sympathizers told them that they had seen a 105-millimeter field gun that the Spartacists had supposedly captured from the Rightists at the end of the Sommerstraße.

"We quickly held a war council in the open street. Should we haul away the cannon with our few men or not?" one of the detachment wrote afterwards. "Our mentor and second-in-command, the old sea dog Sab Sauermilch, rode up on an aging artillery horse with the words 'Children, let's get going, you get nowhere without losses.' We others agreed. Tillessen was our backbone and [*Leutnant Hermann*] Fischer said bluntly 'We're taking the cannon.' After this brief intermezzo we broke the shock troop into two detachments and advanced at double time sheltered by the house fronts, shouting 'Windows shut!' We saw some dubious faces but nobody dared to say a word from the houses.... We came to the branch street from the Sommerstraße and at last, there was the cannon, with a Red flag mounted on the gun carriage. Tillessen, Sauermilch and one or two other men gave us cover to the next street while the rest of the men, led by Fischer, moved the gun. In the middle of the preparations we were disturbed by a wild shot. Tilly had barely gotten to the next street when a sideswipe of gunfire from two machine guns and about 30 rifles opened up on us, and left nothing more to wish for. The cold shower in sparkling garb mostly came in much too high. We answered with Parabellum fire [*Bergmann submachine guns or Luger 08 automatics*] and the crowd that wanted to rush us suddenly changed their minds. In a few moments the cannon had been hooked up and Fischer rode on the gun carriage as the rudder man...." Once the detachment got past the burning newspaper kiosk the company commander passed out cigars with the brisk words "*Sehr gut!*"

Back at the Academy of Paintings, *Leutnant* Berlin's assault company was resting in the corners between the paintings when *Leutnant* Van De Loo rushed in and reported street fighting. Berlin's men were still in their greatcoats—German soldiers used the greatcoat as sleeping apparel in the field. They they stuffed their pockets with fistfuls of rifle or pistol clips and hand grenades and headed for the sound of the shooting. Van de Loo seemed to think he was in charge since he was better dressed. But Berlin retained command. When some of the men held a "soldier's council" and suggested that they take off their greatcoats, *Leutnant* Berlin, no fan of Bavarian probity, pointed out shrewdly that they'd probably never see them again if they took them off in Munich. The greatcoats remained buttoned. The patrol was guided by some "White Guards" who, the reporter said, had probably been wearing red armbands 10 minutes before. Red ambushes had been set up on two side streets but the Reds shot high and no one was hit. The patrol leaders decided to consolidate and headed back. On the way back they found and rescued a wounded man from the Ehrhardt Brigade and got him to the field hospital. A few minutes after they returned, Tillessen's detachment showed up dragging the field cannon.

The 10 hostages murdered by the Reds had been inflated by propaganda to 30 but by 4 p.m. Red resistance in Munich was obviously crumbling: "It looked like our entry to Braunschweig," one Ehrhardt Brigade soldier remembered. "Black-white-red banners on the house fronts had been exchanged for the blue-and-white ones. Flowers were thrown at us. Cigarettes, wine, chocolate and other food was handed out to us left and right. Many older ladies wiped their eyes from tears of joy. The men tipped their hats, their eyes brightened. 'The Prussians are coming!' "It's good that they've come, another day would have been too late!' Young girls cleared the way for us through the crowd. We were among the first. Bands played and hussars, artillery and other troops marched in from the north.

In the middle of the premature victory parade, some White Guardists ran up and told the Ehrhardt Brigade leaders that their men were under attack at the Sendlinger Tor, the southernmost gate of Munich's medieval fortifications and a city landmark.

"Instantly a detachment picked up their rifles and put on their steel helmets. They had already decided. Their eyes flashed—Aha, now let's get at the enemy. The other men seemed almost envious."

Two assault groups and two machine guns were sent to help the White Guardists. As the White Guardists led the Ehrhardt men through the streets, people cried "Hurrah! Long Live the Prussians!" The Ehrhardt men shouted back: "Shut your windows!" Sniper fire was always possible. Energized by the excitement, the men carried the heavy machine guns through the streets until they spotted the landmark Sendlinger Tor. Abruptly they started to take fire from a number of unseen rifles. The Spartacists were under good cover and trees obscured the lines of fire. The Ehrhardt team hauled the machine guns up two flights of stairs and set them up in hotel windows while hotel guests gasped and the women shrieked.

"Open the door or get a hand grenade," a soldier shouted at one apartment door. A terrified tourist shouted back that they couldn't find their door key.... A veteran with infantry experienced told the soldiers to keep the machine gun muzzles well back from the windows to avoid attracting artillery.

"Invariably I though back to the [*Chinese*] Uprising of 1911 where I watched the retreating rebels in Hankou." the naval officer wrote. "They would set their Mauser 88 rifles on their knee, wedge in a clip of cartridges and shoot blindly up into the air. No wonder that so many bullets were falling and so few hits were reported."

The Ehrhardt Brigade contingent saw the Reds gathering for an attack—unaware of their double machine gun position. The machine guns both opened fire and broke up the attack almost instantly.

The Ehrhardt contingent shortly received a message: their capture of the hotel had taken the pressure off the White Guards and they were told to leave the hotel and rejoin their unit. When they got back, the Tillessen contingent had already rolled the captured field cannon back to the bivouac and nobody seemed to talk about anything else.

Rudolf Mann was at a roadblock when two White officers—so they said—showed up at a roadblock in a droshky with two Russian prisoners of war driving for them. One of the officers was wounded and semi-conscious. The two Russians—Mann described them as semi-Asiatic, probably meaning Kalmuks—logically assumed that they were about to be shot, and quietly began to pray. The White officer explained that the Kalmuks were harmless and that he had picked them up to help with his wounded friend. A Prussian officer who spoke some bad Russian explained to the praying Kalmuks that they would not be shot unless somebody brought charges against them. Hearing Russian worse than their own, the two Kalmuks trusted him instantly and smiled with relief at answered prayers.

The Ehrhardt Brigade soon settled up with the Pioneer Barracks, where the telephone truck had been lured into an ambush.

"As our forward patrols got near enough to the area in front of the Pioneer Barracks, the dead-looking building showed some signs of life," Mann wrote. "The Pioneers now understood that their escape was cut off and that things were serious. They opened up an angry fire. Some of them showed that they wanted to fight but others came out the barracks gate waving white cloths in a sign that they wanted to negotiate.

"This gave occasion for some of the gun crew to leave their cover, and their casual attitude proved fatal. More gunfire rattled suddenly from the barracks door and a cannoneer was hit and killed. Nothing remained but for the howitzer to speak and a few shells through the shattering brick wall ended the fight."

The Reds came out of the barracks, first by ones and twos and then in crowds with their hands up. "The female cook for the Pioneers marched out but only held up one hand," Mann noted with amusement. "In the other hand she convulsively held a loaf of bread as marching provisions—you visionary angel, you! I thought, as my own glance drifted from her tear-brimming eyes to the ration loaf. How am I going to feed this many people? We didn't have much for ourselves."

Mann and his soldiers herded the Red prisoners into a couple of cellars and posted guards over them until they could find qualified officers to decide who could be set free and who would be held for further questioning. One angry young man had to be clapped into handcuffs but most of the prisoners simply moaned or wept with fear. People who were obviously harmless were let go almost immediately—two street girls who showed up looking for their pimps were not detained—but children also showed up looking for food. The Starvation Blockade was still in force so Mann began to boil thick noodle soup with whatever vegetables came to hand and to dole out sparing slices of black bread. Some captives drank the noodle soup out of their steel helmets. "I heard about the White Terror but I didn't see any of it," Mann said. He admitted that some of the soldiers seemed to be trading their black bread rations for sex with available women. The German Army maintained inspected brothels for their troops on extended campaigns, but Mann, who had nothing against flirtations, saw seduction based on hunger as a disgrace to soldierly honor.

The hungry children looking for their fathers saddened him most of all—that, and the women in their quaint Bavarian accents wailing over their husbands or sons. *"Xaver, I tol' yuh, don't play in this masquerade! Think of yer family. It's a mess! Now you know I got it right!'* Sadly, she did get it right, because while she was on the telephone trying to get to him, Xaver was lying stiff and still on a slab in the Schwabing morgue."

Mann also interrogated some prisoners for potential release as harmless. "These seven office workers in suits definitely hadn't fired on us. None of us could watch the interrogation without tears in our eyes—tears of laughter. The left flanker of the group was a dwarf who couldn't see over the table, and had a water-head like a pumpkin. They all told us their names under interrogation protocol and were released from the prisoner cellar with friendly words. As I thoughtlessly said *'Auf Wiedersehen'* they played a scene right out of [*Max*] Reinhardt. Fourteen hands went up in self-defense and they screamed *'Nein, nein, nein'* in chorus."

The terrified officer workers were let off, but Mann also interrogated a pair of former Bavarian officers who told him the whole thing was a misunderstanding: Both had obviously signed up with the Reds for the inflated wages. They said nervously that they had served in the transportion corps, not as combat leaders. Mann doesn't mention what happened to these two.

Captain Ehrhardt himself felt bitter and angry over the deaths at the Pioneer Barracks. "If we had made an immediate hard example, I'll testify to the fact that we could have saved the administration troops some bloodshed," he wrote. "Many misled workers would have thought about it and would still be alive today.

"I'm out of sympathy with the compromise system of [*General Ludwig*] Maercker. Anybody who picks up a weapon in a civil war should know that it's not a joke. A few hard, remorseless blows work wonders and save blood. Sparing human life is the highest duty of the soldier. Endless false pity will only be taken as weakness and lead to further losses."

"My staff headquarters lay in the Friedrichstraße, in the house of the architect Max Langheinrich. For two days we took individual pot shots in the house and in the street that must have come from somebody's attic window. At last I ran out of patience. On the third day I had the whole area between Kaiserstraße and

Hohenzollernstraße suddenly surrounded at six o'clock in the morning and searched every house.

"Our behavior aroused opposition, even resistance from the citizens, who felt that they had been deprived of freedom of movement. But we didn't let ourselves be interfered with and luckily we found two young boys in the attic room, who claimed that they were sick. One of them had a freshly used revolver under his pillow and the other was found with a handful of freshly fired cartridges. From that day on, we had peace."

On March 2, other sources record that the Altmann brothers were arrested at 8 Winterstraße in Munich along with a laborer named Josef Seidlmaier. They were taken to the yard of a leather factory in the Pilgerstraße and shot by a firing squad of the 21st Bavarian Guards Regiment led by a *Leutnant* Moeller. No mention is made of the hot revolver and cartridges under the pillows or the pot shots at Ehrhardt's headquarters in the subsequent account by Emil Julius Gumbel, who listed what he claimed were 354 rightist murders between 1918 and 1924. *Leutnant* Moeller was never brought into court to give an account of his involvement, but a captain in the same unit said: "I supervised so many arrests in the first days on May on the grounds of information from the Criminal Police and Army observers that I can't possibly remember all the names. I can't even remember if Seidlmaier and the Altmanns were brought before me personally or if they were shot while trying to escape." Seidlmaier had reportedly never been part of the Red Army—a Red named Seidl had ben implicated in the Luitpold Hostage murders. But Seidlmaier had in fact worked for 14 days in the Workers Defense and Security Service. He had supposedly handed over his rifle to the authorities on April 27—as the *Reichswehr* and *Freikorps* units were closing on on the city. Nobody was ever charged for the executions or murders.

On the night of May 6-7, a genuine outraged occurred. Members of the Saint Joseph Gesellschaft had stayed out past curfew drinking and planning an excursion to the theatre. A detachment from the elite Prussian Guard under orders of a Captain von Alt-Stutterheim, having been tipped off by an informant about the curfew violation, descended on 41 Augustenstraße and arrested about 22 of the young Catholic men, none of them known to be Red Army partisans. A search of documents appeared to have produced something marginally suspicious. The young men claimed to be innocent but the officer in charge said he didn't believe them. A brawl broke out and seven of the young men were killed on the spot. The others were roughed up and released. One unlikely Communist account had the Prussian Guardsmen doing a "drunken Indian dance" on the corpses.

The Saint Joseph Society shootings on May 6-7 were so obviously outrageous that some of the Guardsmen were actually hauled into court. On October 25, *Vizefeldwebel* (Senior Sergeant) Konstantin Makowski and Guardsman Jacob Mueller were sentences to 14 years in prison, another soldier named Granbasch (Grabowski?) got a year in prison for a fatal blow inflicted during the brawl. The former Hussar Stefan Latosi got 10 years for lifting blood-stained watches and money from the dead and injured. Strings had obviously been pulled since *Hauptmann* von Alt-Stutterheim, who had lost three relatives on the Western Front, was never bought upon charges.

"The worst of the fighting we saw was at the Pioneer Barracks," Captain Ehrhardt said in summary. "The Bavarian soldiers there who fought against us were no more Communists than the farmers in the countryside. The resistance stemmed from personal spite. Free food and clothing had convinced them to stay in the army, so they played at Soldiers' Councils and Red Army out of laziness and personal comfort rather than out of any idealism."

The professional revolutionaries born outside Bavaria portrayed the Battle of Munich as an uprising of home-grown Red heroes who fought to the death against the White Terror from Prussia. The professional revolutionaries did this after they escaped from their hiding places in apartment houses while a professional army assisted by loyalist Bavarians destroyed the confused Red Army Bavarian rabble without much serious trouble.

"In the first three days we brought in two cannon, 43 heavy machine guns, 80 light machine guns, more than 4,000 rifles, 12,000 hand grenades, and 80,000 rounds of rifle ammunition," Ehrhardt reported. "In actual fighting we captured two cannons, 12 heavy machine guns, 20 light machine guns, and 600 rifles. The losses to the Brigade were four dead and six wounded.

"Slowly the firebrands of the civil war came out. The revolutionary leaders, above all the student Toller and the Russian Leviné-Nissen—Toller in a nightshirt with a red wig was found behind a hidden door—were captured."

"The Russian, who was responsible above all the others for the spilled blood, had to be shot, and was shot. *Herr* Toller, who had led the Red Army, received a pleasant fortress arrest, where he could write his plays, which are still popular in the eastern quarter of Berlin."

Toller was reportedly betrayed by a woman under pretext of helping him escape: there was a reward of 10,000 marks on his head. He received clemency from the tribunal—to Ehrhardt's quiet disgust—because of his nervous breakdown while serving at the front during the war and because he had refused Rudolf Egelhofer's orders to execute Johannes Hoffmann's captured officers. Toller was a man of great intelligence and imagination and he seemed to have realized that he had begun a terrible process when he took up arms against an elected Socialist government. After the first few days of his five-year sentence in his cell, Toller wrote to a friend: "What will the coming years bring to Europe, to Germany? There are times when I want to scream and run away from the pictures of the horror that haunt me like hallucinations."

Leviné-Nissen was unrepentant. He stood trial, like Toller, on charges of high treason—both Toller and Leviné-Nissen had been German soldiers during the war—and Leviné-Nissen knew that he had no chance of acquittal. He was not tried, however, for the Luitpold Gymnasium murders because these murders had already taken place when he was in hiding. German courts in the pre-Nazi era—Professor Gumbel to the contrary—were generally honest and followed the letter of the law—sometimes tempered with leniency because the vast casualties during the war had reduced the young adult male population of Germany to a perilous level.

"We were compelled to arm the workers to defend ourselves against the onslaught of the dispossessed capitalists," Leviné-Nissen told the court. "We have all of us tried to the best of our knowledge and conscience to do our duty toward the International, the Communist World Revolution."

The fact that he had been in hiding while his rag-tag soldiers were being wiped out by the "Prussians" did not appear to figure in his thinking, but the prosecutor was rude enough to mention it. Axelrod and Levien were not at the trial—they had both hidden out after telling the rank-and-file Reds to sell their lives dearly so that they themselves could bravely escape back to Russia. Axelrod's disappearance was apparently permanent. Levien, sentenced during the Stalin purge of 1936, was executed by the NKVD in 1937.

"[*Eugen*] Leviné faced the Court on the second day of the trial with an indifference to the fate hanging over him which alone could shatter the indictment of cowardice by the Public Prosecutor," the Social Democratic *Münchener Post* reported. "The unstudied posture of the defendant undoubtedly impressed many of those who had not experienced the Leviné of the Second Soviet Republic."

Leviné was sentenced to death on June 3, 1919. He was shot by a firing squad at Stadelheim Prison on July 5. The last of the Mad Lovers told his wife to remarry. She she did so within a year, to another Communist named Ernst Meyer, who died in 1930. Fleeing Germany in 1933 when Hitler came to power, Rosa Broida tried Soviet Russia but discovered that Stalin was not an acceptable alternative on Lenin and Trotsky. She spent her last years in England, glorifying the Revolution and her first lawful husband's role in it.

In the end, it was Ernst Toller, born a German citizen, rather than the Russian-born Mad Lovers like Rosa Luxemburg, Leo Jogiches, Eugen Leviné-Nissen and Rosa Broida Leviné-Meyer, who felt the seismic tragedy about to burst forth.

Jews had lived in the Rhineland since Roman times and in Prussia and other North German states since the Crusades. Some of them served as delegates in the city councils of the "Free Cities" during the later Middle Ages. From the days of the first Prussian kings—Calvinist Protestants due to their French Huguenot tutors—Jews had enjoyed tolerance in Prussia when they were banned from France and Spain, often defamed in England, and lived in Russia at considerable risk to their lives. They reciprocated with an enormous regard for *Ashkenaz*, "the Land of the North"—meaning Germany. Moses Mendelssohn was the co-founder of "Prussian Literature" along with his Christian friend, Gotthold Lessing. Mendelssohn's best-seller, *Phaedon*, used Christian, Jewish, and Greek Classical sources to argue in favor of the immortality of the soul. His personality inspired Lessing to create the literary character "Nathan the Wise," who argued for tolerance and brotherhood between Christians, Jews, and Muslims. Moses Mendelssohn's son Abraham operated a bank so honest that even Richard Wagner—the stereotypical anti-Semite who had Jewish friends—kept his money there. Moses's grandson, Felix, was a musical child prodigy and the darling of the Prussian royals. Once Prussia came to be the founding state of the German Empire, the Prussian Constitution of 1813—virtual full civil rights for Jews—was extended to the entire new nation of Germany. The German Jews reciprocated with a loyalty amounting to outright love. Jews volunteered to fight side by side with Protestants and Catholics to throw Napoleon out of Germany. Some Jews fought and were killed at Waterloo in 1815 while others doctored the sick and wounded. Jewish women donated their jewels to buy muskets and medicine. The traditional Jews of Tsarist Russia, who felt no such allegiance to the Tsar, noted with a wry joke: *"German Jews are not just more German than Jewish—they're more German than the Germans!"* Otto von Bismarck had close Jewish friends and found many Jews admirable—"There is no such thing as an evil race." Kaiser Wilhelm II had Jewish friends all his life. During World War I, more than 100,000 German Jews served in the Kaiser's army: 14,000 were killed in action or died of wounds, 35,000 were decorated for loyalty and valor, and 22,000 were promoted. The German Army medical corps contained 1,200 Jewish physicians and was famous for its excellence—so much so that the French sometimes left their seriously wounded behind for the German doctors, who were often German Jewish doctors. Robert Salomon Weitz, father of author-designer John Weitz, was a Prussian infantry officer and received the Iron Cross for valor. Anne Frank's father, Otto Frank, was an artillery officer who had also served in combat. On the home front, Fritz Haber invented a process to consolidate nitrogen from the atmosphere which prevented the Germans from running out of ammunition and fertilizer. (Haber also invented both mustard gas and—most tragic of all—Zyklon B, originally an insecticide to fight typhus by killing lice.) Walther Rathenau organized the services of supply and resources when the war turned out to be longer than anyone expected. He also argued for a fight to the finish unless Germany received better terms because his business sources told him the Allies were almost as exhausted as the Germans. Albert Ballin, inventor of the ocean summer cruise, tried to head the war off and then tried to arrange a just peace. When Ballin failed, sick and exhausted from overwork and despair, he committed suicide. The Hamburg banking magnate Max Warburg was one of several Jews who put up his own money for hunger relief. And Adolf Hitler, who had spent his years in Vienna supported in part by Jewish charities, ironically owed his Iron Cross, First Class, to his company commander, Captain Hugo Gutmann—a Jewish reserve officer in the Bavarian Army.

Given the chance to live among honest people who appreciated their hard work and intelligence, the German Jews had flourished—perhaps too much so for the under-class of German society, the people had had trouble holding their economic lives together despite the social reforms of Bismarck and Kaiser Wilhelm.

But the underclass envy had been largely held in check—before the demonstrated fact that bloody revolts in Berlin and Bavaria had been stimulated by a handful of rhetorically brilliant but politically deluded Russian Jewish radicals who seemed to think they were still living in Tsarist Russia, after another handful of German Jewish anarchists and eccentrics had ruined Ballin's and Rathenau's efforts for a sensible armistice. The first really fair election in German history—with first-time votes for women as well as for men without property—had shown that nine out of 10 Germans had no desire for Communism or even for Socialism beyond the eight-hour day and some improved workplace benefits. The Reds were not only outvoted by the moderate Social Democrats but by the political Catholics and even by the monarchists who wanted the Kaiser back. The Reds had no hope—except for violence where other people did the fighting while they themselves hid in apartments or escaped back to Russia. And the hatred that they engendered for themselves expanded like a lethal plague inflicted on the whole German Jewish community.

Trouble had been forecast during the war when Matthias Erzberger, a member of the Catholic Center in the Reichstag, had demanded to know how many Jews were on front-line service, how many served behind the lines in staff jobs, and how many had claimed deferments, His constituency was working-class Catholic, people not indisposed to anti-Semitism and his intention was malicious. But the *Judenzählung* [Jewish Count] of November 1916 that came in response showed that German Jews saw front-line service and were killed or wounded in almost exactly the same numbers, based on their population, as German Christians. There was no statistical justification for the propaganda image popular in some circles of the German Jew as a shirker.

But Adolf Hitler, himself something of a shirker, a former Jewish imposter for food and lodging, and a nobody who did nothing during the Battle for Munich, frantically seized on this opportunity offered by the Russian Jewish leadership of the Munich uprising. Experts who have combed through his printed speeches assert that before the first quarter of 1919, Hitler had never made a single anti-Semitic remark. Originally something of a Marxist, he later parroted nationalist phrases about the need to soldier on until the final victory over England and France. Temperamentally unsuited to real fighting, Hitler had sucked up to his Jewish company commander during the war, while a Jewish soldier and sometime buddy, Chief Orderly Karl Lippert, had covered some of Hitler's riskier messenger assignments. But after the war Hitler, always a bit of a sneak, proved to be a capable informer, a *Vertrauensmann* or *V-Mann* under the command of a General Staff officer named Karl Mayr. In simple terms, he squealed on his buddies. One day, Captain Mayr received a letter from a man named Adolf Gemlich, asking if Jews should continue to enjoy full rights of citizenship even after the days of the revolts in Berlin and the Soviet Republics of Bavaria. Captain Mayr gave *V-mann* Hitler the letter to answer. Hitler cribbed some of his ideas from the Protocols of the Learned Elders of Zion, a Tsarist Russian forgery written in 1895 and recently circulated in Germany. Hitler explained to Gemlich that the Jews were engaged in a secret, systematic campaign to take over the world. The ultimate answer, Hitler said, was to deprive the Jews of their rights as German citizens and then evict them from Germany. He described this proposed program with a word that would have a terrible resonance in the future—*Endlösung*—the Final Solution.

Chapter Eighteen
The Shadow of Versailles

On February 24, Dr. Hugo Preuß introduced the proposed draft of the Weimar Constitution. Preuß, who had sent three sons to the war and seen two of them decorated for valor and one of them crippled, was a Democrat—not quite a Socialist, but an educated liberal. He was also a German patriot and an observant Jew. The constitution he drew up was based on both the old constitution of the German Empire and Constitutions of the United States and of Republican France.

Dr. Preuß provided that the German Federation would be a republic, with supreme power emanating from the people. Article 22 and Article 109 stated that men and women were allowed to vote at the age of 20 with no property qualifications, as they had been allowed to vote in the January election where the parties of the moderate Socialists, the political Catholics, and the conservative monarchists individually out-polled the Communists and extreme Socialists. Article 119 placed marriage under the special protection of the constitution and Article 175, a holdover from the Prussian and Imperial German constitutions, made sex between two men or between a man and an animal a criminal act. Under Article 48, the chancellor had the right to rule by emergency decree—a measure that Preuß had intended to protect the rights of temporarily unpopular political minorities. After much debate, the new constitution was adopted on August 11 by a vote of 262 in favor from the Social Democrats, the Democrats, and the political Catholics against 75 negative votes from the conservative Nationalists, the right-wing German People's Party, and the Leftist Independent Social Democrats.

The document that was to define Germany's future, however, was not the Weimar Constitution but the Treaty of Versailles.

The delegates to sign the treaty that would formally end the world war left from the French hotel for the Trianon Palace on May 7, 1919, while the Ehrhardt Brigade and the *Reichswehr* were still rounding up what was left of the revolutionary Red Guards in Munich. The head of the delegation was Count Ulrich von Brockdorff-Rantzau, a jurist, experienced diplomat and former Prussian officer. Three of the delegates were Jewish: Dr. Otto Landsburg, minister of justice; Dr. Walther Schücking, a Democrat and specialist in international law; and Dr. Carl Melchior, a partner in the Warburg Bank and the financial expert. Dr. Walter Simons, ministerial director of the foreign officer, was of French Huguenot descent. Johannes Giesberts, postmaster general, was a political Catholic from the Rhineland and leader of the Christian Workers' Movement. Social Democrat Robert Leinert, the son of a financially troubled potter who had grown up in an almshouse, had become the president of the Prussian Landtag.

When the delegates saw the terms, they were unanimously horrified.

Georges Clémenceau, the 78-year-old Premier of France, had witnesses the German siege of Paris in 1870 and the Communard uprising of Parisian leftists in 1871 which had killed more Parisians than the German siege. An atheist and the estranged husband of the American woman who spent most of her life elsewhere, Clémenceau was obsessed with revenge on Germany. He was widely believed to have said that the world's greatest problem was 20 million Germans too many—a concept to be repeated in the U.S.-sponsored Morgenthau Plan of 1944 – 1945 at the end of the next war.

The Treaty of Versailles was seen by the Germans as an attempt to rectify this perceived imbalance of too many Germans. Germany was to lose 6,000,000 inhabitants and 70,000 square kilometers of territory through annexation. The Germans were to return Alsace and Lorraine to France and give up all their African colonies—and none of this surprised anybody. Upper Silesia would be annexed by the Polish state that the Kaiser and his Germans themselves had authorized in 1916. This was a surprise to the Silesians. Other German land not part of the Kaiser's approved Polish state would also be given away. Danzig—90 percent German since the Middle Ages—would become a "Free City" and Memel would be given to Lithuania—another new country the Germans themselves had approved under their own initiative during the war. Seen from the German perspective, this was the near-exact equivalent of what giving Texas and California back to Mexico would mean to an American. The Rhineland, never anything but German since Roman times, would be demilitarized and The French would be free to extract coal from the coal mines in the German Saar for 15 years.

Kurt Eisner had lifted and published some secret papers from the Bavarian archives which he believed established that Germany bore the sole guilt for the war. He repeated this charge at a meeting of international; Socialists in Berne in February of 1919: "It can no longer be doubted." Eisner also approved the delayed Allied return of German war prisoners. At Versailles, after Kurt Eisner had been assassinated as a traitor, the Germans found that they were expected to sign a statement acknowledging Germany's sole guilt for the war. They found they were also expected to pay the entire cost of the war with the amount to be determined in 1920. The German Army would be reduced to 100,000 men.

The German delegation tried to negotiate. They were told that they could either sign or resume fighting. The Starvation Blockade would remain in force until the Treaty was signed. "What hand would not wither that binds itself and us in these fetters," Philipp Scheidemann said on May 12 before the National Assembly. "This treaty in the opinion of the government cannot be accepted." Frantic telephone calls were exchanged between senior generals to see what kind of resistance might be offered if the French actually invaded. Noske sent out a desperate appeal asserting that he himself had wanted to resist but could get no support. In fact, while Waldemar Pabst wrote that the army would not accept the signing of the treaty, some of the senior generals felt that military action was hopeless.

As the deadline drew near—and with some members of the Catholic Center party joining the Nationalists and the more conservative Democrats and Social Democrats in continued resistance to signing—Matthias Erzberger weighed in and told the Weimar assembly that he would not support any further resistance to signing.

"Our people and morale are such a mess that we *have* to sign," Noske said on June 3. On June 23, a few hours before the expiration of the deadline, the Weimar National Assembly voted to sign. The German Nationalists, the German People's Party, and some members of the Center and the Democrats voted not to sign. The Democrats—the party of patriotic Jews like Walther Rathenau and Hugo Preuß, was also the party of professors and diplomats. Many Democrats doubted that the French would be crass enough to attempt an actual invasion of Germany and that better conditions could be negotiated.

Chancellor Philipp Scheidemann—a Social Democrat—went down with the ship. He was true to his warning about the withered hand and resigned his office as chancellor. Matthias Erzberger, who had made Germany's proposed last stand impossible, stayed on in the government of Gustav Bauer.

The first impact of the Treaty of Versailles struck Hermann Ehrhardt even before the actual signing. The maintenance crew of the German High Seas Fleet, interned at Scapa Flow, was kept informed of the way things were going and Admiral Ludwig von Reuter ordered officers and loyal sailors to prepare for possible boarding by the British. The portholes on the German warships were quietly greased to keep them ready for quick action. On June 21, 1919 at 10:00 a.m., the

German maintenance crews quietly opened the sea valves and left the portholes open. The 10 battleships, five battlecruisers, five cruisers and 32 destroyers were seen to be sinking within an hour. The British were shortly stuck with 400,000 tons of German battle fleet resting on the bottom of Scapa Flow.

"Admiral Reuter was the man who mustered up the decision to rip the mighty power of our fleet away from the enemy," Ehrhardt said. "This action wiped away the dirt that the revolution of the German Red Sailors had left with us. Our hearts swelled. I saw every eye among my people glowing, new pride surged through them that they belonged to the Marine Brigade. I heard them singing *'O Deutschland hoch in Ehren'* on this day."

The melody to the victory song was actually English—*Ye Mariners of England*—but what Ehrhardt heard within the next few days must have soured the source. *Korvettenkapitän* Cordes, who had served with the fleet at Scapa Flow, described sailors who had scuttled their own ships under orders being shot at in the water as they swam away from the sinking ships. If they were hit, he said, the English left them to sink and drown. "The English explanations of the grounds for the shootings were shameless lies," Cordes wrote. "They had already taken away the lifeboats and life jackets.... The English commander ordered that the captains of the sinking ships be shot.... Some of the wounds looked to have been made by dumdums.... An English officer put a pistol to *Leutnant* Lampe's head and told him to stop his ship from sinking. He said he couldn't, so the Englishman pressed the trigger just as the ship lurched and [*Lampe*] got a bullet graze on his head.... Senior Quartermaster Schaeffler from the *Kaiser* protested to the English guards and got four bayonet wounds in the stomach and another in his upper arm.... The Commander of the [HMS] *Resolution* said that German officers had forfeited their right to be treated as officers.... Protests were rejected. All press reports were knowingly falsified...."

Ehrhardt habitually read English newspapers and he noted that some British officers had actually praised the courage of the German scuttling crews. He contrasted this British accolade to fellow sea warriors whose courage he respected to the cowardly newspapers of the German Left, which portrayed the scuttlings at Scapa Flow as a crime and a possible obstacle to peace.

"For the first time the outer world had seen that the bureaucratic character and lack of will was not shared by all of our people," Ehrhardt wrote. "But the men who called themselves the 'German National Assembly' had no feelings for the flag, no feelings for sea-faring, no feeling for honor. They let the colors of the German flag be changed as if such colors could be changed like a dirty shirt.... For the first time I had to involve myself in the political persuasion of the troops. But the flag question was really more than a political question, because the flag would decide the existence or non-existence of national pride among the people." Ehrhardt had served as a sea cadet in the days when the Royal Navy was lionized in the new German Navy to such an extent that German sea cadets were given all seafaring commands in English. The brutality at Scapa Flow and the craven response of the Leftist German press metaphorically burned all his boats for him. He considered himself, for that time, to be the avowed enemy of *"British brutality, French revenge, and Polish greed"*—a phase widely used in the *Freikorps* movement—and he no longer considered the revolutionary Left to be a part of the German community in any way, shape, or form. His new motto was the battle cry of the Frisian pirates of the later Middle Ages: "Better dead than a slave!"

On July 5 at Wilhelmshaven, Captain Ehrhardt spoke about the flag to his assembled men of the Second Marine Brigade.

"The National Assembly has voted by a three-fifths majority to make black red and gold Germany's national colors. This national assembly, born in the confused times of the first revolution, no longer represents the will of the people on many questions. The collapse of morale, the education toward a loss of the will to work, the ruin of our economic life, the seesaw politics in the inner circle have opened many people's eyes to the mistake of November 9. It is inconceivable that men like these who are responsible for the future of the nation should find the time to concern themselves with the color of the flag instead of bread and raw materials for production and the conditions of the treaty.... I find it unheard of that they amuse themselves with a new national flag in a time of deep disgrace, Shall Germany concern itself with the colors of the flag when the Kaiser and the best of our people are taken away from us? The black-white-red flag flew in honor over Germany's people and land for a man's lifetime. We achieved great things under that flag and we were proud to see that flag greeted all over the world. That flag that no enemy has ever hauled down in open combat, shall now vanish without a sound or a song. I say the new flag is nothing, No memories, no famous deeds cling to it. I saw the the black-white-red flags on the trucks of the Brigade with joy. I will not forbid your further progress. Whoever sees the new flag as a symbol for great future times may hoist it. It will never fly over my house.

"I've told the Brigade what the change of flags means to me. I know that I speak for a wide circle of the German people."

Nationalists like Ehrhardt had opposed the signing of the Treaty of Versailles as a matter or honor. The Independent Social Democrats had opposed the signing as a matter of opportunity. Anything that made life more insufferable for the German working classes was an opportunity for a new try at World Revolution orchestrated from Moscow. The same antipodes of Right and Left resisted the proposed Factory Council Law, which had been before the Weimar National Assembly when the Treaty of Versailles came up. The Right—funded by manufacturers—resented the law that would have allowed workers to sit on company boards of directors, as an infringement of the rights of managers to run their own businesses. The Leftists believe that company unions might give the workers enough benefits and satisfaction to wean them away from the international labor movement, which was dominated in Europe, if not in America, by Bolsheviks and their sympathizers.

The cowardice of the National Assembly in voting to sign the Treaty of Versailles—even the German Left resented giving up German territory to France and Poland—prompted a general transportation strike as a possible prelude to a new revolution. The strike was scheduled for the first week in July.

Waldemar Pabst and the Guards Cavalry Division were ready for them. Pabst, as a staff officer, had noticed the extreme Left was able to convince more moderate Socialists and trade union members to take part in potentially paralyzing strikes. He planned ahead in the lull after the Liebknecht-Luxemburg "executions" and the second and bloodier upheaval to blunt the effect of general strikes. *Freikorps* officers and soldiers with a background or interest in technology were trained as telephone operators, locomotive engineers, tram car motormen, electricians, dynamo and plumbing technicians, newspaper printers, and even as bread bakers and food processors. When the general transportation strike was announced, Pabst brought these trained officers and soldiers into action and had them take over most of the transportation and support industries. They had no regard for picket lines and placards. Pabst also sent a directive to Hermann Ehrhardt and to Wilfried von Löwenfeldt, commander of the 3rd Marine Brigade on June 27:

"The General Strike of the railroad workers has been declared....

"The Guards Cavalry Division is authorized to ensure the operation of necessary transportation—if necessary with the use of weapons. The *Reichswehrminister* [Noske] has authorized, under the existing conditions of a siege, that any striker who refuses to work may be put in prison for up to a year. Strike placards are to be openly ripped down....

"The 2nd and 3rd Marine Brigades and the Marine Detachment under command of *Korvettenkapitän* von Löwenfeldt are also authorized to take part in putting down the strike.

"The Marine Brigades under the supervision of Captain Löwenfeldt will occupy the railroad stations, repair shops, locomotive sheds and provision depots. The watches shall be strong enough so that they can escort those workers who are willing to work past the strikers and prevent any violence. Each detachment of soldiers should be led by an experienced officer and include one or two loaded machine guns.... Strike picketers and strikers or hecklers shall be rendered harmless.

"It should be advised that the weapons shall be used ruthlessly to suppress the opposition. It is a point of honor for the *Reichswehr* troops that the railroad strike with its unforeseeable economic consequences should be put down by the military in the shortest time possible.

"Pabst, Captain and First General Staff Officer."

Ehrhardt and the Second Marine Brigade deployed to rail yards all over Berlin to help break the strike shortly after it started. Perhaps incongruously, the Brigade's first mission was an act of kindness.

"With one stroke we occupied a number of railroad stations," Ehrhardt wrote. "We set up emergency aid stations, operated locomotives, and unloaded food from boxcars. Above all we had to care for the animals headed for the slaughter-houses that had been sidelined through the ruthlessness of the strikers. They faced a death by thirst. The confused bellowing of the poor creatures in the cattle cars in the unseasonable heat was terrible to listen to."

The strike began to fall apart as the union workers saw that the city and country could do without them at least on an emergency basis. Pabst's technical teams and Löwenfeldt's and Ehrhardt's iron control over their men had destroyed another weapon of what some called democracy and others saw as an attempt at a Red take-over. But Gustav Noske objected to the fact that soldiers were used as strike-breakers and the implied threat to shoot armed unionists, as opposed to armed communists: "The order must be rescinded. The [*Social Democratic*] Party had declared that they cannot support it. No Social Democratic administration can approve an order to act with violence toward a strike."

Pabst pointed out to Noske—who had never served in the military—that changing the orders when the strike was visibly falling apart would make it impossible to maintain the morale and discipline of their own *Reichswehr* troops and of the *Freikorps* units. He also reminded Noske rather bluntly that Noske was a survivor of the Scheidemann government where Philipp Scheidemann—Noske's fellow Social Democrat—had had the integrity to resign rather than sign the Treaty of Versailles, which had provoked the Reds to make one more try at a take-over.

The story that Pabst threatened to kill Noske on the spot is an unsubstantiated but plausible rumor. Pabst officially offered to resign, but his military superior, General Walther von Lüttwitz, suggested that Pabst was exhausted and that he take a long convalescent leave instead of actually leaving the Army. Pabst agreed and made some confidential telephone calls: he learned that *Reichswehr* generals and *Freikorps* leaders could muster at least 40,000 troops for an attempt to boot Noske out of office. Pabst also pulled some strings for two more acts of dark honor. Otto Runge, the 43-year-old hussar who had clubbed Karl Liebknecht and Rosa Luxemburg, got a two-year prison sentence: *Oberleutnant* Kurt Vogel, who had been in charge of the transport detail that finished them both off, got two years and four months in prison for falsifying a report that said they were killed while trying to escape. Pabst understood that nobody cared much about Runge, who was an automaton, but that in case of a possible Red take-over of Berlin, Vogel would probably be murdered. He saw to it that Vogel had a Dutch passport and some false identification papers. Kurt Vogel shortly snuck away during a family visit and slipped into the Netherlands with a pocketful of Dutch money. The former Hussar Otto Runge got out of prison with the next amnesty.

The USPD—the Independent Social Democrats—smelled victory when the mainstream Social Democrats undermined the *Reichswehr-Freikorps* strike-breaking activities they themselves had approved. Placards went up showing the strikers and the government as friends. Hermann Ehrhard's troops were defending the former

Kaiser's palace with the black-white-red Imperial flag flying when the strike went back into effect.

"On the 21st of July 1919, the USPD took up demonstrating again, despite the declaration of siege conditions and the advanced warning not to resume their demonstrations," Ehrhardt wrote. "A column about 3,000 men strong attempted to enter the Palace Garden. A hundred meters before they reached the machine guns I told them to turn back and to disperse. The people thought they were facing exactly the kind of plum-ripe military they were used to, the kind that they had pushed aside. They erred. Some of those who hadn't learned were hit by a sharp salvo." The volley reportedly killed 42 strikers and wounded a number of others.

"Following my orders, the assault company and some of my other detachments further educated the demonstrators. Every assemblage was scattered. The combined detachments locked on with rifle butts and fists and especially grabbed every 'sailor' who couldn't slip away, along with every heckler, and brought them in.

"It was a welcome wind that blew through the city. The manure smell of the Revolution was blown away, the threat that had opposed the citizens began to soften.

"My youngsters went on with spirit. Perhaps it was naughty to saw off a flagstaff with the black-red-gold flag and throw it into the canal, or to break into a high school and break up a lecture on 'National Unit and Culture' by Herr Rudolf Roker as he was praising *The International* with moving words. I saw rather a re-awakening of self-consciousness that didn't shy away from seeking out the enemy in our own camp.

"It's worth considering that this little shock troop, with ten or at most twelve youngsters, was able to do so much when the Communists wanted to kill anybody who saluted the black-white-red flag."

The Second Marine Brigade, the Third Marine Brigade and the Guards Cavalry Division had won the battle for the streets. But the Social Democrats refused to fight the battle of the flag.

"On the morning of August 1 the black-red-gold banner flew over all the buildings of the Wilhelmstraße," Rudolf Mann wrote. "It was a holiday to celebrate the settlement of the new Constitution, which was having its final reading that day."

The Constitution, in fact, had not yet been voted by the Weimar National Assembly but the new flags were hoisted all around Berlin—to the dismay and disgust of the Second Marine Brigade which, acting under explicit orders, had helped break the transportation strike which itself was seen as a bid for a revolution.

"We watched this holiday of the Democrats with fury, and repeated among ourselves the poem about the black future, the red present, and the golden past, with a giant black-white-red flag of our own that a bank had donated to us, perhaps to get rid of it, flying in defiance over the main railroad station,"

"The word passed among our people that because of hoisting that flag we were being asked to leave Berlin," Rudolf Mann recalled. " That meant nothing to us—we wanted to leave at the first opportunity. As we left, some gentlemen in civilian clothes tipped their hats to us, and some youngsters tried to cry out 'Hurrah!'"

The final draft of the new constitution was called to a vote on August 11—which became Constitution Day. The Social Democrats, the Center Party, and the Democrats voted in favor and adopted the Constitution with 262 votes. The negative votes totaled 75. The Nationalists and the German People's Party voted against the Constitution because they wanted a monarchy, and the Independent Democrats voted against the liberal constitution because it undermined their hopes for a revolution, giving those who believed in democracy everything they wanted.

The day after the constitution was adopted, the Second Marine Brigade entrained for Upper Silesian, where stone-throwing clashes and minor gunfire between the German and Polish populations required disciplined troops to head off actual warfare.

The Weimar constitution had begun in high hopes but soon came under attack, not by the Nationalists or the Communists, but by the Allies. On September 14, 1919, Dr. Hugo Preuß was quoted at length as he took umbrage at an outside attempt to alter the Weimar Constitution: Preuß had allowed that German-speaking Austria—now a small country separated from Hungary and bereft of Czech, Slovak, Polish, and Italian subjects—might be allowed to join Germany on its own initiative. Anything that might strengthen Germany in economic or demographic terms, it appeared, was seen as a threat to the people who purportedly said there were 20 million Germans too many. A professor of law, Hugo Press noted that an attempt to toy with the constitution of a sovereign nation was patently illegal. He separated himself from the Nationalists but concurred with their belief that the attempted editing of his constitution was hostile and vindictive—and ultimately foolish.

"I do not believe in the opinion, widely current among Germans, that the statesmen of the Entente and America are filled with anger and hate and have allowed themselves to be drawn into a planless and purposeless policy. This policy, however, can have but one thought and plan—to make it impossible to consolidate a democratic republic in Germany. The forcible prevention of a union of Austria and Germany is the blow in the face for all the solemnly proclaimed principles of the Entente."

Dr. Preuß pointed out the federation with Austria—to be undertaken only if Austria made the first move—would lead to considerable economic hardship for the Germans, since Austria was now land-locked and had a rich musical and artistic culture but no seaports and comparatively little industry: the Škoda Works that produced armaments for the Austro-Hungarian Army was now located in Czechoslovakia. "[*The federation*] would, however, be of great value and would be of wonderful importance to the young German democracy, as it would bring about a unity which Bismarck's statecraft and Prussian weapons failed to accomplish.

"The policy of the Entente, therefore, can have no conceivable aim other than to prevent forcibly this morale strengthening of the Republic of Germany and to add to the strength of its opponents.... This policy runs counter to all the solemn declarations of President Wilson and the Entente. That should not meant much, for after our experiences since last November the value of these declarations has been sunk deeper than German values. But this policy negatives [*negates*] the obvious interests of the allied and associated powers, which demand the strengthening and the democratic development of Germany.

"It was attempted at first to represent the German democratic republic as a bare-faced deception in order that actions such as were adopted to defeat Kaiserism might be justified. It is harder to maintain the fiction the longer the democratic regime lasts.

"This regime has now concluded a constitution the bases of which are pure democracy and unity.

"Under most tremendous difficulties it has protected the western world as a dam against Bolshevism. It was obvious to every statesman that the German republic cannot be forever represented as a disguised Kaiserism. So new blows must be delivered.

"Such treatment increases the protagonists of the old system and adds to the spirit of revenge in the German people. A desired excuse for mishandling Germany may this be given in future. The goal must be the destruction of German democracy or of Germany herself. If the allied and associated statesmen do not have this aim in view, their policy of making life bitter for the German democracy would be inconsistent, and that must never be assumed in studying the policies of an opponent.

"President Wilson once said that the peace that ended the world war must approximate that made by the Congress of Vienna. The Vienna Congress believed that by measures of force it could make stronger the so-called legitimacy of the monarchical principal. That was a bad mistake, but how light as a feather it weighs against the madness of the Versailles peace, on which, through just those means, democracy and international law were to be attained!

"From the madness of Vienna, America at least kept itself free. But in Versailles...."

Preuß was seen as historically accurate in the days before Hitler. The American author Frederick Lewis Allen, writing in *Only Yesterday*, published in 1931, wrote: "When Germany and other European nations failed to be engulfed by the Bolshevik tide, the idea of its sweeping irresistibly across the Atlantic became a little less plausible." The defeat of the Bolsheviks by the *Freikorps* and *Reichswehr* in 1919, by the Czechs, Romanians and Hungarian counter-revolutionaries in 1919 and by the Poles in 1920, was greeted by most Americans with a vast sigh of relief. But around the borders—as the Germans saw it—malicious opportunism flourished.

Chapter Nineteen
Upper Silesia

"In Upper Silesia the tragicomic play of the Separation Crisis led by *Herr* Korfanty began," Hermann Ehrhardt wrote. "The Bolsheviks had to find themselves a political marionette. When the country was under the grip of terror, the Haller Forces would be welcomed as liberators and rescuers. Even the Germans, so the sly Polish wire-pullers reckoned, would be hopeful, would be relieved as long as their lives and goods would be protected as Polish citizens."

Ehrhardt's explanation may have been simplistic: The Polish Nationalist General Józef Haller von Hallenburg, late of the Austro-Hungarian Polish Legion, late of the French-sponsored Polish Legion, finally of his own Polish legion, was scarcely a Soviet puppet. Haller's most important victories were fought defending Polish territory from the Russians, though Haller had also seized German territory with largely Polish population in the aftermath of the Armistice and with the consent of the Treaty of Versailles. But Ehrhardt's response had to be immediate and forceful. The chaotic and rapidly shifting Polish situation was not amenable to a non-partisan analysis study due to the frequent need for quick military responses.

Kaiser Wilhelm II had begun to consider an independent Polish state as early as 1889, the year after he acceded to the German throne. The Kingdom of Prussia had joined Tsarist Russia and Austria in the three partitions of Poland. The last partition in 1795—inspired in part by an extremely liberal Polish constitution modelled on that of the United States and of Revolutionary France—had eliminated the sovereign state of Poland from the map, just a little over a century after the Polish "Winged Hussars" and their German allies had won the applause of Europe for saving Vienna from the Ottoman Turks. Wilhelm's reasons for sponsoring an independent Poland were a mixture of sentimentality and strategy, especially of an attempt to make friends with his own German Catholic subjects and to ensure the loyalty of those Poles who chose to remain as German citizens. The Polish independence plan was revived by Quartermaster General Erich von Ludendorff, who wanted to recruit Polish soldiers who would fight with spirit in the conflict against Russia—and to undercut the allegiance of Polish Legions fighting against Germany with the Allies.

Kaiser Wilhelm had withheld the plan for a sovereign Polish state during his heyday of friendship with Tsar Nicholas II. But the Kaiser felt betrayed when the Tsar backed Serbia against Austria-Hungary after the assassination of Archduke Franz Ferdinand, a personal friend. The Kaiser was also furious at the brutal murder of Franz Ferdinand's wife, Countess Sophie Chotek, whom Wilhelm had always treated graciously at court, even though she was not a royal. The Tsar's betrayal of Germany for the benefit of the murderous Serbs, as the Kaiser saw it, touched off a war with Russia's unlikely republican ally, France and—"This is our reward for Waterloo!"—with Britain, the homeland of the Kaiser's adored grandmother, Queen Victoria. Even the protest-oriented Social Democrats, the the lone exception of Karl Liebknecht, had voted war credits.

The Tsarist Russian invasion of East Prussia in the first months of the world war was an outrage: Tsarist Russian troops murdered 1,620 German civilians, raped hundreds of women, and burned everything that they couldn't carry away when they were defeated by the quick German counter-stroke that made Paul von Hindenburg and Erich Ludendorff national heroes. In 1915, the Russian Army was back in Poland. The measures that Tsarist General Nikolai Yanushkevich approved to 'cleanse' both the Jewish merchant class and the long-term German settlers who were Russia's best farmers caused outrage even in Russia itself and led to a deeply anti-Russian response in the neutral nations. The Kaiser's Army shortly evicted the Russians from all parts of Poland with the help of thousands of Polish soldiers in Prussian uniform and the covert help of those Polish Jews who took part in the conflict.

"France and Britain's alliance with Russia against Germany and Austria-Hungary is an alliance not only against Germany and Austria-Hungary, but also against the inseparable life interest of all of Europe," a Swedish Socialist named Gustaf Steffen wrote in 1915. Steffen referred to the Russians as a "peasant and national culture... if they are not wholly and purely Asiatic. All social feelings belong to a backward West [Asian] and Central Asian type, not a European one." Antipathy to Russia—and to Russia's allies and the British blockade on Germany that also impacted Sweden's economy—became so pronounced that in November of 1915 King Gustaf V of Sweden considering declaring war on the Allies, though he stopped when he realized the majority of the Swedish people were firmly committed to peace. The other Scandinavian countries, along with Switzerland and the Netherlands, also tilted toward the Kaiser's Germany but ultimately remained neutral. The idea that the German Empire was a terror to its neighbors was a concoction of Anglo-Saxon propaganda aimed at people whose only language was English.

On November 6, 1916, the Kaiser and his government proclaimed the founding of the independent Kingdom of Poland.

"The Polish question is the topic of the day," Admiral Müller wrote in his war-time diary. "According to the foreign press the signs seem to be favorable to us, i.e., rage on the part of the enemy and approval by the neutrals."

The Kingdom of Poland was to be composed almost entirely of that part of Poland, centered on Warsaw, which had been part of the Russian Empire. The Germans had no original intention of divesting themselves of once-Polish lands that had been Prussian or German for more than a century inhabited by a substantial minority of ethnic Germans and also of Jews who would be at considerable risk if the wrong people headed the new Polish state. But the intention that a restored Poland should once again join the family of nations was probably more sincere than cynical.

The new Kingdom of Poland, however popular with neutrals and disruptive to Anglo-French propaganda, was a kingdom without a king. The various German royal houses, the Wittelsbachs of Bavaria and the Wettins of Saxony, tripped over their own feet nominating relatives for the role of King of Poland—but the Kaiser's diplomats understood that putting a lackluster German outsider on the throne would do more harm than good.

The first stumbling-block to the German vision of an independent Kingdom of Poland made from exclusively Russian Poland and land donated from Austrian Poland loomed in the German Reichstag with "an insolent speech by the Pole [*Wojciech*] Korfanty, who wants Danzig for the Poles." Danzig had been a German seaport since the days of the Hanseatic League which was founded in the Middle Ages and was more than 90 percent German by population and culture.

In the autumn of 1918, General Józef Piłsudski, a long-time Polish antagonist of Tsarist Russian rule, an anti-Marxist, and a sometime senior officer of Polish troops in the Austro-Hungarian Army, was locked up in a German prison in Magdeburg after resigning his officer's commission with the Central Powers. Józef Piłsudski understood that the Germans were losing the war but that he himself appeared to be in no personal danger. Being in a German prison might be of some political benefit at the peace talks. He was immensely surprised when he was joined by Kasimir Sosnkowski, another Polish independence advocate and former Austrian officer. The real surprise came when Piłsudski learned that he had been declared Poland's new Minister of War.

On November 8, Count Harry Kessler slithered into Piłsudski's cell and told him that he was free to go. As Piłsudski and Sosnkowski were switching trains in

Berlin, they heard that the Kaiser had abdicated. Prince Max of Baden, the Kaiser's South German cousin, had engineered the succession before he handed the German administration over to the Social Democrat Friedrich Ebert.

Piłsudski was the best man the Germans could find for the job of ruling Poland: he was a lapsed Catholic who had married his former mistress and second wife at a Protestant ceremony. He was not known as a violent anti-Semite, and he patently hated Russia far more than he hated Germany. Piłsudski sometimes shared and sometimes contested power with General Józef Haller von Hallenburg, a freedom advocate from a distinguished German-Polish family and former commander of Austria-Hungary's own Polish Legion. General Haller's new contingent of troops was called "the Blue Legion" because they wore horizon-blue French uniforms. In the intense turbulence that followed, Piłsudski and Haller both treated German citizens decently, though Haller's forces were sometimes accused of anti-Semitic looting. But neither of them shared Wilhelm's ideas of the new Polish borders. Russia—now Bolshevik Russia—remained the standing menace of Polish politics. As long as the Poles were fighting the Bolsheviks, they resisted taking on the tens of thousands of German troops—many of them containing large numbers of ethnic Poles still somewhat loyal to their Prussian officers, if not to the Kaiser— who roamed around western Poland and eastern Germany in small units entirely capable of self-defense. The German policy of basing troops as close to home as possible and maintaining unit cohesion for the entire term of service had stabilized their military: Rather than "Polish Legions," Prussia had proud Polish soldiers in German uniforms, including the pick of the Poles in crack units like the Prussian Guard, whose emblazoned helmet motto—SUUM CUIQUE, Latin for "To Each His Own"—suggests that individual achievement and personal loyalty were more important than serving in your own home town as ordinary German soldiers did.

Kapitän Ehrhardt had to deal with the tactical situation even if the politics were more than confusing.

"Daily there were fire-fights between larger land-owners and bandits, and it was always the Germans who suffered from this Bolshevik plague," Ehrhardt wrote. "And suddenly the insurrection was on, before anybody could believe it, above all before the the Socialist administration in Berlin could believe it, those for whom the new German flag was more important than a threat to the German border.

"But as the confusion in Silesia has risen to its highest point, the Entente threatened our administration with an armed occupation 'in order to hold back Bolshevism.' That made men out of our administration at last, and the 2nd Marine Brigade, with other military units, was sent to secure the threatened area in Upper Silesia."

The troops were divided into small independent sections to provide security for villages and small towns around the countryside, which was about 40 percent German and 60 percent Polish, and had been part of Prussia or later of Germany, for about 150 years.

Star of the Order of the Black Eagle with Latin Motto "Suum Cuique"

"Every village was covered with troops to provide area security. After that we searched for the weapons that had played a great role in the random shootings. We found almost none of them. The giant forests offered ample hiding places for those who knew the territory. The Poles never betrayed people on their own side. The Polish village priests used their authority to good advantage to prevent a slide to the German side of the question.

"The men were happy with the occupation. The soldiers helped with the threshing and they attended the devotional services that the priests held for them."

Oberleutnant Rudolf Mann billeted with the senior priest, whom he described as "German to the core." Mann believed that the priest must have been of noble or aristocratic birth before he took Holy Orders, because the priest loved horseback riding. Mann found a couple of troop horses, fitted them with artillery saddles, and he and the priest took regular horseback rides together—all the better to be seen side by side by the intensely Catholic Polish population. Mann also shared a problem with the priest: remembering good times in Braunschweig and elsewhere, Mann had tried to organize a dance for the local girls, but the girls seemed to be afraid to attend.

"I have simply forbidden them to attend, because I don't know your soldiers, and the girls know that if they go despite my ban they'll have to stand before the church doors in shame," the priest said. "They fear being shut out of the *Jungfrauverein* more than they fear the civil courts. Out of 216 births in this parish we've only had six out of wedlock."

The 2nd Marine Brigade, a 6,000- man Freikorps unit formed and led by Ehrhardt, made its way to Berlin in March 1920 to take part in the Kapp Putsch, which ended in an embarrassing withdrawal.

Oberleutnant Mann put in a good word for his men and promised that the girls would all be home by 10 o'clock with no funny business on the way.

"I could have laughed until I cried as all the *Maruschas* without exception swirled out to the dance with their long gowns and jacket-vests and their high-heeled shoes. Usually they went about bare-foot. The dance conversations caused no problems because all the dancers spoke flawless German. We had the German schools to thank for that."

Ehrhardt attended the dance as a spectator and enjoyed seeing the colorful regional costumes—the *Tracht*, or regional dress, was a hobby of his—and the spirited Polish dancing. He heartily approved.

After the dance, Rudolf Mann sat down to dinner and conversation with the senior priest and the leading men of the town, including a bank director and an industrialist. He ascertained that Upper Silesia was 90 percent Catholic and that this was the key to politics: The Catholics were members of the *Zentrum*, the German Catholic party outside Bavaria, but the Zentrum in Upper Silesia had two factions. The younger priests and the farmers and workers were for merger with Poland while the senior priests and the business owners and professionals wanted to remain part of Germany—unless Germany itself drifted too far to the Left. The abiding fear of both factions was not Germany but Russia, and especially Bolshevism. The Democrats, the leftist Independent Social Democrats, and the Communists were barely represented in Silesia. Mann also thought that the "magnates"—the large land-holders of mixed aristocratic German, Hungarian and Polish ancestry, the mine owners and the financiers—often favored the Polish Nationalists because the Polish Nationalists were strongly anti-Communist and anti-Russian. The magnates were afraid that Germany itself was turning Socialist and might go Communist.

Hermann Ehrhardt was studying the magnates at close range. He boarded with Victor, Duke of Ratibor, Prince of Corvey and Prince of Hohenlohe-Schillingsfürst, a relative of the Hohenzollerns, and his wife Maria, Duchess of Ratibor, born Countess Breunner-Enkevoirth of the Hungarian nobility.

"We thought for a long time that the host of our brigade staff headquarters, the Duke of Ratibor, and a large group of his relations, both gentlemen and ladies, leaned toward the Polish side," Ehrhardt wrote. "In German circles there was a long-standing prejudice against the magnates. But in closer contact with them we learned that they were good Germans.... Gradually we came to feel ourselves very much at home in the huge castle. The day's schedule became a routine. The Duke sent us our breakfast, middays we ate at the *Gulaschkanone*, [*horse-drawn field kitchen*] and everybody had the evening to himself if the Duke hadn't invited us to dinner." "At the table the Duchess held forth along with the Duke. She was born a Hungarian princess. The conversation with the ladies at length strengthened our suspicions of the Poles. There in a row were the Princess Ratibor, Princess Hohenlohe, and guests from all the large estates, notably Count Sierstorpff. The conversation was merry and unrestrained. The Duke was a jolly old gentleman with a great passion for hunting. The whole house was full of trophies he had shot in the field. There were some great pieces that delighted my hunter's heart. Naturally the Duke gladly told hunting stories. We told our war stories and sea stories and noted how isolated from the rest of the world these people were. I always had the feeling that they never got over their amazement at us middle-class officers. But no stiffness came of it and happiness reigned. Gradually the Duke acknowledged how much he sympathized with the Fatherland. In the proper princely manner he asked very little of it for his personal interests."

"Gradually the ladies came to trust us and we took morning rides with them and tried out some jumping exercises with the company horses. I think it was a benefit for them to hear people speak whose noses had grown a bit and to make a judgment about how these men could educate themselves.

"The Princess Hohenlohe became very indignant when I told her one day that I had taken her to be Polish because of her dark appearance. She gave me a look of death for some time afterward."

A German Propaganda poster: "Prayer of the Homeland: Upper Silesia remain German!" Both ethnic Germans and ethnic Poles were mostly Catholic. Silesian Jews voted strongly for Germany.

The nest headquarters was set up at Castle Slawenitz. "Prince Christian Kraft zu Hohenlohe-Öhringen never stinted on hospitality. Every officer had a sitting room and a sleeping room, a bath and a room for his orderly. We organized sports events and harmless little folk festivals to make sure of good relations between the troops and the population.

"The brother of the prince, who shared the residence, showed us an antiquated stiff reserve, even though the other people in the family were happy to deal with us. He had probably been a diplomat and regarded it as painful to deal with mere mortals.

"The Princess herself, who gladly showed herself to be a fresh and accomplished horsewoman, seemed to maintain a hidden distance that sprang from the consciousness of being from a different caste."

"Even as we had at Castle Rauden, we hoisted our proud old black-white-red war flag. From these restful quarters we were moved into the mining country by Myslowitz and Kattowitz in the border area. Here the service became more serious. Many unemployed Polish workers had made common cause with the uprising. Our detachments at Bogutsschütz and Eischewald lay within pistol shot of the Haller troops on the opposite side. The Polish *Soldateska* were well uniformed but they were no heroes. Little shooting incidents on both sides sprang up due to youthful sporting instincts rather than any war-like feelings of hate."

Rudolf Mann noted that the prosperous mining country actually had automobiles and service stations, like Berlin, Hamburg or Munich but unlike most of the rest of Germany. Mann interrupted what appeared to be an enjoyable flirtation with a young lady who had studied pharmacy, but not geology, when he decided to visit the inside of a Silesian coal mine.

"If you know geology, you can look around with enjoyment and attention. The young apothecary girl from Sowocze, who didn't know any geology, but dispensed schnapps for a living, didn't want to hear about it but always laughed when I showed up in my blue linen working smock....

"Now I know why the Upper Silesian coal fields are so important, because the coal strata are sometimes 22 meters thick and not too deep—about 400 and 600 meters—under the surface.... In the Ruhr fields the coal is only about 90 centimeters thick and the strata are full of stones.... All the locomotives in the Empire burn Silesian coal."

Mann won the respect of the Silesian miners by actually venturing underground. He came back with some new perspectives. He expected the mine tunnels to be something out of "*Germinal*" by Émile Zola. The Polish miners told him that the mines had been that way at one time but that electricity for illumination and power tools had changed all that.

"Why do you strike so often and for so long?" Mann, clad in his blue smock, asked some of the miners.

"When you make as much money as we do, you can bring off a small strike," the miners said good-naturedly.

He noted that the miners were paid by the load, not by the hour, and often made 36 to 40 marks a day, with subsidized rent. "A father of a family with a heap of kids often had a heap of troubles but the young gentlemen workers turned out on Sunday afternoons in club clothes to play football. The younger men seemed to spend their evenings eating meat and drinking good liquor. They wore the new wristwatches and smoked English cigarettes planted in silver holders."

Ehrhardt, above ground while *Oberleutnant* Mann was exploring the Silesian coal fields, remembered one Polish attempt at defiance that ended without any violence beyond the verbal.

"One day the Poles started to build a bridge across the river. The company leader of our Brigade asked the Polish commanding officer to come out on the bridge and said to him: 'If this bridge hasn't been broken up by 6 a.m. tomorrow morning we'll open fire on the Polish village. Promptly at 6 a.m. the next morning the bridge was broken up. Our officers and soldiers recognized that a lot of these Poles had been in our own army in 1914."

"The war of ambush was worse. Some guard posts were murdered by the Poles who then disappeared without a trace. It was a notable demonstration of the kindness of the Polish murder organization."

"All the back roads were heavily wired—not so much against military operations as against chronic smuggling.... The smugglers, as a rule, were Polish Jews. We caught a lot of these fellows. These kaftan-wearers were at first absolutely astounded that our soldiers didn't let them go right through like the soldiers had who had been here before. They offered us money in a shameless manner. On my orders, the soldiers would seem to take the bribes. After they had the money, they would take the goods, beat the smugglers on the backs with sticks and send them back over the border.

"One day a Polish peasant drove up in a wagon loaded with composted manure. For security the guard post stuck with bayonets deep in the manure and struck something hard. The many rolls of leather were a welcome booty for the Brigade shoemakers. The black-white-red flag my people hauled up flew for a long time over the *Dreikaiserecke*."

Ehrhardt recalled the Germans of Silesia as quietly hospitable. One night after dinner at the local tavern, proudly known as the *Hotel Stadt Öhringen* he and some of his men, encouraged by the landlady with smoked pike and schnapps and vodka on the house, had such a good time that he admitted they had a tough time finding their way back to their lodgings. When the Brigade was ordered to depart and encamp at Karlshorst outside Berlin, the three Princesses waved good-bye and the men marched off singing happy songs about Polish girls and broken hearts on both sides.

A bilingual Polish Propaganda poster: "Vote for Poland and you will be free"—assures Silesians that the coal mines will continue to operate if the plebiscite goes for Poland.

Chapter Twenty
The Kapp Putsch

When the Second Marine Brigade returned to Brandenburg after a largely peaceful turn in protecting Upper Silesia, they found a cold reception. One company was quartered in drafty barns outside Berlin until, with Captain Ehrhardt's tacit approval, the soldiers took over better quarters on their own initiative and let the previously quartered troops shift for themselves. Ehrhardt's *Freikorps* brigade of former officers, officer cadets, veteran non-coms and patriotic university students had nothing in common with the shirkers who stayed on in the *Reichswehr* for three meals and a warm bed. They made that plain. The Social Democrats saw the black-white-red Imperial flag as a threat. The Brigade was stationed a day's march from Berlin, in Bernau, so they could march to Berlin in case of yet another uprising. Ehrhardt reported that "the food supply was dog-miserable. The Red Brothers used false papers to shift the railroad cars loaded with foodstuffs into sidings, so they never got to the troops. The officers had to assert themselves to make sure the men got the food that was theirs by right."

Another problem came from the *Baltikumer*, separate *Freikorps* units that had served in the confused fighting in the Baltic states—first fighting Russian Bolsheviks with the support of the British Royal Navy, then being expelled by the Baltic peoples with the support of the Royal Navy once the Bolsheviks were defeated. Copies of the 'Protocol of the Learned Elders of Zion'—a forged Tsarist-Russian manual for a Jewish world take-over—had been circulated among the *Baltikumer* by White Russians who professed to admire them.

"The consequences were that these people became politicized. They dispersed gatherings, aroused scandals, beat up Jews. Every day I had to break up some brawl or other."

Ehrhardt's own policy toward the Jews was established and never changed during his control: law-abiding German Jews were German citizens entitled to full civil rights. He protected them from violence. Smugglers got the stick. Armed Revolutionaries or agitators got the same treatment as other Bolsheviks.

Even worse than trying to hold down the *Baltikumer* anti-Semites, Ehrhardt said, were the threats to reduce the German armed forces to 100,000 men. At the end of 1919, the Poles had worked out agreements with the Germans to respect one another's borders—but the Poles themselves were faced by the threat of a Bolshevik attempt to reconquer Poland and to contest control of the Ukraine. No one in Germany—especially the Reds—believed that 100,000 German troops could contain a massive Soviet army with or without the help of Poland—and some Germans believed that 100,000 men would not even contain a Prussian-trained Polish army.

The assault company of the Second Marine Brigade marked the first anniversary of the Revolution with an anti-Revolutionary display of the black-white red flag and wreaths decorated with black-white-red ribbons. As the wreath-bearers marched through the streets of Berlin, people from the crowd demanded sarcastically: "Who's dead?"

"The German government within a year, if you want it that way," the marchers shouted back, ready for a brawl. There were no takers.

Tillessen, who had captured the howitzer in Munich, made a patriotic speech at the Bismarck Memorial as the black-white-red flag was hoisted, and most of the Berliners, Ehrhardt said, seemed to agree with him.

On November 15, Captain Ehrhardt wrote Brigade Order 120:

"I have sent the following telegram to the Chief of the Admiralty serving as a member of the Cabinet on November 9, 1919:

"The 2nd Marine Brigade expects the administration to reject the shameless demands of the Entente. There was in these demands an unmistakable will to exterminate us as their enemies. An agreement with the demands will only incite the Entente to new demands once we are unable to fulfill the conditions. That would place us in the same condition as if we refuse now, except that we would be sucked dry and rendered completely defenseless.

"Commander 2nd Marinebrigade

"Wilhelmshaven

"The 3rd Marine brigade has agreed with this protest."

After the demonstration, the Brigade said farewell to a senior assault commander, *Leutnant* G. Reinhardt, who had been an enlisted man before he was promoted to *Stoßtruppen* commander due to his wartime experience as a infantry squad leader in the trenches of the Western Front. Reinhardt, one of the officers with no independent means, had been offered a job as a policeman, and Ehrhardt told him to take it to support his wife and children. Ehrhardt pointed out that officers and officer cadets taking orders from former enlisted men indicated that caste snobbery was not a factor in the Second Marine Brigade. The fact that an assault group leader, *Leutnant* Berlin, was Jewish and the enlisted men included several Alsatians who wanted no part of France, several Balts, at least two White Russians, a Chinese and a Turk, suggested that Nazi-style racism was not a factor either.

Captain Ehrhardt shortly had his answer to the protest telegram: the Second Marine Brigade was to send its contingents to 20 different locations. Ehrhardt read what he saw as the intention: without his direct command the men would become disruptive like the *Baltikumer* with their brutal attacks on Jews and their rampant brawls among themselves: the regime would have an excuse to break up the Brigade and eliminate it. Ehrhardt took his case to General Walther von Lüttwitz, an early supporter of the *Freikorps* movement. In January, Ehrhardt was ordered to assemble the Brigade at a camp at Döberitz, about 15 miles from Berlin. Rations continued to be sparse and the fodder was so bad that some of the horses sickened and died of colic.

At the beginning of January the question of the reduction of the Army came before the senior generals, "We all stood on the standpoint that a hundred-thousand-man army was an impossibility and that we must have at least two hundred thousand men," Ehrhardt said. General Lüttwitz agreed and said that the reduced army must also have heavy artillery. Lüttwitz also told his subordinates that if the regime refused to demand a 200,000-man army, it would be his duty as the senior serving general in Berlin to take charge of the army and demand an army strong enough to defend Germany, at least from the East—meaning either Poland or Russia. Germany in January of 1920 actually but informally had more than 1,200,000 men under arms. The military commanders refused to accept an occupation of English and French or the Polish *Soldateska* without a fight.

The Noske War Department tried to strike a compromise in January: the army would be reduced to 200,000 men in April and then to 100,000 men in July.

"What do you think of Noske?" another captain asked Ehrhardt.

"I don't trust the 'brother', but he must have learned that he can't get away with 100,000 men," Ehrhardt replied.

"These Red Brothers fear only one thing—having to give up their nice warm administration seats," the other officer agreed. "They're fine with the idea that the arm should be reduced to 100,000 men. For them the Army means opposition and suspicion. We're good enough to take part in a civil war they can drag out from their plush velvet chairs. But when our work is done, we can get lost."

The mood in camp was such, Ehrhardt noted, that the younger men began to buy multiple newspapers—not just the conservative papers that echoed their own opinions but the liberal and leftist papers so they could sample *all* opinions. They also began to paint the white swastika on the front of their helmets for the first time. "How that came about I don't know," Ehrhardt wrote. "Suddenly the swastika was the Nationalist emblem."

Ehrhardt's men were in a fighting mood. When some of them showed up at a dance for what turned out to be a largely Communist wedding in a nearby village, they were rebuffed. The bride and groom ended up hiding in the cellar while several of the invited guests were thrown out of the windows. The village mayor turned up at headquarters and gave Ehrhardt a stern lecture. Ehrhardt managed not to laugh. He noted that the Brigade had given up singing smutty or satiric cabaret songs and gone back to singing folk songs and the old marching songs of the German Empire or the earlier German kingdoms.

"The mood went so far that at one assembly of intelligence personnel, that I held without fail every eight days, an non-commissioned officer called out to me in his excitement," 'Herr *Kapitän,* when do we finally march on Berlin? We want to throw that whole gang to the devil."

Kapp Putschists at lunch—The Ehrhardt Brigade brought their own food for the Kapp Putsch in 1920 and were NOT evicted due to hunger, but to weak outside leadership. Note Ehrhardt's trademark, the Imperial Battle Flag, generally used as a Naval insignia.

Ehrhardt left standing orders to be ready for a 25-kilometer march with full weapons and ammunition, full pack and all officers on duty—and a little arithmetic showed that 25 kilometers was the exact distance between the camp at Döberitz and the center of Berlin. The very same sparrows that had predicted a Red take-over of

Wilhelmshaven before Ehrhardt struck it down were now predicting a Communist uprising in the Ruhr, the key German industrial area not far from the border of France and Belgium. Lenin had reportedly already given Béla Kun his marching orders—based on an exaggerated account of Communuist support. Lüttwitz, meanwhile, dragged his feet and had not yet begun the reduction of the armed forces.

Some Hungarian gentlemen now appeared at headquarters. The Hungarian Communist Béla Kun, Magyar on his mother's side, and his mostly Jewish followers had been routed from Hungary at a cost of about 700 dead Magyar Hungarians. They were known to be active in the Ruhr. The anti-Communist Hungarian administration of Admiral Miklós Horthy wanted to find out whether Germany could be depended on in a general war against Bolshevism. The Poles were engaged in fighting a Bolshevik invasion with the help of French military advisers and weapons with heavy losses on both sides. The Hungarian nationalists—and the German officer corps—were not sure whether the next enemy would be Poland in quest of Upper Silesia, Breslau, Posen and Danzig, or Russia in quest of all of Europe. Ehrhardt had no ready answer, but he took down some names that later proved useful to him.

Then Hermann Ehrhardt met another gentleman who was to have a profound influence on his personal future.

Waldemar Pabst had played out his extended leave after the confrontation where he had (possibly) threatened to shoot Noske, and he never returned to the top General Staff posting at the Guards Cavalry Division in Berlin. He briefly served as a General Staff advisor to the 5th Division in Stuttgart and the 6th Division in Münster and then resigned due to the mandatory reduction of the army.

About the time Pabst resigned, Dr. Wolfgang Kapp, minister of agriculture, founded the '*Nationale Vereinigung*" [National Union] Party. Kapp, a long-time bureaucrat and sometime self-publicist, had founded the *Deutsche Vaterlandspartei* [German Fatherland Party] on December 10, 1918, at the height of the Bolshevik threat to the cities of Northern Germany. The new party mustered 1.25 million members and financial supporters. Kapp's *Vaterlandspartei* strongly opposed any annexations beyond the stipulated return of Alsace-Lorraine to France and the founding of the Kingdom of Poland as approved by the Kaiser's German government a month before. Patriotically Wolfgang Kapp wanted to wave the German flag with pride. Pragmatically he wanted to hold onto the coal fields and ore fields that Germany need to maintain its industrial base. Pabst, who had been immensely impressed with Hermann Ehrhardt's forceful leadership, asked Ehrhardt to spend some time with Dr. Kapp and see if he had any leadership potential as a political player for those *Freikorps* troops who could be trusted to behave responsibly without anti-Semitic outrages or homosexual scandals.

"I won Kapp over to a personal affinity," Ehrhardt later wrote in bemused retrospect. "My grandfather on my mother's side had been, like [Kapp's] own father, a 48er Revolutionary and had had to flee Germany like Kapp's father. Our relationship developed from the stories we told one another. Kapp was a German from his top to his toes, but despite his overwhelming reverence for authority and his sense of *Realpolitik* he was an idealist of a sort, because he believed that a beautiful, powerful thought must arise on its own from the majority of the people. He could captivate people with his speaking, and his temperament won over mass meetings. His huge size gave him a great influence, but he turned out to need better nerves than he actually had. He was charming when he drank Moselle wine as we chatted, and he drank a lot of it.

"One of his finest stories was this: It was winter in the year 1871. The Kapp family then lived on the eastern shore of Lake Michigan [*in the American Middle West*] where the water was frozen over. Deep snow lay thick on the forests and fields, and a cutting wind blew. Kapp's father had traveled to Chicago. When he came back to town after an absence of several weeks, the mother and children went to the railroad tracks to bring him home. At last the ice-bedecked train arrived late. The door of one passenger coach opened itself and a huge figure came out into the snow in a fur coat and cap and tall boots. The children recognized their father and cheered for him. But the giant heaved his hands into the air and the wife and children stepped back. Then the mighty man with a voice that choked back tears said: 'We have had a German Kaiser for three days!'

"As a victory signal just as that moment the locomotive whistled and the whole trainload of travelers burst into cheers in the white wintery forest. Kapp's father soon left America. The Kaiser Movement of 1848 he had fought for and was banned for advocating, was fulfilled. He returned to Germany as soon as he was able.

Dr. Wolfgang Kapp

"This youthful experience remained a stimulating force in Kapp's life," Ehrhardt observed. "Despite some realistic insights he remained an idealist in politics."

The Revolution of 1848—as Ehrhardt knew from his own family history— had been a bid to build a unified Germany able to defend itself against any nation in Europe, and had foundered partly on the demand of workers for job security that went beyond the power of the Medieval and Renaissance Guilds and the Christian principles of their employers. The primary reason was that neither the King of Prussia nor the King of Hanover would accept the role of German Emperor as a constitutional monarch. The King of Prussian king remarked bluntly that accepting a national parliament would be like wearing a dog collar, and the King of Hanover agreed with him. The Austrian dynasty was so rigidly Catholic that it was distrusted by both Protestants and Jews, and Protestants and Jews, no matter how prosperous or eminent, lived in Austria under sufferance. Gustav Mahler and Sigmund Freud both wrote of this discrimination. Freud remembered his grey-bearded father being told by Catholic teenagers to jump off the sidewalk and stand in the street—"Jew, do your office!"—so they could walk past him. The Wittelsbach Dynasty in Bavaria had a reputation for insanity. The kings of Saxony and Baden—whose royal titles dated only from the time of Napoleon—were unacceptable because their countries were small and rather backward. Otto von Bismarck had used the Franco-Prussian War to unify Germany—minus Austria and with some special conditions for Bavaria—into a federation of kingdoms under the King of Prussia as German Emperor. Now Ehrhardt and Kapp faced a future where they saw themselves being dragged back into the past—large chunks of Germany were being sliced away, contrary to the will of the occupants and to the stated policies of Woodrow Wilson's 14 Points. The Prussian Army that had enabled Prussia to unify Germany with Bismarck's "Blood and Iron" and then held a malevolent Europe in stalemate for four long years was being reduced to a police force.

Dr. Kapp gained Ehrhardt's wary approval and began working out a platform of his National Union Party with General Walther von Lüttwitz, the senior commander in Berlin—and, behind the scenes, with General Hans von Seeckt, the supreme commander of the German Army. Seeckt, a World War I general, nicknamed "The Sphinx with a Monocle," was famous for being enigmatic. He was known to be staunchly anti-Communist and generally assumed to be a monarchist should the opportunity arise. Waldemar Pabst, an absolute monarchist and now separated from the *Reichswehr*, worked with Kapp and Lüttwitz to maintain an army strong enough to defend Germany against internal threats and—with the assumed help of a mass popular uprising—to also defend against foreign invasions with no German provocation.

Kapp, Lüttwitz, and Pabst deliberated about what platform would serve their purpose and still be acceptable to the majority of the National Assembly and the voters.

The political goal was to combine all Nationalist parties into a coalition and to publicize their joint policy in such a way as to attract voters and financial support. The military goal was to stop the attempt to reduce the active armed forces below an acceptable level and to publicize the need for an adequate number of troops to protect Germany from international Communism. Pabst said they needed to bring home to the Germans the danger of further outbreaks of Revolution in the Communist manner: there were attempted revolts in July 1919 even after the massacre during the transportation strike. He retained two publicists, Karl Schnitzler and Dr. Fritz Grabowski, to keep producing propaganda with the quiet support of General

Erich von Ludendorff, former quartermaster general under the Kaiser and Paul von Hindenburg.

General von Lüttwitz refined the policy into five key points:

1. The attempt to have so-called war criminals handed over [*to the Allies*] must not succeed under any circumstances.
2. The reduction of the *Reichswehr* must be postponed until the threat of Bolshevism is completely eliminated.
3. The re-entry of the USPD [*Independent Socialists, seen as habitual Communist supporters*] into the administration in any form is unacceptable.
4. The welfare of the troops in service and the welfare of the troops who may eventually be released from service must be better provided for.
5. The unit of the nation must remain protected.

Another demand was shortly added: the further postponement of the Reichstag elections must cease and the changes to the constitution must not be considered.

As the deadline drew near, the National Assembly revealed that both the 2nd Marine Brigade and the now-affiliated 3rd Marine Brigade were to be disbanded. But Seeckt, "The Sphinx with the Monocle" and commander in chief of the *Reichswehr*, let it slip to Lüttwitz that he himself did not approve: "I will not support that we should knock off such a hard-core unit in such stormy political weather." Gustav Noske, who had undercut the 2nd Marine Brigade's drastic actions during the transportation strike, had 42 dead demonstrators to explain away to the Social Democrats. He ordered the chief of the admiralty, Vice-Admiral Adolf von Trotha to demobilize the 2nd Marine Brigade immediately.

Lüttwitz then dropped the other jackboot. He contacted President Friedrich Ebert and renewed the demands he had made to Noske on behalf of the army about six months before.

1. Immediate dissolution of the present National Assembly and new elections to the Reichstag.
2. The dismissal of the Foreign Minister, the Minister of Economics, and the Minister of Finance.
3. The promotion of a supreme commander of the *Reichswehr* and the dismissal of General Reinhardt.
4. The withdrawal of the troop dismissal order."

Noske believed that most of the army's top commanders would only half-heartedly support Lüttwitz and might even sabotage him. He threatened to take Lüttwitz out "with my own fist" and also threatened the customary general strike. More to the point, Noske ordered that Kapp, Pabst, Colonel Max Hermann Bauer and Schnitzler and Grabowski, the two publicists, be arrested. But Capitan Ehrhardt, having been informed of Lüttwitz's demands, sent word that he offered Lütt-witz his full support. An informant tipped Pabst off that there was a warrant out for his arrest. Pabst headed for Ehrhardt's camp at Döberitz with all possible speed. Then he quietly slipped away for what might be his last visit to his sick wife.

Ehrhardt himself had felt a storm warning when Trebitsch-Lincoln "the journalist with many nationalities" showed up at Döberitz. "From the first blink of my eye I felt a strong distrust against him. I didn't have any grounds for it, my instinct was just against him." The incredible and preposterous Ignatz Trebitsch-Lincoln, who seemed to live for the pleasure of deception and betrayal, was the right man to distrust.

General von Seeckt, "The Sphinx with a Monocle"

He was a former member of Britain's Parliament and also a self-styled German Rightist who both supported and denounced Hitler. Trebitsch had started out Jewish but changed his religion five times. Ultimately he claimed to be a Buddhist and worked for the Japanese; at one point claimed he was the real Dalai Lama. Having adopted the name "Lincoln" to espouse honesty, he misled every government he ever worked for from Britain to Japan. The Japanese finally poisoned him in 1943 after he took their money and then betrayed them. But in 1920 Trebitsch-Lincoln missed his chance to betray Hermann Ehrhard, which might have terminated his colorful career more than two decades ahead of schedule. For some instinctive reason, Ehrhardt didn't trust him.

Ehrhardt, having rejected the help of Trebitsch-Lincoln, now set out for Berlin in the armored car. He met General Lüttwitz headed in the opposite direction. They pulled to abrupt halts and they conferred at the roadside.

"The moment for action has come," Lüttwitz told Ehrhardt. "The administration will ruthlessly abolish all volunteer groups and reduce the armed forces to 100,000 men. I can't tolerate that because of my obligation to the people of Germany. I will march on Berlin and achieve the acceptance of my demands."

Lüttwitz astounded Ehrhardt by explaining the political contexts of his demands. The mainstream Social Democrats had established a virtual one-party rule under Friedrich Ebert and Gustav Noske. A wider representation of the electorate including the Catholic parties, Democrats and Nationalists was necessary to save what was left of the German Army and to protect Germany from the crass abuse and plundering of the Allies, border disputes with the Poles, and a possible invasion by the Russian Bolsheviks with the help of internal traitors. Ehrhardt's face was under tight control—but his slanted Hunnic blue eyes showed how he felt. The eyes slanted *up*—the attack signal. Lüttwitz read his eyes.

"Can you march on Berlin tonight?" Lüttwitz asked bluntly.

"Militarily I would need to consider that," Ehrhardt said.

"Can you march tomorrow?"

"*Jawohl!*"

Ehrhardt had long since learned to turn his sometimes formidable bouts of depression into an indifference to his own survival. In battle he was utterly fearless. But both his religious upbringing and his military training had given him an ingrained, almost subconscious awe of authority: He took a last chance and banged out his own ultimatum to the administration if they wanted to avoid a confrontation with 6,000 of the best troops in Europe.

"1. A general will take command of the army
2. A larger number of ministers will be appointed to the cabinet.
3. General von Lüttwitz will be returned to the army.
4. The politicians who took part in the undertaking (Kapp, Pabst, Bauer, Schnitzler, etc.) will be given immunity from prosecution.
5. The troops that take part in the undertaking will be immune from prosecution."

Ehrhardt knew there were spies in his camp. That afternoon, the story was beginning to appear in the newspapers. The result: dozens of fresh volunteers came in and asked to serve with him. The mess sergeants began to prepare marching rations and the motor transport sergeants began to gas up and load the trucks. The orders also went out to arrest anybody who showed up at camp and hold him in custody until the march began. The horses were fed and harnessed, ready to tow the battery of 10.5 howitzers. The assault companies filled their bread-bags with food, their pouches with cartridges, and their belts with hand grenades and checked over their submachine guns and rifles.

"It was arranged that next morning at 6 o'clock sharp I would march my brigade to the Brandenburg Gate. There I would await further orders."

Then Ehrhardt told his men to turn in and get some sleep.

Abruptly, General Bernhard von Hülsen, an associate of General Lüttwitz, showed up with "a tall, slender gentleman with an acne-scarred face and, apparently, a very soft backbone for a man who was over 50. I had already shown some consideration due to his age and released him from custody. He wandered about talking, back and forth, and hoped he could get me to say that I would not march. I explained to him that my promise to General Lüttwitz was binding and not subject to alteration.

"At the end he said that he would join us with some of his troops and leave the others at home."

"Toward evening as the entire Brigade was resting, Admiral von Trotha came to me as a representative of Noske. He wanted to make absolutely sure that we were actually going to march. I now absolutely smelled a rat as far Noske's fabulous intelligence service in Döberitz was concerned.

"I came into a mean-spirited dilemma. I had always respected Admiral von Trotha and would not have misled him for any price. Thank God I was able to turn the conversation in such a direction that I answered all the questions crisply and clearly without giving away any information. I was able to say that at this time the Brigade found itself totally at rest. I could offer to show Admiral von Trotha through his own eyes that this was true.

"After the Admiral left I was myself rather weary. Since I didn't expect much from the first part of the march, I lay down on the sofa in the barracks and gave my armored car driver the order to wake me up at 11 p.m. I unbuckled my holster with the pistol and laid it near me on the chair. Then I fell asleep.

"Suddenly in my sleep I heard the door opening. Through my eyelids I sensed that the light was turned on. Without even thinking I instinctively grabbed the pistol and shouted 'Hands up!' Then the thought ran through my head—you must be completely alone in camp. As these words rang through my head I saw three armed men, straight in the range of my pistol. To my amazement their hands flew up into the air, so quickly that I was't sure whether or not it was all a dream.

"The three men protested that they had only come for the purpose of negotiation. The word 'negotiation' was completely for my pacification because we all knew nothing positive was going to happen.

"My three late-night visitors were two generals and a staff major. They came from the administration and wanted me to stop the march on Berlin.

"I explained to the gentlemen: I would carry out the orders of General Lüttwitz. After communications from the general, other national troops would also march in order to achieve the general's demands. The stone was rolling, and I had no intention of stopping it."

"Both generals asked me to formulate my own demands from the administration.

"Half-asleep as I was, I formulated the demands I had heard from Lüttwitz somewhat carelessly and off the cuff. I said what these gentlemen had to hear; if the administration gave way, Lüttwitz had gotten to where he wanted to be and we had come a step further. What I demanded, however, must have clunked and clattered somewhat, based on the expressions on their faces, when compared to Noske's position. For me, as a soldier, it would have been a simple formula. I didn't become politically active until after the misfortune, when corruption and filth got worse and I myself had to live as an outlaw.

"Now the conversation with the generals went back and forth. They asked me to hold off on any news releases, whether the administration agreed to the demands or not. They asked me to declare I would not march to the Brandenburg Gate itself, but only to the Siegesallee. It would be important to avoid friction with the security police.

"The possibility of stopping the march itself was out of my hands. But I knew the troops who stood behind me and said that as the invader you can always chose where you will make the incursion. And civil war was always conducted more by threats than by weapons. It must be freely conducted with the threat of turbulent wills that did not shy away from violent deeds.

"I would have gladly gotten a little more sleep but my midnight guests came back again. The Brigade, acting under orders, had not let them back on the street. Once again my sleep was spoiled, so I got into my armored car, took the wheel, and gave the order that the gentlemen should be allowed to pass out of camp. Then I marched along with the troops, who greeted me heartily. At the Pichelsdorfer Bridge I allowed the regiment to halt. Here I held a conference and explained to them the sense and purpose of our march on Berlin. The jubilation among the troops was enormous.

"During our march through Charlottenburg we realized that the Berlin administration had run for it in their entirety: the residents were alarmed. My men didn't find the silly Charlotteburgers with their white armbands and old popguns especially frightening. But it was funny when these defenders of the administration burst into cheers and welcomed our orderly march into Berlin. It was like the children's song:

"*Hannemann, go on ahead.*
"*You've got the biggest boots, he said.'*

"We reached the Siegesallee while it was still in darkness. I called a halt and had the field kitchens dole out warm soup. It seemed obvious that the whole Tiergarten neighborhood on all sides was successfully occupied. It was a beautiful picture of maneuvers. Fires were blazing, the men refreshed themselves. Everything was serene and carefree. Around five o'clock in the morning one of the generals I had dealt with from my canapé came out. He brought me the decision of the administration that they would have to refuse my demands. We made nothing of it, since I had knocked out these demands without much thought, and I actually got along splendidly with this gentleman.

"The 'green' security police, with the points on their big stars striking out, pulled back with a good humor. My men called out to them—'Thanks for letting us off, guys!'

"Since the administration had rejected my demands, I felt I was no longer bound by my promise and marched for the Brandenburg Gate. This was the administrative quarter where the *Reichswehr* guard posts stood. I left the parliamentary system intact. The *Reichswehr* people were no greater fire-eaters than the 'green police.' Within a half hour we had occupied the whole sector, but I stayed with the heavy weapons units. I saw in the background some gentlemen in topcoats and top hats, and I assumed they were morning strollers. One of them was General Ludendorff. Only later was it clear to me that this group of 'toffs' were Herr Kapp and his future cabinet."

"At the stroke of seven General Lüttwitz appeared. His face was radiant, and he told me that the administration had fled. The *Reichswehr* and the *Sicherheitspolizei* of Berlin supported him. Then he ordered me to secure the administrative quarter, *Unter den Linden, Friedrichstraße* and the whole vicinity of the *Reichswehr* ministry and to hold them unconditionally. Our point company marched through the Brandenburg Gate to the tune of *'Deutschland, Deutschland über Alles.'* The occupation order was carried out without firing a shot.

"Gradually Berlin awakened. The citizens were astounded to see the battle flag or the pure old black-white-red flag on the administrative buildings. Curious people wandered *Unter den Linden*, The Defense Ministry, and the other administrative buildings. The streets had to be blocked by my people but it went cheerfully and smoothly, because the populace supported the undertaking. Everywhere the black-white-red flag appeared in windows.

"Militarily everything had gone off flawlessly. Now the politicians have to take charge, I said to myself. With this understanding I got a peaceful night's sleep on the first night, although I saw some things that made me feel we were not yet in a safe harbor.

"There was, for example, the relationships between some of the generals. One of them had dealt with Lüttwitz and Kapp and had at first supported them but then grew uneasy as their demands were not met. As he saw on the day of the entry march that Lüttwitz had been thrown out by Noske, and that his place was vacant, he took the side of the administration and took over Lüttwitz's post. On the morning on March 13 Lüttwitz promptly cut him out of his position. But then this general's wife came to Lüttwitz and begged and begged that her dear husband be restored to grace. Lüttwitz trusted him with the senior command in Berlin. I found the relationship of this gentleman with Lüttwitz very distinguished and aristocratic, but I believe I can say that I would not have done it that way. I must immediately point out that the general concerned quit service with the noble Lüttwitz three days later as another act of betrayal.

"In these first hours the previous political traitors—the General Agricultural Director Kapp, Dr. Schnitzler and Dr. Grabowski came to me. They had just been freed from custody, and were lamenting very much that they hadn't been able to do any of the preliminary work for the take-over, and were now left out in the cold. Grabowski asked me if I couldn't do something for him. I went into the Reichs Chancellory Building and said 'Dr. Grabowski would very much like something to do.'

"In my opinion it was a mistake not to make use of the good political sense of Dr. Schnitzler. He immediately saw the outstanding mistake not to take custody of the administration. We could have caught up with Ebert's automobile caravan even with an airplane and held them fast."

President Ebert and Defense Minister Noske fled Berlin, as Ehrhardt said, and called out the general strike as they took shelter first in Dresden in Saxony and later in Stuttgart in Baden-Württemberg. The outright Reds also condemned the take-over and, after some wavering, joined their recent enemies, the moderate Social Democrats—who had supported the shooting of 42 demonstrators during the transportation strike. The Communists and Independent Social Democrats urged all workers to stay home and refuse to support the Kapp-Lüttwitz Putsch.

Waldemar Pabst, back in Berlin, was ready for them. His technical squads from the Guards Cavalry Division—contrary to mythology—were able to keep the electricity and water running with few interruptions, and with the help of many of the regular workers.

"The so-called general strike was not a strong but a defiantly-childish action... like a hunger strike in prison—'now you can see how bad it is for us, now you must finally give in... and so forth," Rudolf Mann wrote. "...The electricity still worked. The streetcar conductors said that tomorrow would be the decision, everything would stop, everything.... The barber said he would work, and the baker where I bought dessert rolls said that if his employees didn't show up, he could order more desserts from Italy.... Not a shot was fired in Berlin.... The Green Police rode around in packed trucks and protected peace and quiet. They worked with the military. Here and there some blood flowed where people were unreasonable....

"Most welcome was the news from East Prussia. Council President [August] Winnig, a well-known Social Democrat, and the senior military commander, had openly acknowledged the Kapp administration. This province showed the old flags.

"If the old administration stayed in Dresden, they sat between two fires, because in Bavaria the turn-out had been completely accepted, the ministry there was thrown out and a new administration with citizen support was set in place. Munich remained completely quiet; the trouble-makers remembered the hosing they got in 1919."

Mann also noted that the 3nd Marine Brigade and the Aulocks and Paulsen *Freikorps* units reported that Silesia was also secure and peaceful. On the trolley he spoke to a metal-worker who belonged to the Independent Social Democrats. Mann learned that the man blamed the Rightists for another bout of unpaid unemployment. Mann found himself sympathizing with the man, if not his politics. He now felt that some extremists were ready for an armed clash with the Ehrhardt Brigade.

About midnight, Mann went to his assigned shelter at the old Leopoldspalais in the Wilhelmstraße, looked around for a place on the floor where a man who was five-foot-nine could stretch out, and never found one. He curled up for a few hours. Overnight, he later learned, crowds of student volunteers turned up at Döberitz ready to serve with Captain Ehrhardt.

The next day, Captain Ehrhardt tried to sort some sense out of the situation. General Reinhardt, who had been targeted for an ouster, was now in Württemberg in South Germany urging resistance. Ehrhardt said, as Pabst had, that while Reinhardt was a brave enough man in battle, he was none too bright.

Korvettenkapitän Hermann Ehrhardt

"He was always a brave solder, but a bit of a Swabian slit-ear," Ehrhardt said, referring to a breed of semi-feral South German pig.

General Hans von Seeckt, "The Sphinx With The Monocle," had reportedly refused to order the *Reichswehr* regulars to fire on the 2nd Marine Brigade —"troops don't fire on troops." But Seeckt did not publicly support the take-over either. "General von Seeckt... reckoned with other possibilities than the military ones. He had always shown good diplomatic sense. He embodied the General Staff mentality in every situation. This man had perhaps the coldest judgement of the situation. He was against Kapp. But he never said so out loud.... As a counterpoise to Seeckt, Noske had as big a mouth as ever. According to reports he said 'With a few machine guns we could make this specter disappear.' It remains noteworthy that the Supreme Commander never had the courage to let his machine guns speak.

"We officers didn't take the calling of a general strike by the fugitive administration as tragic. It was no more dangerous than an ox without horns, if our firm will to fight had not been thrown away. The fate of the Kapp Putsch was that the new administration had an instrument of power in the troops in Berlin, but lacked a political soul and the nerve of statesmen.

Ehrhardt noted that Vice-Chancellor Eugen Schiffer had stayed on in Berlin when Ebert and Noske fled. Born in 1860 in Breslau, Schiffer was an attorney, a convert from Judaism to the Lutheran faith in 1896, a loyal monarchist through the Kaiser's time who never supported the Kaiser's overthrow. After the Kaiser fled, Eugen Schiffer, like Hugo Preuß and Walther Rathenau, became a liberal but not an extreme Socialist. As the ranking member of the Weimar Republic administration still in Berlin, Schiff negotiated with the invading Kapp administration in what was probably a humanitarian way in attempt to win them amnesty if and when the Putsch failed—or in Ehrhardt's opinion to weaken their resolution and avoid conflict.

"The fact that *Herr* Minister Schiffer could roam around the capital stirring up the bureaucracy also stemmed from political idiocy, " Ehrhardt said cynically. "When the supporters of the Republic boast that the bureaucrats were loyal to the new republican system, they err. The take-over wasn't deflected by the persuasion of the bureaucrats, but by the helpless posture of the new administration. The famous resistance of the bureaucrats against Kapp can be explained in their feelings for their own security: the new meal ticket failed to win their trust. If the old bosses came back and they had compromised themselves, then they and their families would be turned out hungry into the streets." Eugen Schiffer, in fact, may have simply been trying to avoid having more Germans killed by other Germans—Schiffer had supported the Kaiser to the very end—and he himself was in fact turned out of office when the power shifted again.

"When the general strike began I told my officers—'To me this strike is sausage! Let the people general-strike until they have nothing more to eat, then they'll stop by themselves,'" Ehrhardt said. "We in any case won't be starved out!

"But in the councils of the administration the word 'General Strike' led to bad nerves. They didn't show it outwardly but they had an inner fear of revolutionary gestures. I marked that well, but I trusted that soldiers stood near this administration. Ludendorff—so I thought—would be the connection that held the movement together." Ehrhardt, like any professional officer, had brought in tons of food by truck and the soldiers themselves carried a week's rations in their packs. Wagons carried fodder for their artillery and field kitchen horses. But the civilians running the Kapp Putsch were geared toward winning elections and not battles.

"I got a hand-clenching proof of the minimal authority of the new men. I gave an order to take 10 million marks out of the *Reichsbank*. I trusted an officer with the transaction, as is customary. But the *Reichsbank* sent the officer away without the money. I later received the order, as things got worse, to take this money with violence. It went against my principles to function as a safe-cracker. Officers should function as trusted persons, never as executives against civil institutions, such as the *Reichsbank*. The administration later sent the police to get the money."

"On the third day I went to the *Reichskanzlei,* because I had the feeling that nobody was running the store. In the antechamber I saw a gentleman I definitely didn't recognize. He led me without comment into the cabinet conference room.

"Here I saw a frightful picture. At the first glance I saw that Kapp had completely crumbled, body and soul alike. He had the head seat at the table. His eyes were swollen. His voice was heavy as he said mechanically: 'I want to inform you...' He was no longer in charge, but was departing. He didn't seem to know what he was saying. I left immediately and was very much devastated by this."

Kapp beat it out of Berlin in disguise. Like the Communist amateurs in Braunschweig and like Ludendorff himself, he hopped a plane for Sweden. One of his last orders—though it took some time to make its force felt—would have fatal implications not just for the crumbling Kapp Putsch but for the future of the monarchist Right in Germany.

"In the afternoon a commander's conference was called by General von Oven in the Reichschancellery," Ehrhardt wrote. "I still completely trusted the influence of Ludendorff. But after the morning session I smelled something bad coming and took as many of my officers as could be spared from duty along with me. I also had a full company of soldiers standing by outside.

"General von Oven had his chief of staff explain the situation to us, said that the men were no longer standing behind their officers and they were not standing behind Lüttwitz; Consequently of this, *Excellenz* Lüttwitz had to resign.

"I spoke out against this in the sharpest manner possible. I declared: 'This meeting is not a Soldier's Council. It would be a sad sign if the soldiers no longer stood behind their officers. But for me this whole speech is a nothing but the jammering and croaking of spineless officers trying to blame things on their soldiers.

"But my words didn't do any good with the support of General Ludendorff and the gentlemen had seized him by his sword-belt.

"Ludendorff went into another room. Now the question came up: Who stands behind Lüttwitz? It turned out that outside of two officers from small units and myself, nobody did. I strode fuming with rage against this faithlessness and treachery into the next room, told *Excellenz* Lüttwitz the tactical position and urged him to have a number of generals and commanders taken into custody. Unfortunately Ludendorff and General von der Goltz had already left. Lüttwitz ordered me not to take any action against these cavaliers. So I went back into the conference room and declared to General von Oven: 'I'm pulling my Brigade out of Berlin!'

"This sent a healing fright through these gentlemen's limbs. They knew that the detachments they had expected were not arriving and if I didn't stay to hold Berlin for God's sake, the city would fall into the hands of the Communists. I promised to help them against the Communists.

"As I was leaving this conference, an officer came up to me whom I recognized as such in spite of his civilian clothes, He presented himself to me as *Oberst*

Hoffmann. He said to me 'Ehrhardt, you were the only decent human being at the entire meeting.'

Members of the Marinebrigade Ehrhardt, with swastikas on their helmets, distributing leaflets on March 13

"Everything that played out at this sit-down got around Berlin and the troops like a flash fire. Kapp and Lüttwitz stepped back. Now I sat with my men abandoned by all the world in the middle of Berlin."

Rudolf Mann had been covering the street and reading the papers on the second and third day of the take-over. Mann took great pride in his ability to speak French and read English and he was delighted—after answering questions from a reporter from Havas, the French news agency—to see that France appeared to be neutral. Britain also appeared to be neutral. The general strike had raised concerns in Poland that Berlin and Germany might be taken over by the hated Communists. Mann saw Poland as not only neutral but somewhat supportive.

The problem started with the swastika.

"The swastika..." Mann observed. "Many of our soldiers who had it on the front of their helmets thought it was Ehrhardt's monogram, others thought it was the Latvian people's coat of arms that had found its way there from the Baltic. Some people had no idea at all. The Jewish population of Berlin had seen it before without any anger. On Sunday, the Jewish ladies and their children came up to the guards posted with swastikas on their helmets, cautiously, as one might approach the lions in the Berlin Zoo, and gave them flowers, cigarettes and chocolates. They simply considered the invaders to be reactionaries or guardians of Capitalism.

"Then Kapp issued a proclamation that he had confiscated the white flour that the old administration had set aside for the Jews for Easter and was about to dole it out to the workers.

"The *Deutschvölkischer Schutz-und-Trutzbund* scattered explanatory flyers among the people. The danger of *pogroms* lay just ahead.

"Judah was in danger, and they drew themselves together and fought back by every means possible.... With one stroke the Nationalist movement had been painted as an anti-Semitic movement and the opposition to the Nationalists took on a Jewish coat of paint.... I can say without fear of contradiction that every street agitator we took in from that time forward except one was Jewish, and that most of them were attorneys at law, students, business owners or senior managers."

Mann immediately understood the implications of the blunder: "'Kapp must be capped' the people of Berlin demanded, and the security police demanded his resignation."

Leutnant zur See Berlin—pointedly wearing his Imperial officer's cap and not a steel helmet with a swastika painted on the front—loaded up a truck with leaflets printed by Waldemar Pabst's technical printing-press team and some of the youngest and most boyish members of the Brigade and distributed them to anyone bold enough to accept them. He looked rather leery. The damage was done: outrages by the *Baltikumers*—including the beating of Jewish citizens that Ehrhardt had interrupted with his own men—had permanently associated the swastika with violent anti-Semitism. Kapp's panicky and stupid confiscation of the *matzoh* flour made it official as far as Berlin's large and influential Jewish community was concerned. The financial and journalistic consequences were disastrous.

After Kapp left, Lüttwitz also began to panic, in a somewhat more dignified manner. He called a conference of *Reichswehr* officers and asked for their advice. Most of them urged him to resign, which might save their pensions. Captain Ehrhardt—Pabst was at the conference and remembered this—disagreed.

"We should arrest the whole association of labor leaders with a company I have at my disposal and if necessary shoot the ring-leaders," Ehrhardt said bluntly. Lüttwitz blamed the officers who had broken their word to send more troops to his support—but not Ehrhardt—for the failure. He resigned.

"I had only one duty at this point—to keep the troops intact, because I sensed that Communist propaganda would work to our destruction by every means," Ehrhardt wrote. "Orators turned out, leaflets were put in people's hands. Wild rumors popped up. This gave us only one way out: I had to give up the security of the whole city and and assemble the Brigade in the Wilhelm Quarter. Here I climbed on an artillery gun limber and spoke sharply:

"We have not reached our goal. The blame for this belongs with the cowardice of the civic population and the spinelessness of some of the leading military personalities. The Bolsheviks believe their moment has come. Our Brigade is the only reliable military unit in Berlin. Therefore we must be the first and strongest to take up the fight against the Communists. I expect as before unconditional discipline and obedience. I march as one with the Brigade and with every individual man, as the Brigade stays locked in behind me. Loyalty for loyalty."

The speech ended with three thunderous cheers.

Chapter Twenty-One
The Captain Vanishes

While Ehrhardt was trying to hold the Kapp Putsch together without Kapp, Adolf Hitler was flying from Munich to Berlin in a biplane with Dietrich Eckart. Hitler, who had recently decided to become an anti-Semite, was still a V-Mann—a spy inside the *Reichswehr*—and his flight was unofficially official. He and Eckhardt, also a V-Mann, had been ordered to find out what was actually happening in Berlin and report back to headquarters in Munich. In Berlin, however, the V-Mann Adolf Hitler encountered Ignatz Trebitsch-Lincoln.

"Who was Lincoln?" Rudolf Mann asked after have made his acquaintance during the Putsch. "He was, speaking honestly, an extra collaborator we Kappists couldn't shake off. He had energetically helped us through his connections.... When I first met him I took him for Swedish.... As I later read from newspaper articles, he was a member of the English parliament, born of Jewish origins in Hungary, with the actual name Trebitsch—if it was reported from London, it must have been true: most stuff out of London is true.

"He must have carried out the dark plans of the English administration through his collaboration with us. That says just about everything..."

"England, of all places, had undermined the movement through this man, in order to set aside the Reactionaries in Germany: first perception. Through Lincoln, London had worked up a Putsch to bring the old system into further mistrust:

second perception: the English sent him to spy when there was almost nothing to spy on.... I had the impression he was an imposter who took part in politics as a form of sport.... He should have helped Horthy found his administration in Hungary—that would have been quite a sports event for him."

Trebitsch-Lincoln, whatever his assignment, may have done the world a huge disservice: he may have saved Adolf Hitler's life. Trebitsch told Hitler and Eckhardt to make themselves inconspicuous and get out of town if they didn't want to be arrested—and Trebitsch, whose political instincts were sharp, clearly understood that neither the *Freikorps* nor the Reds were likely to be keeping prisoners if things got any more exciting.

At the time he met Trebitsch, Hitler, a lapsed Catholic fascinated by Darwinism and the occult, had just come off a 10-year cycle of pretending to be Jewish to take advantage of Jewish charities in Vienna and in the Bavarian Army, where two of his company commanders had been Jewish. Trebitsch may also have recognized that Hitler was of mixed Jewish ancestry. He was himself a skilled imposter. Centuries of persecution all over Eastern Europe had ingrained an abiding principal in most Jews: No Jew was ever handed over to the executioner if he might possibly be saved. Trebitsch had seen Captain Ehrhardt and knew that Ehrhardt would not think twice about executing a spy or informer—which Hitler essentially was. Trebitsch could also see that the betrayal by most of the generals had put Ehrhardt in a very dangerous mood. Hitler knew who to trust and he and Eckhardt played it smart and got out of town. They later blamed their speedy departure from Berlin on unruly Leftists at the airport, not on a formerly Jewish spy who may have worked for England.

Trebitsch-Lincoln had a competitor for the strangest person to turn out for the Kapp Putsch: B. Traven was also said to have turned up in Berlin. Traven—one of the many claimants to have been a natural Hohenzollern, in his case the natural son of Kaiser Wilhelm II—was the author of 10 books, including "The Treasure of Sierra Madre." He was a lapsed Socialist and later a sentimental Communist of the pre-Bolshevik variety. The one profile police photograph of B. Traven, taken in London in 1923, bears no resemblance to the Kaiser, and if B. Traven was actually in Berlin he was probably not there to support the Kaiser, his purported natural father.

The opponents of the Putsch, however, had started a rumor that the Putsch was an attempt by monarchist officers to bring back the Kaiser—and re-start the war. Mann remembered seeing a truck plastered with a big yellow placard that read: "The Lie about the Monarchist Putsch." Leftist newspapers also printed stories that made Ludendorff the actual leader. Mann said that when Ludendorff showed up at the Reichschancellery, the guard post turned him away because he had no credentials. An officer slipped up and whispered *"Mensch, Ludendorff!"* to the guards. They finally let him in to get a set of identity papers processed. Ludendorff responded to the flop of the Putsch as he had to the failure of the German Army's 1918 *Friedensturm*. He fled a second time.

The former Kaiser called for champagne at his chateau in the Netherlands when he was told that *Freikorps* soldiers had taken over Berlin. Wilhelm was as surprised as anyone in Germany and felt that it served the Social Democrats right for running him out of Germany two years before. No one had asked him to climb back on the throne. He told his loyal advisors that the situation in Germany had become too turbulent to support a constitutional monarch and that the next step would be a dictator. The Allies had implausibly demanded that Wilhelm be put on trial as a war criminal along with his scoundrel son Crown Prince Wilhelm of Prussia, throwing in Crown Prince Rupprecht of Bavaria for good measure, but all these worthies spoke fluent English and French. A trial would probably have embarrassed the British and the French, since the Allies were about equally guilty of starting the war and had unilaterally rejected German and Austrian peace offers, sponsored by the Pope, that could have saved Britain, France and the United States a million lives but still returned Alsace and Lorraine to France and fostered Polish independence.

Princess Viktoria Luise concurred that the whole Kapp Putsch came as a complete surprise to the whole dynasty. She was visiting relatives in Germany at the time and described the take-over as happening "in Munich."

"We were beset by questions as to what had happened but we knew nothing," she said. "...We ourselves were once more on the move, this time to Stralsund and Rügen, and once came came face to face with fighting at the railroad stations. Before Rostock we were surprised by the Reds at a small station and I had to take cover behind a wall and next to some Rostock professors who had fled from the terror. Passing backwards and forwards in trucks and armed with machine-guns, they were really in search of the professors and were checking from station to station. The situation was anything but pleasant. In Stralsund we encountered more street fighting and had to lie flat on the floor of our hotel room while fighting was going on in the station opposite... the journey was gruesome, for the revolt had spread westward and I thought I would never see my husband and children again."

As the Putsch continued to slowly disintegrate, Ehrhardt was ordered to confer with General von Seeckt.

"Seeckt asked me earnestly: 'Can I count on the Brigade to support me in battle against the Bolshevik threat?' I assured him that he could."

"I have not approved the actions of the Brigade, but I acknowledge the flawless discipline of this unit and hope that I can firmly rely on the 2nd Marine Brigade in the forthcoming conflict," Seeckt said.

Ehrhardt noted that his assault company with *Leutnant* Berlin and *Leutnant* Tillessen had come with him for the interview with General Seeckt and had seemed relieved at his expression when he left Seeckt's office. Seeckt—"The Sphinx With the Monocle"—caught the mood instantly and had a written order prepared:

"I give the 2nd Marine Brigade the assurance that no arrest order against its commander will be issued as long as it remains under my command."

Ehrhardt ordered the 2nd Marine Brigade to march out of the Wilhelmsviertel with the band playing "*Deutschland, Deutschland über Alles*"—and the crackle of sporadic gunfire behind them.

"Suddenly we heard the cracks—a half-minute long. Then it was silent and we could hear the marching music again," said Rudolf Mann, who was in the rear of the column. "Profiteers and riff-raff in their new spring coats felt themselves very strong. They dared to step on a dead lion when the lion wasn't really dead, but could spring up alive and a little bit bloodthirsty.... Some of the men wanted to jump into the crowd and punch out the hecklers. Some of the crowd tried to wrestle away rifles....

"There were warning shots, fired into the air, over the hecklers' heads. Terror shots.

"A man in a dark blue suit fired a pistol out of a window. The bullets struck. Soldiers were wounded.

"The threatening crowd tried to storm the Hotel Adlon. We later learned that some Swedish ladies had waved to our soldiers and that this sign of approval threw the whole mob into a rage.

"Enough. Some rifles were pointed at the center of the mob and the men picked out targets. Men were hit and fell. The officer in front of the company, riding a horse as only a sailor can, tried to turn his horse around. The horse slipped on the wet pavement and fell. The younger soldiers thought the company commander had been shot. They aimed shots into the crowd. Dead men lay in the vestibule of the Adlon Hotel. Panic!

"Some of the soldiers also panicked and the company set up firing lines to the left and right without orders. Horses in the gun teams shyed. A hand grenade fell on the pacement and blew up. Wounded horses dragged a *Minenwerfer* down the street.

"Then as last the command rang out—'cease fire.'

"In the dead silence you could hear the raindrops and the tapping of many fleeing feet.

"The troops got back in order and the march continued.

"Behind the Tiergarten a short volley fire struck another threatening swarm of riff-raff from the Moabit neighborhood."

The second firing reportedly occurred when a "boy" of indeterminate age either laughed or mocked the marching soldiers and two young soldiers broke ranks, decked him with their rifle butts and kicked him. The mob surged forward the soldiers fired another volley that killed a dozen men and wounded another 30 to 60 more.

Ehrhardt had no apologies to make, though he never gave the order to fire. The men knew that two days before, the world war flying ace Rudolf Berthold—the

Iron Knight—had been murdered by a Red mob in Hamburg, Berthold listed 44 aerial victories and multiple gunshot wounds. He was a fighter pilot who flew with one hand after the other arm was amputated. His motto had similar to Ehrhardt's —"Better to die like a man than live like a coward." As the leader of a small *Freikorps* unit, Berthold had been recognized because he always wore his *Pour le Mérite* visible around his neck. Rumor had it that Berthold had been strangled with the ribbon from his own medal. In fact, Berthold's enfeebled injured arm had been ripped from its socket and he had been shot six times with his own pistol and his face stomped beyond recognition. The Reds also stole his overcoat and shoes. Bechtold was an outspoken anti-Communist and the vulgar brutality with which a crippled war hero had been murdered enraged the *Freikorps* soldiers even more than the death itself.

Ehrhardt also noted, in defense of his own soldiers, that a detachment of the Greater Berlin Defense Regiment had been cornered in the *Rathaus Schöneberg*, (a district municipal building) and had called him by telephone to ask for help. He asked if they had weapons and ammunition and then told them to fire at will, but they were reluctant to kill other Germans. Ehrhardt sent one of his own companies to the rescue. When his men got there they found that the aging Berlin militiamen who had laid down their weapons had been overwhelmed and beaten half to death by their Red neighbors. In different districts, other *Freikorps* units were attacked by the Red Guards, whose battle cry was "Moscow is ready."

Not everybody felt that way. "I had one nice experience in Charlottenburg during the *Ausmarsch*," Ehrhardt remembered. "During a rest break for the troops I rode up and down the street, back and forth, and a lady came out with her seven-year-old son, looked me over and said to her little sprout: 'See, this is Ehrhardt!' The little guy waved his cap and shouted out 'Hurra!'"

Back at Döberitz, Ehrhardt set up guard posts, not knowing what to expect, or from whom. Ehrhardt was not sure whether shooting his way out of Berlin had voided his contract with General Seeckt, whom he saw as a brilliant military organizer but an even better political survivor.

A few days later the 3rd Kurland Infantry Regiment arrived at Döberitz. "The Kurlanders, to make a joke, completed the complement of *Baltikumer* who were always being confused with my men," Ehrhardt said dryly. (The charter of the 2nd Marine Brigade had excluded the Brigade from combat outside the borders of the pre-war German Empire, and none of the 2nd Marine Brigade soldiers fought in the confused Baltic campaigns.) "They were brave men. On the 21st through the 23rd of March they had a serious fight against Communist fighting units. On the 21st, their attack with by a company armed with only 60 rifles had been too weak. On the 23rd, they attacked with 200 rifles, 10 light, and 5 heavy machine guns, against 450 Communists in a regular battle, supported by a half-battery from the Spandau Arsenal. They captured 26 machine guns, 200 rifles and 2 light armored cars that belonged to the magistrate at Spandau. Their losses were two privates and an acting sergeant dead, 7 men seriously wounded and 1 officer slightly wounded. I throw these numbers up to show how show how heavy the conditions were on the outskirts of Berlin and also how good and effective the troops who had supported the new administration were."

The return of Friedrich Ebert and Gustav Noske was followed by a major outbreak of Red revolution in the Ruhr industrial area. Some 60,000 Red workers staged a confused uprising quietly sponsored by Moscow. The Social Democrats blamed the Kapp Putsch for triggering the revolt due to the shootings on the way out of Berlin and the continued fighting with Communists. The Right blamed the Social Democrats for their political weakness. Ehrhardt was dismayed when the Ebert-Noske administration ruled out the use of any *Freikorps* troops and sent *Reichswehr* troops to the Ruhr instead.

Red Guard in front of the state parliament, Munich

"So it was actually Kapp who brought brought about the outbreak of the Ruhr uprising!" Ehrhardt wrote sarcastically. "Revolutionary movements don't stay firmly in the hands of their leaders, because they aren't possible without the agreement of the masses. The old Socialist regime in Berlin demanded a general strike against Kapp: Out of this strike and its atmosphere grew the Ruhr uprising: the avalanche from the heights rolled down into the valley."

Ehrhardt was further appalled by the tameness with which the senior officers knuckled under to the Ebert-Noske regime after they had asked him for help when their demand for a full-sized army led to their arrest orders. "A general strike of the officers against the Revolution would have been the right thing to do."

Some of Ehrhardt's officers advocated compromise with the Social Democrats, and even with the semi-Communist Independent Social Democrats—to form a new power block in Germany, an alliance of the laborers and the soldiers against the old power of the industrialists, the bankers and the landed aristocracy.

"Their demands were nonsensical," Ehrhardt said. "There were in the German Empire two power factors: the socialist workers, whose only weapon was the strike, and the military, who could put the final question to the community: strike and you die. We had the military power firmly in our hands.... I was not interested in their proposal." Ehrhardt suspected that the whole thing might be a plot to discredit him and embitter his loyal troops against him.

"Then at last Noske stepped down from his post as *Reichswehr* Minister," Ehrhardt noted with satisfaction. "His own party hunted him out with insults and disgrace. A civilian *Reichswehr* Minister sat in his seat—it was Herr [*Otto*] Gessler. That was a kind of success."

Otto Gessler, a Democrat born in Wurttemberg and former mayor of Nuremberg in Bavaria, described himself as a "reasonable" supporter of the Republic. His party—the party of Hugo Preuß and Walther Rathenau—wanted to keep Germany strong and was far more concerned about rampant chaos and the Red uprisings than about the fearsome but unlikely return of the Kaiser.

"The new minister displayed a sense of humor without a doubt," Ehrhardt noted. "As the representatives in the National Assembly requested the disarmament of my Brigade in Döberitz, his answer was: 'I'd like to see the gentlemen go there and try it.' The fact that Gessler shared the same name as the Austrian tyrant Hermann Gessler, whom the home-spun hero Wilhelm Tell assassinates in Friedrich Schiller's 1804 play to the applause of his fellow Swiss peasants, made the joke even more amusing.

"I'd never needed to explain that the mass of my troops were unshaken," Ehrhardt noted. "At the first disturbance guns would have gone off."

Realizing that the confiscation of ammunition was one of the Allied demands in the enforcement of the Treaty of Versailles, the 2nd Marine Brigade and some of the other *Freikorps* units ordered huge inventories of ammunition. They conducted battle drills with live ammunition—but most of the live ammunition went into hidden stockpiles.

Orders came through to remove the Brigade from Döberitz and set up operations at Münster, a Prussian city in Westphalia and close to Hannover, where an active *Welf* independence movement was advocating secession from the Prussianized part of Germany. The Welf Dynasty—*Guelph* in Anglo-Saxon usage and in Dante's Italian—was the parent dynasty of the House of Windsor in Britain, whose monarchs had ruled Hannover from 1714 until 1837. Princess Viktoria of Prussia was married to the Welf heir, Duke Ernest Augustus of Brunswick. Ehrhardt was not unhappy to be leaving Döberitz, where some of the neighbors were hostile and rations were sometimes interrupted. But he suspected that the Brigade might be confronted in transit. He arranged to move the troops in three different trains with three independent elements—each element including infantry, machine guns, and artillery in case of a fight with the national troops.

"I was now rather unconcerned about an arrest order," Ehrhardt said bluntly. "Nobody dared to take possession of me. We arrived in Münster without incident and felt better as we got farther from the pestilential air of Berlin."

"During the first week I recalled the Berlin experience in which our troops were to be used for political purposes, in a very charming form. A deputation of Welfs came to me. The names of these people no longer reside in my memory. They said to me decently and honestly: 'You have now flopped with the Berliners. That must be why your Brigade is here. Would you not now place your troops on the side of an independent Hannover? We can be sure that the movement is coming like a flood, and we will come smoothly through a plebiscite.' These good people were astounded that I laughed at them, but it was a bitter experience for me."

Laughter stopped when Ehrhardt abruptly got the anticipated bad news: The Weimar government was about to serve him with a warrant for his arrest.

"A howl of wrath rose from the troops and the over-all mood was: 'March on Berlin!' I had the duty to dissuade my people from any unreasonable action.... Wild rumors swirled through the camp. The Security Police were on the march from Hannover! The Criminal Police from Hamburg were lurking outside the camp! And anything else a military mind could think of. My people took extreme measures to block any surprise again my person. Tree trunks were rolled over the streets, watchers were posted in the fields, and anyone who came near the camp was confronted in the sharpest manner. I let the angry mood cease raging and then removed the massive security precautions because I knew that nobody would actively take on the Brigade."

On May 1, the Socialists from the Münster region asked the camp commander to hoist the red flag over the camp—which meant hauling down the black-white-red flag of the German Empire and of the Ehrhardt Brigade. The camp commander asked Ehrhardt to haul down the flag temporarily to reduce friction between the camp and the handful of Socialists from the countryside.

"I would hold it for a bottomless shame above all else to haul up a red rag and I would never even consider hauling down the flag of Skagerrak to appease a Red mob," Ehrhardt told the district commander. The black-white-red flag stayed at the top of the flagpole. On May 1, some soldiers from the Brigade saw a train flying the red flag. They stopped the train, hauled the red flag down, trampled it, and beat up anyone who tried to interfere. "After that the threatened specter was set aside for while in Münster," Ehrhardt said.

Princess Victoria Louise of Prussia, the only daughter of Kaiser Wilhelm II of Germany, with her fiancee Prince Ernst Augustus after the announcement of their engagement

More bad news followed. Ehrhardt received a telegram from Hamburg that his mother was seriously ill. "A strong bond had always held us together. She had lived my life with me, thought my thoughts and felt my feelings. She had always been completely healthy until a few days before the Kapp Putsch she wrote in a letter: 'Now I have become sick and tired of life.'

"I knew after the telegram from Hamburg that it was hopeless with her. To see her once again, I must, as her son, set everything aside and defy an arrest. That evening I made my decision. I set out with three of my officers in the car, and drove to Hamburg. By the pure grace of God I was able to see my mother alive. I was able to talk to her for some long hours. She couldn't hear what was going on all the time. I was able to get back to Münster that same night without being stopped.

"Two days later I received her death notice. Immediately I headed back for Hamburg, to take my final leave of my dead mother.

"As I attempted to make a third trip, for her burial, the troops objected. Officers and enlisted men told me they would not allow me to take the car out of camp again.

"And the troops were right. The delegation who had taken part in the burial reported me to the management of the cemetery. Arrest was threatened as I left my mother's grave. Our political system had become so depraved that they had to impose on a son's mourning because the weaklings couldn't take him any other way. In the wildest times, in the Seven Years' War, in the Thirty Years' War, in the Middle Ages and during the Germanic tribal migrations, the peace of the cemetery was always respected. We highly civilized people have forgotten the honorable right of asylum, and no longer respect that ancient custom. The simple people of the troop felt for what was done to me. Their participation in my personal loss showed me how tight and beyond the call of duty the bond was that linked me to the Brigade."

Ehrhardt's men refused to allow him to leave the camp when he received a summons to Berlin. "We will not allow the *Herr Kapitän* to appear before the Berliners," one deputation of enlisted men told him. Ehrhardt shook his head and one sergeant said: "Then hang us all, Captain, so that we can be where you will be."

"I was the prisoner of my own people," Ehrhardt said.

When he arranged what might be a farewell party with his wife and three children—he recognized no distinction between his Gilsa step-daughter and his two Ehrhardt sons—Ehrhardt took a room at a small inn. The landlady's daughter told him after the first night of a three-day stay that five men she had never seen before had been lurking outside the inn all night.

"They could have been vagabond bums, who were very popular during this era," Ehrhardt said. "But I admit I felt a little bit uneasy. In any case, I slept with my Browning on the night table."

The next morning the girl knocked on the door while he was still asleep: "The five guys from yesterday are sitting in front of the inn again. It appears they've been laying in wait for Captain Ehrhardt all night."

"I thought it over. I had strong ties to the Brigade. If these guys wanted my pelt, they could have it, if they were tough enough. I buckled on my pistol and stomped into the dining room of the inn to get a look at these guys. How astounded I was and how I laughed when I found five of my own men in the dining room who hadn't expected their commander to come downstairs this early. They had stayed up all night watching over me while I rested with my family.

"Another time, I rented a carriage with a gentle old horse to take my family on a country ride in the surrounding forests and sent the coachman home. I drove the horse myself to be alone with my wife and my children. It was beautiful in the great, sunny forest. I felt myself free from prying eyes. I turned the horse down a side road to have a picnic with my loved ones. As we chatted comfortably, I saw some riders through the budding trees. At first I was edgy because we were quite far from the camp. Then somebody I knew waved to me and I saw I once again had five loyal men set on my heels to watch over me.

"In the vicinity of the Münster camp there were weapons dumps and ammunition dumps. The main road to these depots led through the camp. One day an automobile came down the big street, then turned, as no other truck had done, on the street to my guest house. The guard who was at the road crossing saw this right away and immediately gave an alarm that a fully-occupied automobile was driving in the direction of my house. A few moments later a second truck, loaded with armed men, was roaring after the suspicious vehicle: probably criminal police. They took it for granted that this was an attack on my person. The first automobile came back and my men assumed the attack had been successful. Riflemen were placed across the road, in the attack position, and the first automobile pulled up to a crooked stop. The driver hit the brakes quickly, all hands flew up in the air, but the investigation showed that the automobile was full of Allied officers, whose job was to control ammunition. They had turned onto my street by mistake. Somewhat annoyed, but very happy to be out of the hands of such dangerous guys, the gentlemen drove away."

Ehrhardt was touched by his men's deep loyalty—so much so that his conscience troubled him. He took two of his oldest soldiers aside while they were voluntarily standing guard over his quarters and questioned them.

"Gentlemen, consider for a moment, what do you think would happen if you take part in opposition to the state?"

"One of them said very quietly:

"*Herr Kapitän*, it's all the same to us, we have our muskets and cartridges and know what we owe the *Herr Kapitän*; You have been loyal to us, and we too will be loyal. What happens to us is sausage.'"

Ehrhardt had warned his men when they first arrived at Münster that the Brigade would be disbanded within a few weeks. He also promised to try to find military or naval enlistments for any members unable to go back to their previous jobs, universities or technical schools. The other option was the *Arbeitsgemeinschaft*, a scheme in which whole sections of men would become laborers on the same landed estates or the same factories: "On the great estates and the liquor distilleries the owners had the security of knowing that these men, with weapons in their hands, would be able to defend them at need.... But many of my people feel themselves not to be fit for any career but soldiering. They had become *Landsknechts* and could not find their way back to their turf. An honest judgement has to take that into consideration. I succeeded before my departure, to make sure that no one who was unable to support himself outside the service was turned out without some sort of arrangement."

The *Reichswehr* officers who handled the dissolution of the 2nd Marine Brigade with Ehrhardt's cooperation were General Gustav Behrend and General Staff Major Alexander von Falkenhausen. "The achievement was a thankless one for both of these gentlemen. The troops were outraged and saw in the generals only agents of Ebert. It required my whole authority over my men to prevent any stupidity and to press through and make sure my men's discipline was as flawless as it would have been with their own officers."

Ehrhardt was especially impressed with Falkenhausen,

Alexander von Falkenhausen, archetypal Prussian militarist

an exceptionally intelligent officer with remote natural ties to the Prussian royalty. Falkenhausen had learned to speak fluent Japanese as a military attaché and had been decorated with the *Pour le Mérite* for successfully leading Turkish troops against British troops in Palestine in 1915. He had also prevented the Turks from mistreating local Jews as suspected British agents.

"He took over as my chief of staff at my request.... He carried out his task not only with consideration, but with his whole heart, and provided a delightful contrast to other staff officers I've brushed up against.... He expressed his dismay that he could not stop the reduction of the army to 100,000 men, and the thanks for his integrity was that he later had to explain this statement in court."

Alexander von Falkenhausen with horsehair helmet in Japan

"I didn't want to part from my people without music and song. I held a last, big parade for the Münster camp so I drew up the men and took my leave with a short speech:

"I anticipate that the seed corn of national consciousness and love of the Fatherland, that I have planted in you, now, when we part from one another, will not slumber within you, but will come up and bear a thousandfold. For that reason I say not 'farewell' but rather '*Auf Wiedersehen!* [*Until we meet again.*]'"

Ehrhardt saw many of the officers and men trying to hide their tears, and he had some of his own to hide. "At this moment, where I saw the bond that had bound the men to me was cut through, and I felt a sudden emptiness inside. I came away strange and lost as I set out on my flight."

Ehrhardt later penned a more detailed note in which he explained that the Brigade's attempt to shield him from arrest could have led to a battle that destroyed many lives and the Brigade's reputation. His letter and his further comments elsewhere showed that he himself was filled with disgust for the Ebert government, which meant to have him in custody at all costs but lacked the courage to storm the encampment—and utterly lacked the political influence to convince his own men to hand him over. Ehrhardt also wondered what his best escape plan would be: Most German officers in trouble that didn't require suicide tended to seek the protection of a senior general—almost an instinct in a paternalistic and authoritarian society like the German Empire—but Ehrhardt had seen enough of the senior generals at the Kapp Putsch to consider them worthless. Major von Falkenhausen, a man two years older than Ehrhardt, a monarchist with distant ties to Prussian royalty and a strong anti-Communist, had shown Ehrhardt that there were actually some senior officers worthy of respect. Ehrhardt also accorded some respect to Waldemar Pabst, who had swiftly seized on an invitation to lead the Austrian *Heimwehr*, the home defense force, and had gotten out of Germany as the Kapp Putsch flopped. The fact that Pabst had ordered the murder of a woman—even a Rosa Luxemburg, whose battered body had recently been found in the Landwehr Canal—seemed to unsettle Ehrhardt, a pastor's son who thought in very rigid patterns of chivalry. But Major Pabst too had been betrayed, and Pabst had never betrayed Ehrhardt, whom he lionized as the greatest natural leader of men he had ever seen. In Ehrhardt's book, the older officers had shown themselves to be careerists: they might face a second-rate foreign army on the battlefield, but they would never face the possible loss of a pension, not even to save the German nation from fatal weakness before Communism. Perhaps more to the point, Ehrhardt no longer trusted the Ebert government for a fair trial or a safe place in prison while he waited for a trial. He was now an outlaw both by fate and by choice.

"The next morning very early I left the camp in the armored truck in the direction of Hannover. After an hour's drive was half done, I whipped out my straight razor—a lieutenant held a mirror before me—and in a few minutes my proud beard had vanished. One of the lieutenants laughed and said "*Herr Kapitän* no longer looks so *chic*." As I passed through the railroad station in Hannover my heart was beating urgently in my throat. My two companions and I got into the D-train for Munich. We got there without incident. Obviously due to the skills of people in my Brigade I had a good set of false papers. In Munich I took a studio apartment on a quiet street. The landlord, a former officer of the good old type, bid me a warm greeting to my first sanctuary and gave me a heap of provender. But these were the hardest days of my life. Without activity, without responsibility for others, with a burdensome feeling for a duty that no longer existed, I lapsed into a severe nervous condition."

Ehrhardt had strong tendencies toward depression in moments where he felt he had failed others. This may have started when he got kicked out of school for hitting the teacher and his father sadly but wisely decided that Hermann was not cut out to be a clergyman. The defeat in the world war almost destroyed him. The mutiny on the rust-bucket freighter taking him back from Scapa Flow may have saved his life because he defied circumstances to save others and himself. He enjoyed boxing but he was also adept in *judo*—and he had learned to turn overwhelming force against itself. Once again, as on the freighter, he turned the black, almost suicidal depression that engulfed him after the arrest warrant at his mother's burial, the dismissal of the Brigade, and his own unknown fate into the sort of indifference to his own survival that went beyond ordinary courage and turned ito utter fearlessness in the face of visible enemies.

Chapter Twenty-Two

Organization Consul

Shortly after the 2nd Marine Brigade was disbanded at Münster, arms-cache informers and politicians who had opposed the Kapp Putsch and supported the reduction of the German Army to 100,000 men began to turn up dead around Munich and in other parts of Germany.

The first murder was one of the most spectacular. On May 21, 1920, a group of 60 *Freikorps* members stormed the country estate of Hans Paasche, a former officer in the Imperial Navy and shot him to death while he was reportedly attempting to escape. Leftist sources have Paasche being murdered in front of his children.

Paasche had volunteered to serve in German South West Africa at the same time as Hermann Ehrhardt, and like Ehrhardt he was appalled by the huge number of Africans who died of typhus, along with the several hundred German soldiers, in the aftermath of the fighting. The epidemic was a life-changing experience: Paasche changed from a big-game trophy hunter to a vegetarian and from a Darwinian racist to a sympathetic observer of tribal culture. Unfortunately his intense desire for social reform took him so far to the Left that he was charged with high treason during the later stages of World War I. A negotiated plea got him into a mental health institution instead of a military prison. His wife's death seems to have unhinged him completely. His own father disowned him as a traitor. Paasche then lined up with the Workers and Soldiers Council in Berlin and turned his country estate at

Waldfrieden into a shelter for wanted Leftist insurgents. His murder, whether in flight or in front of his children, was attributed to Operation Consul.

On October 6, 1920, Marie Sandmayer, an attractive 19-year-old servant girl recently discharged from a country estate, was found dead under a tree in Munich's Forstenrieder Park. She had been strangled. A note was attached to the tree. "You slut, you have betrayed your fatherland. You will be judged by the Black Hand."

Sandmayer had spotted an arms cache while working on a manor in Upper Bavaria where she was shortly discharged. She wandered around Munich seeking employment when she noticed placards printed for the Allied Control Commission ordering all weapons to be turned in. Apparently she hoped to make trouble for her former employer. But she reported the weapons she had seen at the printing shop rather than to the Allied officers. Somebody at the printing shop told a member of a local Rightist group. On October 5, a man who said he was from the Allied Control Commission called at her lodging and asked her to come with him and give her report. She turned up dead in the park the following morning.

On October 20, a former *Reichswehr* soldier named Hans Dober survived a murder attempt during an automobile excursion to Landshut. He had reportedly offered to sell information about a cache of hidden weapons to the German authorities—who may not have appreciated his lack of patriotism.

Hans Hartung, a waiter and former spy sent to keep an eye on the Communists, turned up in Munich and expressed dissatisfaction with the wages for his free-lance espionage activities. "You'd better be careful, I know plenty," Hartung told the Bavarian Rightists as he walked out. The next day, he was reportedly heard boasting that he could get a truck and clean up a whole arms cache he knew about. On March 21, 1921, he was found underwater in a stream with 11 pistol bullets in his head and his body weighted down with stones.

The Munich police found that the stones used to weight Hartung's body came from a parade ground and that an army truck had been seen the night before near the stream where his body was found. Suspicion in both the Hartung murder and the Sandmayer murder focused on two young lieutenants. But Munich Police Chief Ernst Pöhner interfered with the investigation and no charges were filed in either case.

One story had Pöhner being indignantly asked: "Do you know that political murders are taking place within your jurisdiction?"

"Yes," Pöhner replied. "But not enough of them!"

Terrified Leftists, some of whom got out of Germany then and some later, conjured up a name for the faceless assassination team they accused of 354 political murders: Organization Consul. The Consul was said to be Hermann Ehrhardt.

Ehrhardt had his own explanation for the initials O.C. that had become indelibly attached to his name: it was an employment agency for former officers. His once-gusty sense of humor had darkened since the betrayals of the Kapp Putsch and its aftermath.

Shortly after he reached Munich, Captain Ehrhardt found out that some of his officers needed help in finding employment.

"It was impossible for all my unemployed officers to come to Munich, so I had to take the risk and get back on the train to Northern Germany." he wrote. "The automobile met me at the little railroad station and took me through wonderful old oak forests to the meeting place at Prignitz. The owner of the estate greeted me heartily, a regular old Germanic giant, and 30 to 40 officers of the Brigade were waiting for me."

Under the protection of the regular old Germanic giant, Ehrhardt said, the officers told him how hard it was for sea-faring officers to find work in a country with almost no navy and a greatly reduced merchant fleet—as per the terms of the Treaty of Versailles. Ehrhardt used whatever influence he had as a war hero and former anti-Communist to find maritime jobs for his senior officers. But the officers reported back that the indiscipline and the sloppiness of naval service under the Weimar Republic was a far cry from the Kaiser's Navy. To cheer them up—he said —he founded an association of former Ehrhardt officers under the initials O.C., a sort of veteran officers' group which provided job counseling and comradeship.

"In the time after this the O.C. appeared," he wrote. "The know-it-alls, the tower watchmen of the '*Vorwärts*' [*Social Democratic newspaper*] and the '*Berliner Tageblatt*' [*Upscale liberal newspaper*] made this out to be 'Organization Consul' because I sometimes called myself Consul Eichmann in those days." He said he invented the new name because being called "*Herr Kapitän*" could have led to the arrest he was still trying to avoid. Titles like "*Doktor*" and "*Geheimrat* [Privy Counsel] sounded too unmilitary.

"The name O.C. was later given to the organization itself, probably first in the summer of 1921 during conferences with leaders of other groups. The frequent question was 'What do you actually call your organization?' I told them 'Call it whatever you want, O.A. of whatever. The abbreviations were the mode then, so the '*Orgesch*' was generally called O.E. for Organization Escherich. One of then gentlemen said 'Why don't we go with the letters of the alphabet and just say O.C.?' That went down well with me, and so the name O.C. originated to stand for Organization C. The C had nothing at all to do with Consul Eichmann."

Whether Ehrhardt expected anybody who read "Adventure and Fate" in 1924 to believe all this is anybody's guess. The Bavarian authorities seemed to be deliberately blind to his illegal presence. He was still a hero in Munich after the destruction of the Soviet Republic of Bavaria in May of 1919. But he often had to leave Munich, and sometimes Bavaria, for various missions that he never describes in any detail. He was sometimes followed, he says, by the Red Crows, spies from Berlin and from 'Red' Leipzig, where the new federal court system and prison were located: "You could always recognize the Red Crows by their big beaks and their red cheeks."

While Ehrhardt wandered Munich "like a ghost" running what he described as a veterans' social group and employment service, bodies attributed to Organization Consul kept piling up. Karl Gareis, a USPD [*Independent Social Democrat Party*] delegate to the Bavarian *Landtag*, or state parliament, had demanded an investigation of the Rightist groups operating in Munich. The USPD was regarded as a Communist front and a hot-house for Red Guardists by the nationalists, not without some demographic evidence. On June 9, 1921, Gareis was returning to his home in the Bohemian Schwabing district of Munich. He had just made a speech at a political meeting, again attacking the Rightists in Munich. As he said good-bye to a friend, four shots rang out from the darkness and Gareis was hit behind the ear. He fell to the pavement and was dead within 45 minutes. The killers were never arrested. Ernst Pöhner, the police chief in Munich, probably made sure that they weren't arrested.

Pöhner, according to Ehrhardt, was a supreme political realist: "The weapons will rule," Pöhner told Ehrhardt. "Five percent of humanity alone has the will to use them. If I have three percent out of this five percent on my side, I can render the other two percent harmless or strike them dead and rule the apathetic masses." He once told a court, speaking as an honest man from top to toe "What you call high treason is what I've been doing for the past five years."

Ehrhardt wrote that in the Kaiser's day high treason had been a heinous crime because the will of the honest people supported their government. The German courts—tribunals of judges with legal training rather than juries of laymen—were widely known as the most honest in Europe out of sheer professional pride. Every trial took place with a Crucifix in plain sight to remind everyone present, Christian or otherwise, never to mix justice with partisan politics. In Weimar, by contrast, Ehrhardt said that the trials were so utterly political that no opponent of the administration stood a chance of a fair verdict. Ehrhardt made it a point to stay out of court.

"The political excitement was a new source of unrest for us," Ehrhardt wrote of his secret residence in Munich. "As I sat peacefully writing in my business room one day, I got a telephone call. An excited acquaintance asked me: 'Have you already heard? Erzberger has been shot!'

"I answered: 'That interests me. But I can't help the man any more. '*First do the thing, then drink and sing!*'

"I didn't think about the consequences any further, and the course of our work went smoothly. Relatives came two days later to visit, and I drank a cup of coffee with them. At three o'clock I always went out of the firm first, and I always came back to the office at four o'clock. Then suddenly I saw one of the youngsters who had served with me bending over his mug in a beer garden.

"I thought to myself scornfully: 'By Thunder! Why is this lousy dork in a beer garden instead of at his office?' I walked over to him to put him in his place. The kid threw his arms around my neck and said: 'Thank God I've found Herr Consul! The heads of the office have all been arrested and the office is under surveillance.' 'Where did you find that out?' I asked. The lousy dork—otherwise *Leutnant* Franz-Maria Liedig reported: 'I went to the office without any suspicion at all. When they didn't open up after I knocked, I was startled because the work day had already begun. A stranger opened the door. Something doesn't seem right, I said to myself, so I asked, 'Are you Doctor Schuster?' The criminal investigator didn't really know anything. He said, 'Doctor Schuster isn't at home.' I thanked him politely and took off, without even throwing a glance into the office, where most of the people were gathered under strict observation. I have to thank the naiveté of the investigator that I came away unharmed.'

"Well, Someone clearly had His Sheltering Hand over me, because it was a miracle that I got clean away when all my Munich people were dragged before the Offenburger Tribunal. Some of them spent nine months in custody before they were found not guilty."

"At the same time, a feast table under the black-white-red colors with Munich journalists, artists, politicians and some sea officers had been planned in advance. A former captain of the Kapp Putsch had gotten wind of something, went to them individually, and convinced them to hold the feast table in another location. The only guy the police caught was an eccentric publisher from Hamburg."

Ehrhardt says he disguised himself as a Bavarian peasant, *lederhosen* and all, and slipped off to Hungary, where the murderous Bolshevik regime of Béla Kun in 1919 had made Communism extremely unpopular. Colonel Bauer, a comrade from the Kapp Putsch, was already working there and found Ehrhardt an office job—and some false papers made out in the name of Eschwege, the Alsatian loyalist Ehrhardt later impersonated in court to shield Princess Magrethe.

Erhardt may have not been guilty, as he said, but he was a plausible suspect. Matthias Erzberger had been strolling with his political friend Carl Diez in the Black Forest during a stay at a health spa—he had managed to cultivate a serious weight problem despite the Starvation Blockade—when he was confronted by two young men who asked his name. The two young men, Heinrich Tillessen, , and Heinrich Schulz, 27, were both former members of the disbanded 2nd Marine Brigade. Heinrich Tillessen had served with Ehrhardt's torpedo boat half-flotilla during the war. His older brother Karl had sometimes served as Ehrhardt's adjutant. Heinrich Schulz had volunteered for the Army at the beginning of the war and had been wounded three times in infantry combat. Schulz had been decorated for valor or service multiple times, and was discharged as an *Oberleutnant*. When Erzberger identified himself, the two young men pulled pistols, shot him six times—he pathetically tried to shield himself with his umbrella—and then reloaded and shot Erzberger twice in the head. The horrified Carl Diez was also seriously wounded.

Heinrich Tillessen and Heinrich Schulz fled to anti-Communist Hungary, which refused to extradite them.

Matthias Erzberger's tragic life convinced many liberals and Leftists that he was the innocent victim of Rightist maniacs. The Nationalists took a nuanced view. To his credit, Erzberger—originally a war supporter in the Reichstag—denounced the Turkish attacks on Armenian and Aramean Christians in 1915. To his discredit, Erzberger had touched off the "Jew Count" with a speech in the Reichstag that sounded like an attempt to prove that German Jews were slackers: The anti-Semitic ploy backfired when the German Army's own meticulous statistics showed that German Jews had been enlisted, decorated, promoted and killed in almost exactly the same ratio to their numbers as Lutherans or Catholics. Matthias Ehrhardt obviously played to the gallery—his supporters in the *Zentrum*, the Catholic Center and in the Bavarian Peoples Party were appalled at the massacre of Christians anywhere. Erzberger assumed that they wanted to hear that Jews were over-privileged slackers—though the evidence that followed his charges proved that a great many Jews were actually good soldiers for the Kaiser.

In July of 1917, Erzberger began the cycle that led to his doom. He demanded peace in the Reichstag and also pulled strings to promote the dismissal of Theobald Bethmann-Hollweg, Kaiser Wilhelm's intensely loyal chancellor. Princess Viktoria, the Kaiser's only daughter, described what happened next: "On July 17, 1917, the majority parties in the Reichstag—the Centre, the Social Democrats, and the Progressive People's Party—adopted the Peace Resolution. It immediately harmed the German war leadership considerably. My father had already taken steps toward peace in 1916, in consultation with Bethmann-Hollweg and other responsible statesmen...."

"Knowing the huge opposition to the war, the Allied governments continued to reject peace offers, including those that offered to evacuate conquered territory and return Alsace-Lorraine to France: an initiative by Pope Benedict XV was also rejected.

"The contents of the Pope's proposals show that the Vatican had worked for the good of both sides, but it was all in vain—Erzberger's defeatist 'Peace Resolution' knocked the wind out of the sails of those in the Allied camp who inclined to peace. And publicly, as the United States Ambassador in Berlin, Mr. J.W. Gerard was to say later, 'It would have been easier for Germany to make peace with Bethmann-Hollweg at the helm. The whole world knows him and honors him for his honesty.' The Papal Nuncio Pacelli [*the future Pope Pius XII*] said, though in confidence, that if Bethmann-Hollweg had not resigned, the prospects for peace would have been good. Regarding Erzberger, the initiator of the Peace Resolution and leader of the conspiracy against Bethmann-Hollweg, the Nuncio added—'Now everything is lost, even your poor Fatherland.'"

Erzberger probably got the surprise of his life when, summoned to Compiègne to sign the Armistice, he found not Woodrow Wilson's 14 Points but the advance work for the Treaty of Versailles. His 'Peace Resolution' had led to mutiny in the German Navy and mass desertions in second-string units of the German Army, so the military commanders told him to sign whatever the Allies put in front of him. When the full details of Versailles were released, he became the most hated politician in Germany. When Erzberger replaced the more conservative Dr. Karl Helfferich as Weimar's finance minister, Helfferich insinuated that Erzberger was engaged in income tax evasion. A libel suit followed. On January 26, 1920 when leaving the court during a session of the trial, Erzberger was shot by a former naval ensign, Oltwig von Hirschfeld. A metal obstruction in Erzberger's pocket deflected the pistol bullet. Erzberger was only slightly wounded. But when the boyish assassin was sentenced to a mere 18 months in prison, Erzberger seemed to sense he was doomed. "The bullet that will kill me has already been cast," he told his daughter Maria. The court's libel verdict came in and Helfferich was punished with a minimal fine. The trial ended the day before the Kapp Putsch.

Perhaps defiant, perhaps vengeful, perhaps resigned to fate, Matthias Erzberger began to prepare a new scheme of taxes that would fall most heavily on the landed gentry, the aristocracy, and the nobility: the people who tended to support the Nationalist cause with both their money and their sons. His critics put out the rumor that his envy was based on his secret ancestry: they claimed he was the son of an employer, possibly even a Jewish employer, and a servant girl. Erzberger's actual parents were respectable lower-middle-class Catholics and the rumor was a fabrication. But Rightists hated him for all the other reasons—especially for his ability to maintain a weight problem through the blockade which gave other Germans nutritional problems.

"He's as fat as a bullet," Nationalists said, meaning a musket ball. "But he may not be bullet-proof."

Josef Wirth, also a Catholic politician and Erzberger's successor, made whatever political hay he could out of the Erzberger murder but the Nationals made no bones about their delight.

All sides—including the Communists—had begun to make political hay out of anti-Semitism. The events of 1919 in Berlin and in Munich had shaken the Germans. The extremists of both the Right and—more surprisingly—of the Left had tried to blame everything not on a few hundred of the 80,0000 Russian Jewish immigrants but on the entire German Jewish community of about 1 million, most of whom were either patriotic or humanely pacifistic.

In 1919—the first year Hitler began to make anti-Semitic statements, after years of subsisting partly on Jewish charities—a National Socialist publication was published under the title: *"Judas, The World Enemy; What Everyone Must Know About the Jews"* Hitler's attempt to capitalize on anti-Semitism because, unlike most of the *Freikorps* leaders, he had done no real fighting in 1919, was predictable. Less predictably, on March 20, 1920—in the immediate aftermath of the Kapp Putsch— an article in which the Communist newspaper *Die Rote Fahne* [The Red Banner] inferred that Gustav Noske, one of the most hated men in Germany by Left and Right alike, was actually of Jewish ancestry. Gustav Noske was compared to Judas, but the comparison expanded into anthropological terms: "Noske must have had a criminal or a voluptuary among his forefathers. And this ancestor acted through him. You only need look at his skull in order to recognize his criminality."

Die Rote Fahne also suggested that the Hunnic-looking Hermann Ehrhardt was secretly a Jewish tool. "Ehrhardt and his henchmen draw towards Lockstedt. They remain stuck in transit, supposedly because workers do not want to let them through; in reality, because they, in open and secret agreement with the Mueller government, want to crush the general strike in Berlin. Now on April 1 they are once again to migrate from Döberitz to Lockstedt, as officially reported, as soon as the agreements between the workers and the officials of the railroad district of Altona are settled. Ehrhardt thus wanders here and there through Germany, an *Ahasver* with armored cars and swastikas on his helmets." *Ahasver* is German for "Ahasuerus," the name of the Persian king who destroyed a plot to exterminate the Jews with the help of his Jewish queen, Hadassah, as told in the Old Testament Book of Esther. The fact that *Die Rote Fahne* tried to undermine their most dangerous Nationalist enemy—a Hunnic atavism if ever there was one and the son of four generations of Lutheran clergy—by portraying him as either a Jew or a rescuer of the Jews is more than a little revealing about the integrity of Hermann Ehrhardt in rescuing law-abiding Jews from thugs, and about the venality of the Communist press in making up any lie that served their interests.

Even the Kaiser was secretly Jewish, according to propagandists who didn't want Wilhelm II back on the throne. In 1922 the German author Max I. L. Voss re-discovered a royal scandal and published *"England als Erzieher"* [*England as Educator*] in which he traced the immediate ancestry of Prince Albert, Queen Victoria's German-born husband and the Kaiser's grandfather, to a Jewish court banker, Baron von Mayern, who supposedly conspired to help the impotent duke produce an heir—the future Prince Albert. "[*Albert*] is to be described without contradiction as a half Jew, so that since his time, Jewish blood has been circulating in the veins of the Hohenzollerns," Voss wrote. Kurt Tucholsky, a Jewish war veteran and a left-wing Democrat, chimed in with a description of the Kaiser in his Dutch exile surrounded by "men in kaftans with their ringlets curly" and "a little Jew from Krakow" among the Kaiser's imaginary dwarf garden of toadies and freaks.

Satire may also have been involved in one of the strangest assassination attempts that Organization Consul was blamed for. On June 4, 1921, Philipp Scheidemann, the first chancellor of the Weimar Republic before Friedrich Ebert nudged him out, was strolling through the park in the city of Kassel, where he was lord mayor, with his daughter and his granddaughter. When they left his side to pick some flowers, a young man with wrap-around leggings rushed up and used a syringe with a red rubber bulb to spray a weak solution of prussic acid into Scheidemann's face and neatly trimmed silver beard. Scheidemann, temporarily blinded, pulled out a concealed pistol and fired two shots. He missed. The assailant and a companion ran for it but they were arrested later that year: Hanns Hustert and Karl Oehlschläger were the actual attackers. But Erwin Kern, a former naval officer and long-time member of the 2nd Marine Brigade, was cited as the instigator. Kern would soon show that he knew how to use a gun better than a syringe. The choice of prussic acid on a Prussian statesmen sounds like a very drastic pun. Scheidemann recovered. He was not Jewish but his house was shortly defaced with swastikas.

Suspicion and frequent hatred of Jews was to claim a notable victim—and Hermann Ehrhardt received collateral damage from the emotional ricochet.

"Our Organization was forbidden and disintegrating," Ehrhardt wrote. "Anything that happened in Germany was shoved into the shoes of the fabulous O.C. Both the Republican and the Marxist spirit pursued us with deadly hate. But our own spirit was even stronger, our bond of loyalty never broke.

"Then a new bomb exploded. It was in the late morning, about 11 o'clock. I was sitting in my office. Opposite sat a gentleman of the *Großdeutsche Partei*, trying mightily to win me over. A newspaper extra edition fluttered onto the table.

"Rathenau was no more."

"The face of the gentlemen before me went pale. He got up and left immediately.

Wilhelm II, German emperor (1859–1941) at his house in Doorn, Holland after his exile, at the age of 70

Walther Rathenau said himself that he was a German first and Jewish only afterwards. "I am a German of Jewish descent. My people is the German people, my fatherland is Germany, my religion is the Germanic faith which is above all religions." Walther Rathenau, whose father became a multi-millionaire by buying the German rights to manufacture Thomas Edison's new electric light bulb, believed in assimilation: Rathenau had served in the Guards Cuirassiers Regiment, the ultimate snob regiment of the Kaiser's army, but he served as a sergeant because Jewish officers were then snubbed, if not banned, except as medical officers, in the elite regiments. When the Western Front began to collapse and the Kaiser's generals wanted to sue for peace, Rathenau, like Albert Ballin, had urged a fight to the finish with a mass conscription of every able-bodied German. He believed that Socialism might be inevitable but that Marxism was out of date: letting Bolshevism in the Russian mode into Germany would be a disaster in humane as well as economic terms. He compared the Russian form of Bolshevism to German Socialism as the performance of a masterpiece by a third-rate provincial opera company. He also said that killing 10 million people to liberate 10 milion other people made no sense to him. Quietly and without fanfare, Rathenau had donated over a million dollars of his own money to those *Freikorps* groups which avoided violent anti-Semitism.

Erwin Kern, one of the two Rathenau assassins

Rathenau, however, was probably too intelligent and too deliberately enigmatic to be understood by people in desperation. He was also too arrogant to waste a lot of time explaining himself. His friend Count Harry Kessler assumed Rathenau was homosexual, but since Kessler himself was homosexual, this could have been a wistful projection. Rathenau had exchanged passionate letters with women and gruff friendships with men for 20 years, but not one instance of consummation with either gender has been confirmed. Rathenau was notably close to his mother, and he accepted the post as foreign minister at the urgent request of his friend Chancellor Joseph Wirth and against his mother's advice.

"Walther, why have you done this to me?" she asked.

"I really had to, Mama, because they couldn't find anyone else," he replied.

What he meant to say was that nobody else could envision his next tactic, which may have saved Germany at least temporarily at the cost of his own life.

In the spring of 1922, Britain and France had convoked a protracted conference at Genoa, in Italy, to try to fix the amount of German's mandatory reparations for the war. Rathenau broadly supported reparations to maintain the national credit but wanted the amount kept as small as possible. The British and the French delegates—having seen how popular the idea of soaking Germany with a huge indemnity was with their voters at home, especially the veterans, the millions of war widows and the soldiers' bereft parents—wanted the maximum payments. In the middle of arguing that heavy reparations would wreck the German economy and foster a depression across Europe, Rathenau received a telephone call from the Soviets, who asked him to travel from Genoa to Rapallo, on the Riviera. On April 16, 1922—Easter Sunday—Weimar Germany abruptly became the first major country to recognize the Soviet Union with the Treaty of Rapallo. The British and the French were thunderstruck. The Russians were delighted, since the treaty developed into an economic alliance where Russia traded raw materials and training space for German industrial and technical advice. Britain and France had to contend with the fact that Germany, no longer isolated, was now backed by the largest country in Europe with the largest population and by far the greatest natural resources.

Walther Rathenau had saved Germany from overwhelming reparations—the reparations issue gradually fell apart—but he had signed his own death warrant at Rapallo. Chancellor Joseph Wirth, also an advocate of the treaty, had helped arrange it, but Joseph Wirth was a Catholic from a humble background, not a Jewish multi-millionaire. Nationalists in Germany put Rathenau's Jewish background together with Russia, the homeland of Rosa Luxemburg and her coterie of lovers and of Eugen Leviné and his coterie of thugs. The Nationalists decided that Rathenau had sold them out to the people who had taken over Munich in 1919 and who had three tries at taking over Berlin before Hermann Ehrhardt arrived.

On Saturday, June 24, 1922, Walther Rathenau was being chauffeured from his home in Berlin to his office. Rathenau's open car was overtaken by a six-seater driven by Ernst Werner Teschow with two young men in new leather coats in the back seat. As Teschow passed Rathenau's car, Erwin Kern—a suspect in the attack on Scheidemann—stood up in the back seat with a submachine gun and sprayed Rathenau with bullets. Kern hit Rathenau five times. Hermann Fischer then stood up and threw a stick hand grenade into the back seat. The blast lifted Rathenau into the air and broke his back but his body stayed in the car. Helene Kaiser, a middle-aged nurse, jumped into the Rathenau car. She was a tiny woman and Rathenau was over six feet tall and sturdily build but she was able to examine him enough to see the wounds were fatal. His backbone was shattered and his jaw was broken. "He was bleeding heavily.... He seemed to be already unconscious."

Teschow drove away at high speed, pulled to a stop, and lifted up the car's hood to simulate a breakdown. The police drove right past him. Kern and Fischer dumped their new leather coats and fled on foot.

Rathenau's chauffeur drove the car to the nearest police station 30 yards away to report and then drove Rathenau, already dead or dying, back to his house.

The murder was a shock to the Germans in general. On Sunday, 200,000 people gathered for oratory in which the Left attacked the Right and blamed them for the murder. On July 27, the following Tuesday, Berlin's workers received a day off and a million workers and their families marched in a parade to honor the murdered foreign minister. Six other German cities saw demonstrations with more than

100,000 marchers. The German Jewish community was especially hard hit by the death of a man they saw, with some justice, as a genius and a hero: the biographer Emil Ludwig and the composer Arnold Schoenberg both recanted their essentially patriotic conversions to the Lutheran church in protest.

The extreme right was delighted. Rathenau's mother received thousands of postcards calling him a traitor because of the Treaty of Rapallo and rejoicing in his death. The moderate Right was not happy. When Teschow showed up at his uncle's country estate just outside Berlin, asking for protection, his uncle turned him in to the police. Teschow claimed that he himself had misgivings about the assassination but that Erwin Kern had told him that Rathenau was a Bolshevik who wanted to bring all German under the domination of the Jews. Kern also told Teschow that if he was apprehended he could tell the court that he had believed Rathenau was one of the Learned Elders of Zion —which sounds like the key to an insanity defense. Rathenau had once actually said that 300 men who all knew one another controlled the destiny of Europe, and to a paranoid sensibility this was close enough. The Russian forgery known as the Protocol of the Learned Elders of Zion had an extending work-out by proxy as the young men who had failed to inform on the actual assassins were put on trial.

Hermann Ehrhardt, then hiding in Hungary, was not part of the group that decided to kill Rathenau and was never consulted. The meeting, according to an informant who could only have been Ernst von Salomon,

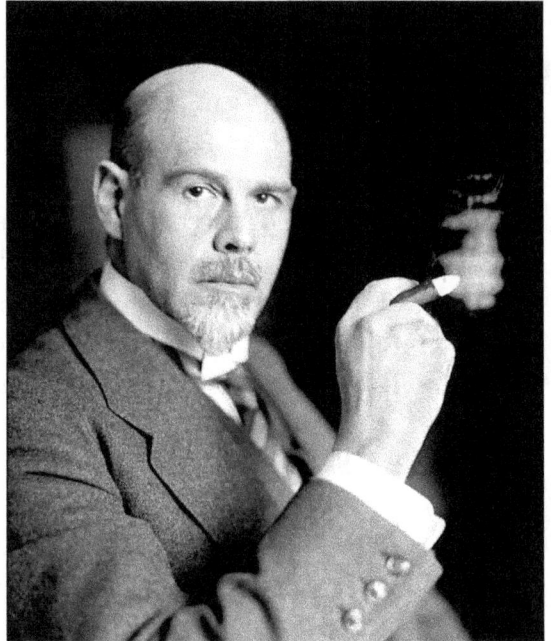

Ehrhardt was detained on suspicion in the June 1922 assassination of Foreign Minister Walter Rathenau, a Jewish industrialist and German nationalist.

took place among junior officers in a room in Berlin filled with cached rifles and hand grenades. The conspirators wrote and then scribbled over names on a single sheet of butcher's paper. Rathenau was the only Jew on the list, and the confused former officers determined to kill him because they wanted to avoid a Communist take-over of Germany and, if possible, to foster a return of the monarchy—not because of Rathenau's religion. Salomon himself, a Quixotic and dangerous Romantic of remote French ancestry, got five years in prison for obtaining the car that Teschow drove and that Kern and Fischer used as a gun platform. Salomon wrote in prison and later (illegally) married his Jewish girlfriend, Ille Gotthelft, and protected her all through the Hitler era by citing his credentials as a former *Freikorps* soldier and Right Wing death squad member. He wrote anti-British propaganda screenplays during World War II but never joined the Nazi Party and was in fact arrested several times during the Hitler years.

Kern and Fischer had expected that they would become heroes, at least of the Nationalists. Instead, they were denounced and disowned by just about everybody. A reward of 1 million marks was posted for them. They escaped in disguise by bicycle but just missed a boat for Sweden. They were pursued by an angry posse when they hired a boatman to ferry them across the Elbe. They bicycled through the dark forests of Thuringia, half-starved and sometimes fed by isolated farmers who had no idea who they were. Suspicious farmers, however, called the police and Kern and Fischer barely escaped a two-hundred-man police dragnet by slipping through thickets.

Kern and Fischer took shelter at the gloomy, deserted castle of Saaleck. A light in the east tower alerted residents and someone in a nearby village recognized Kern's face from a photograph. The police arrived on the night of July 17. A gunfight broke out. Kern and Fischer shouted defiance and refused to surrender. The police broke down one door and seized the western tower. Bullets flew back and forth in the middle of a thunderstorm and heavy rain. Kern was hit just below the right temple and killed almost instantly. Fischer dragged Kern's body to one of the beds. Then Fischer sat down on the other bed and placed his own pistol to his forehead. He shouted out *Es lebe hoch Kapitän Ehrhardt!*—"Long Live Captain Ehrhardt"—and pulled the trigger.

Hermann Ehrhardt's astonishment—which may have sounded convenient in 1924—was confirmed when the American historian and author Otto Friedrich spoke with the surviving conspirators in the 1960s. But while Ehrhardt had not approved the assassination, he refused to condem the assassins, who were former comrades in arms, first in Kaiser's navy and then in the 2nd Marine Brigade. To soldiers of this generation and some others, dead comrades were a palpable presence.

"The justification of such a deed is that the perpetrator is willing to give his life for his country. At any time, when the fatherland is in danger, it's no longer 'Love your enemy' but rather 'Eye for eye, tooth for tooth, blood for blood.' "

"As Kern and Fischer, the two men who shot Rathenau, were hunted through Germany like wolves, and at last came to their manly end at Saaleck, we said to ourselves: 'Such a heated love of the fatherland must be honored in some way.'...right or wrong, success or failure—they were heroes who gave their lives in the trenches sustained by the thought—'Help your people. Help your fatherland.' The noble man shows that he can die for this thought. The ignoble man stigmatizes the thought so that he doesn't have to die for it.... On their grave we put the words:

"Tu, was du mußt, sieg oder stirb
Und laß Gott die Entscheidung!

"Do what you must, win or die
And leave the decision to God.
—E.M. Arndt"

"Without my beard and under the name of Herr von Eschwege I came back out of Hungary to Germany."

A week after Walther Rathenau had been killed, two former *Freikorps* members not affiliated with Ehrhardt, Bert Weichardt and Albert Wilhelm Grenz, stalked Maximilian Harden, the journalist who had exposed the purported homosexual camarilla around the Kaiser 20 years before. On July 3 they broke his arms and legs with crowbars. Harden required critical care for two weeks. He had long since lost most of his credibility, since he had supported both the German invasion of Belgium in 1914 and the Treaty of Versailles. Harden's savage beating was a footnote compared to the Rathenau murder, which had shaken all of Germany, but this violence too helped stimulate political action. On July 22, a week after Kern and Fischer fought to the death and three weeks after Harden was beaten into a crippled wreck, the Reichstag adopted the *Republikschutzgesetz*, the law for the defense of the Republic, which endorsed stricter enforcement and heavier surveillance against militant groups. Rathenau's death promoted the law for sentimental reasons but also for practical reasons: once Germany had a treaty with Russia that benefitted both sides, the Soviet government ramped back on funding or encouraging violent take-over attempts and fell back on advancing Communism mostly by political means. The supervised plebiscite that had given Upper Silesia to Germany by a margin of 60 to 40 percent on March 20 of 1921 also diminished the threat of a Polish invasion.

Emil Julius Gumbel, the Independent Social Democrat who told U.S. Immigration he was a pacifist when fleeing Germany after Hitler came to power, and long after many of his colleagues voted to go Communist, had published his classic study *Four Years of Political Murder* in 1922. Professor Gumbel claimed that there had been 376 political murders in Germany since the war ended: He said that 354 murders were committed by right-wing groups and 22 murders committed by left-wing groups. Gumbel said that 17 of the left-wing murderers received heavy sentences, including 10 executions, while 326 of the 354 ring-wing murderers went unpunished. The average prison sentence per murder, Gumbel said, was 15 years for a leftist and four months for a rightist. Gumbel was a statistician and his figures were taken very seriously after the rise of Hitler, after Hitler's own internal political murders, and especially after the discovery of the Nazi death camps. But the American academic David Felix, examining the Gumbel figures in 1971, after Gumbel was dead, pointed out that the figures were lurid hokum. Emil Gumbel, Felix said, had been a Marxist and an apologist for the Soviet regime—Gumbel portrayed Russia as an idyllic worker's paradise in the winter of 1925—1926 during a period of deadly forced labor campaigns. Professor Felix responsibly noted that the Red groups in Germany had touched off the revolts in Berlin and Bavaria where hundreds of Germans had been killed on both sides or by mistake, and that the *Reichswehr* and the accepted *Freikorps* units were acting on behalf of a democratically elected German government. "The various radical groups had contributed their share to the destruction of life in the Berlin fighting of December 1918, and in January and March of 1919; the anarchist-Communist putsch in Munich in April, 1919, and the rising of the Red Army in the Ruhr in the spring of 1920. It is a fact, furthermore, that the radicals had started these actions. Gumbel's conclusions were clearly influenced by his political orientation, and his book has failed to assist a fair judgment."

Hermann Ehrhardt's thoughts were of a similar drift. With the help of informers the Weimar government had cobbled together a purported recruiting poster of Organization Consul for release to the Socialist newspapers which described the 5,000-man assassination gang in colorful prose supposedly written, or at least endorsed, by Ehrhardt himself: The charter, which may have been based on geographic reality, showed Organization Consul divided into seven main regions: Hamburg, Hannover, Berlin, Frankfurt am Main, Dresden, Breslau, and Tübingen. Each group had one, two, three or four sub-groups, and control—this part sounds accurate—was maintained out of Munich, the Robbers Roost of anti-Communist in Germany.

The mission statement assembled from various informers for this organization said:

"Spiritual aims: The cultivation and dissemination of nationalist thinking; Warfare against all anti-nationalists and internationalists; warfare against Judah, Social Democracy, and Leftist-radicalism; fomentation of internal unrest in order to attain the overthrow of the anti-nationalist Weimar constitution....

"Material aims: The organization of determined, nationalist-minded men... local shock troops for breaking up meetings of an anti-nationalist nature; maintenance of arms and the preservation of military ability; the education of youth in the use of arms.

"Notice: Only those men who have determination, who obey unconditionally and who are without scruples... will be accepted.... The organization is a secret organization."

Someone who read the article with these melodramatic statements asked Hermann Ehrhardt, during an amnesty, what the actual facts were.

"Personally I'm very much against Bolshevism and only a little bit against Judah," Ehrhardt winked, remembering that *Die Roter Fahne* had already proclaimed him Ahasuerus in an armored car. This sinister verbal shorthand suggested that Jews who avoided Bolshevism or criminal activities were safe—as they had been during the days of the 2nd Marine Brigade and of *Leutnant* Berlin, assault group leader and art connoisseur.

Having read Gumbel's book and its 354 right-wing murders, the interviewer asked Ehrhardt how many men he had actually murdered.

Ehrhardt rolled his blue-grey eyes around behind the slanted lids as if he were calculated.

"*Acht*," he said not so simply.

The German word "*acht*" with a small 'a' means eight. *Acht* with a large A, as a noun, means "respect" or even "caution." This may have been one of the puns Ehrhardt enjoyed. Conversely, the number eight may have been an actual estimate. Ehrhardt never pretended any remorse at the death of Erzberger. He was a Swabian, and spraying a Prussian statesman—Philipp Scheidemann—with Prussic acid was his kind of joke. Tillessen and Kern, the leading attackers in these cases, were Ehrhardt loyalists and personal friends. A number of informers must have disappeared, as they always do among underground operations where money or nervous fear prevails over loyalty and Hard Honor—the meaning of Ehrhardt's name, as Rudolf Mann had noted with approval when he signed on. Rudolf Mann himself had needed work and had resigned his commission when the Brigade was dissolved after the Kapp Putsch—"I held onto my officer's coat and kept it in the closet—there were flecks of blood on it, but no stains." But many of the other officers, the naval officers in particular, were in touch with one another. They would soon prove their continued loyalty to Captain Ehrhardt.

Chapter Twenty-Three
Guest of the State

Ehrhardt seemed to react to having the steel door of the prison cell slam behind him as a wolf does to a frontiersman's pit trap—with complete physical apathy. But his mind remained active. He wrote later that he felt the Weimar "pseudo-democracy" had put him behind bars on charges of "militarism" that made no sense at all after he stuck his neck out to defend Princess Margrethe, a helpless young lady he saw as part of Germany's lawful ruling class. His own conscience was absolutely clear. He knew he had been arrested primarily because of the Rathenau murder and he knew he had not ordered the Rathenau murder—but expected that no one on the Weimar bench would believe this. The next move was to escape and continue the war against Bolshevism—and now against Weimar.

"Their political investigation imprisonment was a type of torture related to the sort of thing that had developed under the Russian despotism," he wrote of the Weimar Republic.

"A soldier's virtue is to find a way to to deal with circumstances of the sort that now confronted me, body and soul. Every advantage was worth something. Good. My quarters were over-heated. Thus the window frame had to be left open to the breadth of two fingers whenever possible. But this was not officially allowed. Things of all sorts were always forbidden and people were didn't care.

"The heating system streamed heat out. Beautiful! I made myself a heat shield out of my gradually growing heap of newspapers. It was a great advantage for me that the watch-master, who covered my corridor, was a splendid fellow, who freely displayed the pedantry of the old-fashioned sergeant or bureaucrat. It was impossible to get him to go one millimeter beyond the regulations required in his position. But within his circumscribed playing field, he would do whatever he could without giving up his responsibility. He was a true son of Saxony and spoke that cheerful dialect that made my insides curdle with amusement. Bur he covered his section with cleanliness and order and made life as easy as it could for the prisoners. He was polite and decent and had none of the Saxon *Schadenfreude*, like most citizens of his kingdom. He was a talkative fellow and made no secret of the fact that he had a large number of friends and was outstandingly well-read for a man of his station. He had been a soldier at the Front and even made it to an officer's rank. Then he was heavily wounded and was given a place of rest as a prison warder.

"He won my heart on the first day. As he inspected my cell he looked at my water jug and said '*Ye're a captain, seems ta me, ya need more water. Ye're used to it, and there ain't no rule agin it.*'" He immediately brought a second water jug and made sure it was always kept re-filled with fresh water.

"...You must have been a front soldier..." Ehrhardt told him.

"'*Ye got it,*' he said, and I believe he was flattered."

Ehrhardt found a sort of grudging amusement in sitting at a prison table to reclaim the contents of his pockets that had been confiscated on the day he was arrested. One of the warders went over the list and the packet he had received from the Bavarian Criminal Police. As the warder bent over the list on the table "...I took the opportunity to take everything I could without being seen, especially my pocket knife and my little manicure set."

"From the first days I was entirely dependent on prison food, but as prison fever heated up in me, what I ate became sausage to me. In the morning we always had porridge with heavy lumps in it, the terror from my childhood that I could never overcome. I let it sit there untouched and made my prison breakfast a slice of dry black bread and a glass of water. About 10 o'clock we had black bread again, but with marmalade or margarine. Around 12 o'clock came the midday meal. With the exception of Sunday it was a dish of things all cooked together, only once in a week with meat cooked in it. Otherwise the food consisted most of peas, carrots, beans, rice, barley, and noodles. It was quite appetizing, at least in my opinion. But I'm by no means a fastidious eater. In the afternoon around 3 o'clock we got another entry with coffee and two butter rolls that were brought from a bakery in the neighborhood of the prison. That was the only treat of the day, as I saw it. In the evening we had peeled potatoes with or without herring, or bread with cheese, or soup or oatmeal porridge.

"Each of these meal times announced itself with a great crash. A giant food container was wheeled through by a trusted convict, from cell to cell. Then the door was opened. In the blink of an eye, we had to be standing there with our little basins in our hands. The turn-key dipped a huge ladle in the cooking water and put a measured portion in the little basin. In the Navy we called that a "*Schlag*."

"The food itself was not a sheer delight, because the ladle was the only utensil. Now we had to eat peeled potatoes or herring from the ladle without spoons, with nothing but the help of our fingers. And the herring smell on our own fingers was added to all the other good smells of the place."

Ehrhardt discussed the problem with his new friend, the former sergeant and reserve lieutenant with experience at the Front.

"'*Herr Kamerad*,' I said, 'You wouldn't be able to answer for it in our barracks if the men had to eat peeled potatoes with their fingers.'

"'*I din't never think about that*,' the *Wachtmeister* answered in his heavy Saxon accent. '*Whatta we do in this case? A puh-tition! Paper is real important. You write up three of 'em, real fast, Herr Kap'taen, and I'll run 'em to the offices. I'll put 'em in all the places where they put puh-titions that they already got that are furious. If the writer is furious, they pull all the stops out of the way an' then they go through dam' quick.'*"

Ehrhardt was impressed with the former sergeant's understanding of the bureaucratic process. He wrote the three letters. In a short time, the prison authorities issued him a knife and fork. He also got a visit from the prison director.

"He was an old gentleman, who asked me kindly, 'How do you find your sad accommodations? Is there anything I can do to make things easier for you? It's not that I have too much to do, and the mistrust of the courts toward the political prisoners is very extreme.'"

"An old torpedo-boat man is accustomed to cramped quarters, and if the story is only going to last another couple of weeks, I can put up with it very easily," Ehrhardt replied—seeking information while pretending to be cooperative.

"Have you already thought of subscribing to the newspapers?" the prison director asked. "You have that right. Make good use of it. It's really good to know what's playing out beyond the walls."

"How would I go about that?"

"Just make a petition," the prison director said dryly. Ehrhardt thought of the friendly *Wachtmeister*. He had the new petition ready before the end of the day.

"The prison director took his friendly leave. After him, gradually, I had visits from every inspector in the prison. I had the feeling that I was becoming a regular topic of dinner conversation among all these men and their wives."

Ehrhardt said that the bureaucrats were decent, somewhat conservative people themselves but they appeared to stand in awe of the Weimar government.

"The first real help I got was from my dear cousin, Karl Ehrhardt, who was a merchant in Hamburg. He personally traveled to Leipzig, and he realized as a practical man that a fellow in prison needed some money, and put a considerable amount of it at my disposal. He was also concerned about my nutrition, but I assured them that a person who's locked up in custody didn't need to eat a lot."

Shortly, Ehrhardt was receiving food packets from his relatives, from his business partner in Munich, and from people he had never heard of. "I sent most of the packets along to my family right away," he noted. "I would have had to have 20 stomachs to eat everything I got in the mail... These gifts of friendship and love brought me a double joy, because I could improve the lives of my loved ones."

Ehrhardt also received communications that were not exactly food packets. One of his former comrades, *Kapitän* von Hase, set up a mail drop in Leipzig where he received letters to Ehrhardt that might have alarmed the authorities. *Kapitän* von Hase visited Ehrhardt once a week to pass along information he had memorized verbally and to receive instructions. Ehrhardt was still running whatever Organization he was Consul of, even while he was behind bars. He steadfastly maintained that it was actually an employment agency. He also used his cousin's donation of cash to subscribe to a wide spectrum of newspapers—conservative, liberal, and Leftist: "This was fodder for my political instinct." He learned from the newspapers that the German inflation, which began with the announcement of the terms of Versailles and increased with Rathenau's murder, was now raging out of control as printing presses churned out money that lost its value between the factory pay window and the food store. The prison authorities offered him the chance to make some pocket money by folding and glueing paper bags but he turned them down: his wife was still an heiress who now got plenty of food in the mail and the earnings would have become worthless overnight. Instead, he decided to develop a skill that could one day produce daily wages so that he could support his family when he got out of prison. He sent home for the electrical textbook of Professor Graetzen. "In the course of eight months I worked almost all the way through it."

One day the cell door opened and in walked Judge Metz with two officers of the Criminal Police. As Ehrhardt described it, he was led down a long corridor to a hearing room where he and Judge Metz played a game of question and answer for about two hours. At one point, Metz asked him, with some very one-sided humor: "Do you have a little more time?" Ehrhardt got the joke but failed to laugh.

Ehrhardt asked Judge Metz if any charges would be made against the princess and Metz told him it wasn't likely.

"Don't worry about it. In any case the lady will be set free. It's possible she may not even have to appear in Leipzig. Her case doesn't actually fall under the jurisdiction of the state's court system. At the most she may be subject to a cash fine for being an accessory... *Herr Kapitän*, don't make yourself uneasy if the investigation is drawn out somewhat longer. This time will be considered as part of your punishment when the sentence eventually comes into question. In any case, you'll soon be quite free.... If you're still worried about the princess, remember that I call it to your kind attention that I don't have her in custody."

Ehrhardt pretended to believe this but he no longer trusted Judge Metz. Ehrhardt's perspective placed him on the opposite side of Professor Gumbel. To Ehrhardt, the Leftists were given the soft sentences and the Rightists were given heavier punishment. In the Rathenau murder case, Teschow beat a death sentence by prating on about the Learned Elders of Zion, as the late Erwin Kern had advised him, which may have convinced the judges that he was mentally unbalanced: he got 25 years for driving the murder car. Salomon got eight years for providing the car. Conversely the Leftist Ernst Toller had gotten five years in prison for leading Communists troops in the battle at Dachau where a number of regular Bavarian soldiers were killed in combat.

"Now that I read the judgments against the men of the Right, my trust for the workings of the nation's court grew less and less all the time..." He said that some people seemed to feel that Karl Liebknecht and Rosa Luxemburg were martyrs while he and others felt that they were destroyers of the nation.

"One prosecutor in the Rathenau case theatrically barked out 'Death! Death! Death!' Anybody who saw him would have been reminded of a sea lion that had just bobbed to the surface—but that would be too good for him."

Ehrhardt was not alone in his contempt for the court system. Herr Metz himself once confidentially described one of his colleagues to Ehrhardt as "an outstanding wash-cloth." "Naturally that was not my judgment since I didn't know the man personally," Ehrhardt wrote dryly.

In January of 1923, Ehrhardt's wife visited him in prison. He was delighted to see her but deplored the lack of privacy—and the fact that a respectable lady came into even marginal contact with criminals.

"We spoke a lot about the amnesty, which is where my case really belonged," he wrote. "Even in my absence, it could be judged that I was not the leader of the Kapp Putsch. I may have achieved something as one of the supporting leaders, but this accomplishment was that the Kapp Putsch take-over shaped up to be bloodless...."

"In her belief that I was certainly entitled to an amnesty, my wife wanted to seek out Federal Court President Schmidt. I tried to talk her out of it, because I knew based on experience that such visits are never edifying to either side. But she wouldn't be held back.

"The next day she came back to me. She was highly indignant because the court president had dealt with her in an unfriendly manner. As my wife had begun to cry in her excitement, he said to her 'That won't be of any use to you, *gnädige Frau*, even if you force out a tear. That won't influence us at all.'

"I got into such a stinking rage that on the same day I wrote to my friends and comrades and told them what to expect if a wife who came to visit her imprisoned husband could be treated in such a manner.

"All my mail was censored, and the good fellow whose job it was to read my letters usually did a decent job. But this time he laid my letters before the Senate President, who came to me in person and threw the tone of the letters up before me and reprimanded me because of them.

"I must contradict what your wife had to say. I can only excuse it due to her extreme excitement. I myself was almost moved to tears with sympathy for your wife. And incidentally, *Herr Kapitän*, I can't conceive how your wife could write me up as 'Schmidt' when to her and everyone else I am 'Senate President Schmidt!'"

Ehrhardt admitted that it took a great deal of self-control not to act on his obvious impulse: "Rage that cannot be expressed extinguishes a great deal within a man. But I had to keep saying to myself 'Just be calm and kept wearing a sugary mask of trust on your upside-down face. Only that will give you the chance to be able to use your fist later.... Sometimes I had to grit my teeth to get control over myself: 'Don't go slack. Don't show these people the picture of a man who loses his nerve after a few months in prison."

The chimney that passed through his cell rendered it extremely hot in January. He slept badly, and the extreme heat in the cell left him so weakened that some mornings he could barely get up.

"I couldn't discipline myself to keep working my way through Graetz. Serious reading of any kind left me immediately exhausted. So instead I read the collected stories about Sherlock Holmes, they were just the thing for my weakened body and my frame of mind under the circumstances. The activities that were depicted in these books were also appropriate to my short-term situation."

One night, when Ehrhardt stood up to cool himself from the "disgusting heat," he slipped and broke the third finger of his right hand. He also knocked himself out. When he woke up the next morning he saw the broken finger was bent almost backwards. A physician came and put a splint and a bandage on it right in his cell.

"I tried to use this accident to convince the physician that my condition wouldn't permit a longer stay in prison. But my appeal was unsuccessful. I think it was fear of the '*Vorwärts*' that none of these people dared to decide that *Kapitän* Ehrhardt belonged in the hospital. If a Communist physician had a Communist patient before him, it would have been quite otherwise. I've often been struck by the absolute cowardice and lack of conscience in the ordinary citizen."

Ehrhardt also complained that the confinement was aggravating the rheumatism he had picked serving on the torpedo boats. But this appeal got him nowhere either.

"The Herr Professor gave no factual answer. He mocked me and said 'Do you picture yourself getting out of there for something like that?'"

Ehrhardt got a shock when he read an editorial in '*Vorwärts*' demanding that his trial take place not in the state court but the federal court: "We want this process to take place in the Federal Court because only with Superior Court Attorney Ebermayer will we know that things are in good hands."

Ludwig Ebermayer, son of a Protestant deacon and a moderate conservative, had admitted to the "undoubtedly noble motives" that led Traugott von Jagow, the son of a Prussian officer and a countess, and brother of a Prussian general, to briefly take on the role of Secretary of the Interior during the Kapp Putsch. The court reduced the charge against Jagow, a former Guards Cuirassier officer, from "high treason" to "condoning treason." The court then gave Jagow five years in prison.

"On the day I read that, the game changed with the thought: 'You must get out of here, it's a matter of life and death.'"

Chapter Twenty-Four
Keys to the Future

While Hermann Ehrhardt was locked in his cell reading Sir Arthur Conan Doyle's complete Sherlock Holmes stories, three of his young officers, perhaps less familiar with the real-life world that inspired Arthur Conan Doyle, were plotting to get him out of prison.

"I knew exactly that the brave boys that had followed me unconditionally would find their way to me," Ehrhardt wrote. "I only had to get the word to them what my reality was: I was sick and if my custody lasted much longer I would go to the dogs—and I was sure they would find their way to me.

"I knew for sure that my people on the outside would try to set up communications with me. In my fantasies I worked out possibilities of how this would happen. When I walked back on forth in my 17-foot cell, my attention was always focused on seeing some sign of life from the outside."

One evening Ehrhardt heard something clatter against the closed exterior window of his cell. He chinned himself on the grate to lift up the window-frame—officially forbidden but informally tolerated due to the extreme heat from the chimney. He saw a man he didn't recognize outside on the street standing in a dark corner.

"*Achtung!*" the man called with a suppressed voice.

"I saw a shooting-iron in his hand and understood in a flash that he was going to shoot something through the window. I pulled my head back in and a dart from an air pistol hissed into my cell. I shut the window right away. I picked up the dart and found that it had been wrapped, in a very clever way, with a strip of paper.

"On the paper was written: 'We have observed you precisely. On your exercise walk information will be left in a round hole about a meter high.'"

Ehrhardt had always skipped the exercise walks around the prison court, to the satisfaction of Dr. Metz, who didn't want him talking with the other prisoners. Ehrhardt himself said he preferred to avoid the other prisoners: he himself washed as often as possible and most of the inmates didn't wash at all. But when Ehrhardt pointed out this objection, Dr. Metz insisted that Ehrhardt take his exercise walks along with everyone else.

"Dr. Metz himself accomplished the greatest service for my eventual liberation," Ehrhardt wrote. "My cousin, my wife, and the prison chaplain took no role in my escape and didn't need to."

"I can say that on the next day I went on the exercise walk with a certain heart-thumping, because I knew that one or the other of my loyal men had put himself in danger. Nothing is more painful for a leader than when he doesn't have an exact insight into the situation of his people."

"I started my exercise walk and walked to keep up a keen observation.

"But one of the convicts, a newcomer with a limp, wanted to talk to me: '*G'day, g'day, Guv'ner! Whut tricks hav' ya turned?*'"

Ehrhardt wanted to get rid of the man. Talking was against the rules. He also needed to keep his attention focused on looking for the message—and to keep the guards' attention away from his search. But the man just had to talk.

"*Just lemme know who ya are,*" the limping convict asked.

"...a three-time armed robber with murder," Ehrhardt said bluntly.

"A wholesome terror seized him and he left off on the silly questions."

When Ehrhardt reached the hole mentioned in the note, a meter high off the floor, he bent over and pretended to tie his shoe, peered in, pulled out the hidden note. He concealed it in his pocket. Back in his cell he read it.

"'Constantly watch the hole and the crack under the door. Shove own notes under crack. They will be picked up and put back. In threatening cases messages will be slipped or whispered on the exercise walks.'

"This was the first glimpse of light in these grey days. I now had a way of sending and receiving messages. The spirit re-wakened in me and a new spring of life bubbled up. There were loyal men outside. And that protected me from the prison stupidity around me. I saw the deed before me."

The next pick-up was a written code key so that any intercepted messages Ehrhardt wrote out and left for pick-up could not be read by the authorities if they found them. He also obtained grey paper of the type used for writing overseas mail to blend with the color of the walls and make the notes less visible. The messages kept asking for descriptions of every detail of the prison, every corner and every lock. Ehrhardt's work in torpedo research had rendered him a fair draftsman. One morning Ehrhardt was astounded to find a crude-looking pass-key wrapped up in one of the notes—but the pass key didn't fit the lock of his cell.

After some weeks went by, Ehrhardt found a note: 'Tonight and tomorrow morning be ready. The patrol will be attacked by surprise. Keys are ready. The watch will be overwhelmed. You'll be taken out."

Ehrhardt had kept track of the two-man patrol through many sleepless nights and knew they came through at midnight, and again between 2 a.m. and 4 a.m. He got dressed and ready to escape without being obvious. He hoped his hand grip was strong enough as the broken finger healed. Then he lay in bed waiting.

"Around 12 o'clock I heard the door to the courtyard rattle. I heard the paces of the guards. I had a heart-pounding like that only once before in my life—after the trip to India fell through when I was a boy and my father walked over to the bed while I pretended to be asleep... but the patrol came and went, and nothing disturbed them."

Ehrhardt waited all night with his heart beating so hard he felt as if he were choking. The patrol made a second sweep with the court. Nothing happened.

"'You're turning into an hysterical old spinster,' I growled to myself. 'How often have you been attacked on the torpedo boat with a near miss, and when it was nothing, you passed it all off with a shot of *Schnapps*?'"

He wished he could punch it out with an enemy he could see and fight.

"I said to myself: 'All is lost. Your friends must have been caught by the pack of hounds that runs the prison and tomorrow they'll be in here *with you*.'"

Ehrhardt spent the next three days without an appetite. He verged back into depression. He ignored his black bread with margarine and even his favorite butter rolls. No messages showed up on his exercise walks.

"At last on Thursday I saw a piece of paper under the door. I picked it up.

"In the cell I read: 'Misfire. Don't lose courage. Another way will be tried.'"

Plans began to circulate under the cell door for all kinds of escapes: lock-picking, beating up guards, full-scale prison invasions. But nothing actually happened. Ehrhardt's nerves, always excellent under enemy gunfire, began to bother him. As Ehrhardt had told General Lüttwitz during the Kapp Putsch, he was no safe-cracker. He began to fear that his junior officers weren't either. At one point a note in the one-meter hole on the exercise route told him that one of the barred gates to the outer court would be open from 2:30 to 5:30 a.m. The note tacitly suggested he might want to have a try at the other doors. The note under the cell door was wrapped around a clumsily made skeleton key—but he now had only ten days left before his trial and probable sentencing.

Ehrhardt noted that July 13, 1923 fell on Friday the 13th—just like the march-out for the Kapp Putsch almost three years before. He had been scheduled for his weekly tub bath and in the July heat "....I've never sweated in my life like that, and the *Wachtmeister* was watching my every move. The ticket now was to make sure the key never slipped out of my pocket. Now, with God's grace, back to my cell. And this time I was lucky. I didn't see anybody. But as soon as I got to the door the *Wachtmeister* came out of the door of the little washroom and asked right away, '*Wul, wul, whadaya want then?*'

"Some good spirit gave me the inspiration to say: 'I'm completely exhausted and delirious from that hot bath. I have to take a nap for a couple of minutes. I can't get my breath.'

"The sweat was flowing down my brow in rivers, and the *Wachtmeister* believed what I said: '*Dammitall, it's way too much, inn't it?*,' he agreed."

"Lord God, I forgot my soap in the tub." Ehrhardt improvised. "Can I go back fast and pick it up?"

"The good fellow wasn't about to try any more stairs so he let me go alone; '*Yeah, g' down but come back real quick.*'"

Ehrhardt padded quietly down the steps. He pretended to be searching for his soap. He got to the iron gate and—the key didn't fit.

"I said to myself—now it's all sausage!"

He rubbed the key with soap. The key slipped into the lock. As he jiggled it, Ehrhardt found out that the gate lock was already open....

"I ran down the passage but when I came to the third iron gate, a terror came over me: the gate seemed to be locked. I looked around, glanced at the wooden door behind the grate, and hissed. The wooden door started to open itself and I saw the two flushed, ruddy faces of my two friends who both said: 'Lean on it!' I pressed on it, the latch sprang open, and they pulled me through."

Ehrhardt never gave the names of the three younger officers who helped him escape because they were now wanted accessories: he called them Frix, Frax, and Jumbo. He offered no physical descriptions or regional accents. Frix and Frax may well have been the Tillessen brothers and Jumbo may have been Friedrich Wilhelm Heinz, a big muscular decorated infantry veteran with a rather large nose and ears: no absolute confirmation is possible. While Frix led Ehrhardt out of the prison to the street, Frax ran back in briefly to lock some of the doors behind them to give them time to make their get-away. The three fugitives ran to the outside of the hotel "*Deutscher Kaiser*" to pick up their ride. They walked around pretending to be ordinary strollers until Jumbo showed up with Sachs, another conspirator—somebody with a Saxon accent—and the rented truck, cranked up and ready. Ehrhardt slipped inside.

"Well, we did it!—but I expressly forbade you to take the point role!" Ehrhardt brusquely told 'Frax.'

"I said '*Jawohl*' and shut up," Frax wrote in an addendum to 'Adventure and Fate'. "Inwardly I rejoiced mightily. He was the real Ehrhardt again. Once again he was the same man who, years ago, piped up the rescue and fished out the crew from the lead torpedo boat after the loss in the minefield. Even after this act of daring, the Chief objected when others risked their lives to do the same *for him*."

Ehrhardt wanted to get as far away from the prison as possible as fast as possible. But the four younger men convinced him that high speed would only attract unwanted attention. When they spotted a heavy truck filled with young men in uncertain uniforms coming from the opposite direction, everybody slipped their concealed Brownings off safety. The heavy truck they spotted turned out to be an all-boys school excursion. Everybody on both sides cheered and waved as the trucks passed one another. The fugitives ate cherries from a basket rather than dismount for food until they were safe in Bavaria and headed for Austria, where Waldemar Pabst maintained some influence as chief of the *Heimwehr*. At midday, the truck came in sight of the Alps.

"God be thanked, the escape is successful," one of the young officers said.

Ehrhardt shortly looked at himself in a full mirror at the first safe house and was appalled. "My nails had gotten soft and, and the hair came out with my comb. Thick shadows lay under my eyes, and under them I saw a couple of colorless cheeks." That night, his host stood guard over Ehrhardt with a pipe in his mouth and a rifle tucked under his arm so Ehrhardt could get some sleep. After a night's rest Ehrhardt got dressed up in Bavarian peasant costume and hiked through the Alps to cross the Alm River into Austria with considerable effort. He breathed in the cooler mountain air with intense relief and satisfaction despite the unaccustomed exercise.

"But this deep breathing couldn't last long," Ehrhardt wrote. "Up there in the mountains I abruptly received the news that the Princess had been arrested and taken to Leipzig. That was the heaviest blow that could have fallen on me. Even though I knew Metz, it had never entered into my head this would happen because the man had so often assured me 'Don't worry about the Princess, nothing will happen to her.'

"I immediately made the decision: you must go back to Leipzig and offer to give yourself up if they let the Princess go.

"But my two loyal comrades, who had hiked with me into Austria, grabbed me with the same resistance my men had shown at the camp at Münster and wouldn't let me go. They said they would use force to stop me if I tried to go back to Leipzig. After all they had risked for me, I would not be allowed to leave safety.

"Messages began to reach me: 'For God's sake, don't put yourself there. That's a very false conception of chivalry. You yourself will be exterminated and you won't help the Princess one bit.'"

Princess Margrethe "a noted beauty and reactionary," had been arrested at her home in Munich on July 16, 1923, according to the New York Times, which reported that the German police were also looked for Ehrhardt's wife, who was suspected of slipping him the keys he used to unlock himself from the inside. The keys, according to Frix, Frax, and Jumbo, had actually been made for them by a grizzled

Hungarian safe-cracker. The three young amateur jail-breakers had consulted a spiritualistic medium, routine procedure among German staff officers. The medium told them that a grey-haired man with a foreign accent held the keys to what they wanted—or so they said later. Previous escape plans involved attempts to throw a grappling hook over the wall—nobody could make it reach—or hiring a couple of disaffected policemen who said they were experts in *jiu jitsu* but never showed up for the attempt to overpower the guards.

The Ehrhardt faction apparently got word to Princess Margrethe that Ehrhardt was safe and he could safely be blamed for everything without endangering him. Prosecutor Schmidt was forced to admit, once he picked the other Ehrhardts for questioning, that Ehrhardt's wife and his cousin had done nothing to assist his escape. Neither of them was convicted. But it was obvious to the court that the Princess had known who Herr Eschwege actually was. She was charged as an accessory to his escape.

"A mountain has shuddered, and now we see a little mouse," the prosecutor said to the Princess—which made Ehrhardt every bit as furious as the same prosecutor's insulting comments to *Frau* Ehrhardt, widow of a nobleman. Under the Kaiser's Empire, either of these remarks could have led to a challenge to a grown-up's duel with rifled pistols—illegal but very popular and very apt to be fatal.

"Loyalty in this political climate of confusion and danger is the single proof of personal worth," Ehrhardt wrote in 1924 at the conclusion of '*Adventure And Fate*.' "Loyalty is the mark of honor, loyalty is the last feeling for morality. Once it disappears, it is no longer possible to renew a people or to found a new state."

Observers told him that the Princess, under great stress, had conducted herself with dignity and integrity: This suggests that she had been told that Ehrhardt himself was safe from apprehension: blaming him for everything that had happened was the best way out of the troubles he himself had gotten her into.

The New York Times reported what sounded like a somewhat guarded judicial attempt to destroy Captain Ehrhardt's reputation by trying Princess Margrethe for perjury and for "abetting the flight of Captain Ehrhardt."

"A court in Leipzig, Germany, today convicted Princess Margrethe of Hohenlohe-Öhringen of perjury and abetting the flight of Captain Ehrhardt. She was sentenced to six months in prison. The sentence[d] was reduced to the irreducible minimum in view of extenuating circumstances, chiefly that the Princess had been victimized and tricked into perjury by Captain Ehrhardt.

"The Imperial Court, from a political point of view, 'practically convicted the absent Captain Ehrhardt of treason.' He was referred to as 'no gentleman,' because he 'induced the Princess to commit perjury to try and save him.' He also escaped from the Leipzig jail, leaving the Princess to stand trial alone.

"Before the judge imposed the sentence, he asked Princess Margrethe why 'she hesitated to take back her perjured statement and had acted with such flapperish defiance.'

"'I was merely silly,' the Princess responded.

"The Chief Justice asked her: 'You were, then, victimized and misled?'

"'Yes, I see that now.' This was the Princess's final comment before she was sentenced to six months in prison.

"The Judge referred to Ehrhardt as a leader of the Kapp Putsch. He added that there would be no amnesty. 'The Princess must be convicted of perjury and giving a traitor aid and comfort. I realize Ehrhardt's splendid qualities of courage, bravery, and military discipline, but his bright escutcheon received a blot with his perjury. But he did not stop at that. The worst was his behavior to the Princess. Through his powers of suggestion he made her commit the most wretched swindle, then during her preliminary interrogation, instead of warning her against it, he directly drove her into committing perjury. He capped his infamy by fleeing jail ten days before the trial started, leaving the Princess in the lurch. Perhaps his followers and admirers, too, will now recognize Ehrhardt's true character."

Chapter Twenty-Five
Return Engagement

When Captain Truman Smith had interviewed Adolf Hitler a week before Hermann Ehrhardt's arrest, he had no doubt as to who was in charge in Munich and in Bavaria: "The *Regiments Verein* have reserve, *Ersatz* (replacement) and *Landwehr* (reservist) battalions," Captain Smith wrote in late November of 1922: "In all, they number 600,000 men.... Ludendorff is believed to be the head of the organization. The actual head... is Ehrhardt..."

Ehrhardt's slow-motion arrest and his mad act of chivalry in pretending to be an Alsatian in the courtroom to shield Princess Margrethe shocked the Rightists in Munich. Some of Ehrhardt's loyalists spent the next six months helping him break out of prison in Leipzig. But the Germans got a worse shock on January 11, 1923, about six weeks after Ehrhardt was imprisoned: citing the German failure to deliver 135,000 telegraph poles and a quantity of coal, the French and the Belgians occupied the Ruhr industrial and coal-mining region. Shocked by this seismic stupidity just as things seemed to be calming down, the German National Assembly encouraged the workers in the Ruhr, recently in armed revolt against the National Assembly, to stage yet another general strike, this one against a recent former enemy. The remainder of the *Reichswehr* and the *Freikorps* groups clustering in Munich reviled the German government for its cowardice. Many of the former soldiers argued for an immediate war with France while Germany still had its artillery and large numbers of trained soldiers. The occupation of the Ruhr led to a consolidation of every armed faction in Germany: the *Reichswehr*, the *Freikorps* groups, the Nazi movement, local militia, and even the Communists, now committed to winning elections rather than insurrections. Walter Krivitsky, a Soviet agent born as Samuel Ginsberg in Austria-Hungary, boasted that he headed a team of seven Soviet gunmen who specialized in assassinating policemen. German patriotic saboteurs, Right and Left, blew up railroad tracks, sabotaged factories and urged workers to stay off the job. Railroad workers who took part in the general strike were banished by the French Army. Railroad accidents became frequent and the train schedules, once meticulous, were no longer reliable. Transportation and production faltered. Political hatred of France became endemic. The Germans actually stopped fighting among themselves as long as the French and Belgians were in the the Ruhr. The Ruhr resistance gave the National Socialist German Workers' Party their first martyr: Albert Leo Schlageter. A decorated and injured war veteran, a *Baltikumer*, veteran of Upper Silesia and a member of ultra-conservative Catholic groups, Schlageter blew up railroad tracks in the Ruhr until he was denounced to the French and shot by a firing squad on May 26. His letters to his mother and sister and his calm composure in facing his own death turned Schlageter into a popular hero all over Germany. Some sources claim the French tortured him after his capture. The fact that Schlageter had signed a Nazi Party card after hearing Hitler speak gave the Nazi propaganda machine the first claim on his memory. Friends said that Schlageter had lived a clean life and once hoped to become a Catholic priest: he was not really Nazi material. Schlageter became so popular after his death that Karl Radek, the leading German Communist, wrote an essay called "Leo Schlageter: The Wanderer into the Void." Radek wallowed in his own guilt for not being able to win Schlageter over to the Communist cause, which is where Radek felt Schlageter, a fighter for justice and against oppression, really belonged. The Nazi movement proved more effective in claiming Schlageter's posthumous glory. A man named Walther Kadow was cited as the informer who handed Schlageter over to the French Army. Two Nazi Party members who understood the values of their movement paid Kadow a visit and murdered him. One of the murderers was Rudolf Höss, a decorated war veteran from the Turkish front and future commandant of Auschwitz. The other murderer was Martin Bormann, later Hitler's personal secretary. Bad money had clearly driven out good.

The Ruhr Occupation while Hermann Ehrhardt was locked up in prison in Leipzig was, in a sense, the making of Adolf Hitler and the Nazi Party. Those of Ehrhardt's officers from the 2nd Marine Brigade who had gravitated to Organization Consult were skilled in conventional small-unit tactics and also at assassination and sabotage. All they lacked was a leader. Truman Smith had predicted who the leader might be. He was not alone in his vision. Eric Warburg, a young Jewish veteran who had fought the Bolsheviks in Berlin, wrote of hearing Hitler speak in a Munich beer hall at the beginning of the Ruhr Occupation in January of 1923. He told his brother that the man was obviously insane but that if the Allies didn't soften the terms of the Treaty of Versailles, this beer hall maniac or somebody like him would someday rule Germany.

Portrait of Corporal Adolf Hitler (far right) during his stay in a military hospital, 1918.

Truman Smith had inadvertently seeded his own prediction when he asked Ernst Hanfstaengl, a Munich intellectual, German-American art dealer and Harvard graduate, to keep on eye on Hitler because he himself was due back in Berlin. Ernst Hanfstaengl and Truman Smith went to the Munich railroad station "where we met a singularly ill-favored individual who was waiting on the platform, a sallow, untidy fellow who looked half-Jewish in an unpleasant sort of way…" Hanfstaengl wrote "'This is Herr Rosenberg. He's Hitler's press chief and gave me the ticket for the evening,' (Truman Smith said) I was far from impressed."

Once Hanfstaengl heard Hitler speak—sitting about eight feet away—he was more than impressed: he was hypnotized. So was everybody else Hanfstaengl saw in the audience.

"The audience responded with a final outburst of frenzied cheering, hand-clapping and table-pounding…. I had really been impressed beyond measure by Hitler."

Hanfstaengl, who was six-foot-four and was always impeccably dressed, approached Hitler, who was about five-foot-ten, sweating profusely after his bravura performance.

"Herr Hitler, my name is Hanfstaengl…. Captain Truman-Smith asked me to give you his best wishes.

"Ah, yes, the big American."

"I agree with 95 percent of what you said and would very much like to talk to you about the rest some time."

"Why yes, of course…. I'm sure we will not have to quarrel about the odd five percent."

Hanfstaengl later remembered some advice he had received from a couple of his Jewish friends,

"That is a dirty business your monarchist friends have organized," the Austrian writer Rudolf Kommer told him after a spate of swastika paintings on Jewish sites. "This race romanticism of theirs will get them nowhere. There is only one danger. If any political party emerges with an anti-Semitic program directed by Jewish or half-Jewish fanatics we shall have to watch out. They would be the only people who could put it over."

Hitler was no monarchist—he had been an anti-monarchist long before he became an anti-Semite. Hitler dismissed royals as a worthless anachronism. But Hanfstaengl noticed, as he became a regular at party functions, that a startling number of Hitler's inner circle appeared to be anti-Semites of mixed Jewish ancestry. Alfred Rosenberg was given the scornful nickname "the only one that isn't" by envious rivals for Hitler's attention. Hanfstaengl, a trained art connoisseur, said that Rosenberg, one of the harshest anti-Semites in the Nazi Party, looked unmistakably Jewish to him and had Jewish advisors. Julius Streicher, whose pornographic fantasies of rich oily Jews stealing blonde women from hard-working blond men, hppened to have relatives named "Weiss" and "Roth." Streicher, a lapsed Catholic, supposedly understood Hebrew and kept track of Jewish liturgical holidays, but his anti-Semitism was so obsessive that even regular Party members thought he must be a bit crazy.

"I suspected the Aryan background of many of the others. [*Gregor*] Strasser and Streicher looked Jewish to me as well as later arrivals like Ley, Frank, and even Goebbels [*who*] would have had difficulty in proving their pedigree."

Hanfstaengl also puzzled over Hitler's sex life: there wasn't any. Hitler made such a fuss over Hanfstaengl's beautiful blonde wife Helene Niemeyer, born in the United States, that Hanfstaengl began to take protective measures. But Helene frankly reassured Hanfstaengl that he had nothing to worry about: "Putzi, I tell you, he is a neuter."

Helene Niemeyer was, in fact, right on target. The Mend Protocol, later taken down from the dictation of Hans Mend, one of Hitler's fellow front-line messengers, provided particulars of *Gefreite* Hitler of the Bavarian List Regiment as an active homosexual pervert during his years at the Western Front and for some months afterwards. Hitler had deliberately neutered himself out in 1919 for the same reason he stopped playing up to kindly or prosperous Jews so he could use them. He became both a nominal straight bachelor and a vicious anti-Semite due to political opportunism.

Hans Mend, one of Hitler's buddies from the Bavarian List Regiment during the war who fell afoul of him later was a petty criminal, sometime jockey and circus rider, and unwed father. Mend had lived on the lower fringes of German society until his ghost-written book "*Adolf Hitler im Felde 1914 – 1918*," published in 1930 during Hitler's comeback bid, helped stifle rumors that Hitler was slightly irregular if not crazy and that Hitler's war heroics were somewhat imaginary. The contemporary German historian Lothar Machtan, who wrote extensively about Hans Mend in "The Hidden Hitler," traces "The Mend Protocol" to Mend's feeling that Hitler had slighted him and perhaps to Mend's dismay that another war had broken out that Mend knew Germany could not win. Mend was not a great intellectual but he knew he was signing his own death warrant when he was interviewed by Friedrich Alfred Schmid Noerr, a German anti-Nazi, in December of 1939, after the defeat of Poland but before the defeat of Britain, France and Belgium drove Germany into euphoria.

"...Hitler never had anything to do with guns from the time he joined us at the front as a regimental orderly. He was never anything other than a runner based behind the lines at regimental headquarters. Every two or three days he would have to deliver a message; the rest of the time he spent 'in back,' painting, talking politics, and having altercations. He was very soon nicknamed 'crazy Adolf' by all the men he came in contact with. He struck me as a psychopath from the start. He often flew into a range when contradicted, throwing himself on the ground and frothing at the mouth. Private Ernst Schmidt (now a master builder at Garching, near Munich) with whom Hitler had been friendly early on, because he sometimes worked on building sites with him, was his special pal. The others he was friendliest with were Privates Tiefenbock (now the owner of a coal merchant's in Munich) and Wimmer (now working as a Munich street car employee). All three were runners at regimental headquarters. The only one who volunteered for combat duty was the Jew [*Karl*] Lippert (a commercial traveler by profession; he later became a clerk at the *Braunes Haus* Nazi Party headquarters), where he worked from 1934 on—and still does, so far as I know, not being subject to the Jewish laws). The List Regiment's battalion adjutant was Lieutenant Gutmann, a Jewish typewriter manufacturer from Nuremberg (now emigrated) whom Hitler made up to whenever he wanted preferential treatment of some kind. It was also Lieutenant Gutmann who got him the Iron Cross 2nd Class at Christmas 1914. That was a Bezaillere... near Ypres. Colonel Engelhardt of the List Regiment was wounded in this engagement. When he was carried to the rear, Hitler and Bachmann tended him behind the lines. Hitler contrived to make a big fuss about this exploit of his, so he managed to gain Lieutenant Gutmann's backing in the aforesaid matter.

"Meanwhile, we had gotten to know Hitler better. We noticed that he never looked at a woman. We suspected him of homosexuality right away, because he was known to be abnormal in any case. He was extremely eccentric and displayed womanish characteristics which tended in that direction. He never had a firm objective, nor any kind of firm beliefs. In 1915 we were billeted in the Le Febre brewery at Fournès. We slept in the hay. Hitler was bedded down with 'Schmidt', his male whore. We heard a rustling in the day. Then someone switched on his electric flashlight and growled 'Take a look at those two nancy boys.' I myself took no further interest in the matter."

"Hitler could never forbear to deliver inflammatory political speeches to his comrades. He always described himself as a representative of the 'class-conscious proletariat.' Whenever he thought he was safe, he referred to his superiors as an 'arrogant bunch of officers' and called them 'robber knights,' 'highwaymen of the nobility,' or 'a clique of bourgeois exploiters.' His oft repeated tirades included remarks like the following: 'Those swine live on horsehair mattresses, while we eat horseflesh soup.'

"I met Adolf Hitler again at the end of 1918. I bumped into him on the Marienplatz in Munich, where he was standing with his friend 'Schmidt.' He greeted me as follows: "Well, Ghost Rider, where did you spring from? Thank God the kings have topped off their perch. Now, we proletarians also have a say. Hitler was then living in a hostel for the homeless at 20 Lothstraße, Munich. Soon afterward, having camped at my apartment for several days, he took refuge at Traunstein Barracks because he was hungry. He managed to get by, as he often did in the future, with the help of his Iron Cross 1st Class and his gift of the gab. He laid less stress on the fact that in 1915, when the List Regiment was terribly mauled, he had been promoted to lance corporal like every last one of the other survivors. It was striking, after all, that a man who had served throughout the World War from

October 1914 to the very end should not have received any further promotions. In January 1919 I again ran into Hitler at the newsstand on Marienplatz. I couldn't help feeling ashamed for 'Red Hitler,' he looked so down at the heels.... Then, one evening, while I was sitting in the Rathaus Café with a girl, 'Adi' and his friend Ernst Schmidt came in. 'Hello, Ghost Rider,' Hitler said to me, "do you know some lodgings for the two of us?' I offered to put him up for the night out of charity. Afterward my girl friend me 'If you're friendly with people like that, I'm not going out with you anymore.' ....

"Then one day in January 1920 Hitler came to my apartment on Schleiß- heimer Straße and complained that he couldn't go home. When I asked why, he didn't answer. I didn't care in any case. 'Alright,' I told him, 'you can sleep here....' He stayed in my place for a day or two.... But Hitler couldn't make out in Munich. He went to see Jakob Weiss at abends in the Holledau... who took him to his par- ents' house and fed him. It was this erratic roaming that finally brought Hitler into contact with General Epp.

"My impression of Adolf Hitler in those early postwar days in Munich thoroughly conformed to my countless experiences with him in the field. Hitler struck me as a book with a thousand pages. He had always been two-faced. He was hypocrisy personified. One of his faces was that of the self-important busybody he impersonated to his superiors, and, if need be, to his comrades. When Hitler was off duty behind the lines or at headquarters and he heard that some success had been gained at the front, it was quite usual for him to burst in on the other men waving his arms and shouting. 'We've won! We've given the French (or British) another bloody nose!' But with his superiors he was always playing the ingratiating telltale as soon as he saw it might benefit him in some way. That's why his comrades were wary of him. ...Hitler's other face was that of a secret, sinister criminal. His whole attitude was that of a ruthless person who knows how to wrap himself in a halo. He had always, ever since I've known him, been... a great actor. Not a word he uttered could be trusted. He lied whenever he opened his mouth, always did the op- posite of what he said....

"When Hitler returned to Munich in the winter of 1918 he made persistent efforts to obtain a senior position with the Communists, but he couldn't get into the Munich directorate of the Communist Party though he posed as an ultra-radical. Since he promptly requested a senior party post that would have exempted him from the need to work—his perpetual aim—the Communists distrusted him despite his mortal hate of all property owners. They stalled him, and he may have thought they were spying on him from a certain stage onward. At all events, he took his re- venge by joining the *Freikorps* Epp and gained Epp's confidence because of his Iron Cross 1st Class. Epp made it Hitler's first job to boost the troops' morale and paid him for it. He was soon able to call himself 'an officer instructor.' In that capacity he visited all kinds of hostelries at night and came across Anton Drexler.... Hitler thereupon joined Anton Drexler's party and was assigned Party Membership No.

5012. But he promptly set about splitting the party by accusing Drexler's secretary, a man named Harrer, of complete incompetence and thrusting him aside. Drexler, who hated disputes of any kind, gave way to Hitler out of weakness. Hitler immediately made use of the burglar's tactic he later employed with such success, which was sticking his foot in the door and refusing to yield until he was on the inside. That was how he was able to smash Drexler's party. And then he opened his own shop with seven men."

Mend may have had some help pasting this account together in terms of spelling and punctuation. Once it was published, he was arrested for sexual child abuse—Mend's prior police record involved only mature women. Mend reportedly died in prison of tertiary syphilis, a disease known to induce delusions. But Mend's description of Hitler's apparent homosexuality is substantiated by Hanfstaengl, who met Hitler about the time that Mend's friendship was breaking off.

"I felt Hitler was a man who was neither fish, flesh, nor fowl, neither fully homosexual nor fully heterosexual," Hanfstaengl wrote. "...He had these people of unsavory habits around him, from [Ernst] Röhm and [Edmund] Heines on the one side to Rosenberg on the other, and seemed to have no sense of moral displeasure at their behavior. [Karl] Ernst, another homosexual S.A. leader, hinted in the thirties that it would only need a few words from Hitler to silence him when, for political reasons, he started to complain about Röhm's behavior....

"...From the time I knew him, I do not suppose he had orthodox sexual relations with any woman. He was probably incapable of a normal reaction to their physical proximity."

Hanfstaengl's near-swooning admiration of Hitler's oratory sounds as if he too were a candidate for membership in Hitler's gay entourage: Oddly, the reverse is true. Hanfstaengl's son Egon described his father Ernst Hanfstaengl as having "polygamous impulses" which made him "eminently unsuitable" for monogamous marriage. Putzi's marriage to Helene Niemeyer appears to have broken up due to his cheating. Two of the madcap Mitford sisters, Diana and Unity, roaming Germany to get away from "chinless wonders" in upper-class England, summed Hanfstaengl up even more bluntly, as a groper: "Putzi gives handies." Yet this educated, artistic, snobbish, grotesquely elegant man, perhaps too normal for his own good, sponsored Hitler into polite society, advised him on his wardrobe, paid some of the early Party's debts out of his own pocket, and gave the Nazi Party a $1,000 interest-free loan to buy two used American-made rotary printing presses so that the Nazi Party newspaper could become a daily rather than a four-page weekly. Hanfstaengl's social connections introduced Hitler to women twice his age, who delighted in mothering him, feeding him, and giving him large amounts of money—and who sometimes, to Hitler's quiet terror, trying to arrange marriages for him. Hanfstaengl, who frequently spoke to Hitler as a close friend and entertained him at his own apartment, learned the secret of Hitler's oratory. Hitler understood that the man on the street was none too bright and liked simple ideas pounded into his head with a lot of drama—and Hitler tried to pick simplistic concepts that his audience would agree with and repeated them endlessly: he had a knack for making dumb people feel smart. Hanfstaengl and his smart set wanted a constitutional monarchy or a liberal republic. Hitler wanted a dictatorship with himself as dictator. He appealed to violent people willing for fight for a piece of the action and a chance to beat up people they didn't like—Jews, intellectuals, and educated snobs with inherited money. Not all Germans felt that way, but enough violent people—part of Ernest Pöhner's 5 percent—were often able to give Hitler frequent control of the street, even against the Communists who recruited from socially marginal people filled with resentment the same way Hitler himself did.

Hanfstaengl took too long to be afraid of Hitler, who flattered him and seemed to respect his artistic background and rich friends. One man Hanfstaengl was instinctively afraid of was "...a slightly mysterious man named Lieutenant Klintzsch, who was one of the stormtrooper leaders and had been and probably still was one a member of Organization Consul, which had been associated with Captain Ehrhardt in the abortive Kapp Putsch in Berlin in 1920 and had a hand in the murders of Erzberger and Rathenau."

Hans Ulrich Klintzsch, born in the Spreewald section of Berlin in 1898, had been a *Leutnant zur See* during the war and had joined the 2nd Marine Brigade during the first recruiting in Wilhelmshaven. Klintzsch was a hard-core Ehrhardt man and had been arrested after the assassination of Matthias Erzberger, but had been released after some months for lack of evidence. His official job, after the Ehrhardt Brigade had disbanded, was athletic director and training officer for the Munich Nazi *Sturmabteilung*—but he was essentially an Ehrhardt 'plant' keeping an eye on the Nazi movement and hoping for the return of "the Chief."

The Chief was returning at his own speed. Hermann Ehrhardt had been alarmed at his own poor physical condition after all those months in prison and was gradually recovering his health with mountain air and hiking. Fortunately from him, the *Steckbrief* [arrest warrant] run off the presses the day after he broke out of prison and posted all over Germany used a dark photograph taken during the Kapp Putsch: Ehrhardt on the poster looked like a Japanese samurai with a heavy sun-tan and the visor of his officer's cap obscured his eye color. The reward offered—25 million marks—was a reflection of the break-away Weimar Inflation. By the autumn of 1923, a mug of good beer cost a billion marks.

Steckbrief.

Der Angeklagte Korvettenkapitän a. D. Hermann Ehrhardt, gegen den die Untersuchung wegen Hochverrats, Meineids und Beihilfe zum Meineid verhängt ist, ist am 13. Juli 1923, nachm. 5 Uhr, aus dem Untersuchungsgefängnis in Leipzig entwichen.

Auf Grund des gerichtlich gegen ihn erlassenen Steckbriefs ersuche ich, den Flüchtigen zu verhaften, in das nächste Gefängnis abzuliefern und zu den hiesigen Akten C 21/20 telegraphische Mitteilung zu machen.

Die Reichsregierung hat eine

Belohnung von 25 Millionen Mark

für denjenigen ausgesetzt, der durch sachdienliche Angaben oder andere geeignete Mitwirkung zur Ergreifung des flüchtigen Angeklagten beiträgt.

Personalbeschreibung des flüchtigen Angeklagten Ehrhardt:

1. Familienname: Ehrhardt.
2. Vorname: Georg Hellmut Hermann.
3. Geburtsort: Diersburg in Baden.
4. Alter: geboren 29. 11. 81.
5. Größe: 1 Meter 70 bis 72 Zentimeter.
6. Haare: dunkelblond, grau meliert, kurz geschnitten, wenig Scheitel, etwas gelichtet.
7. Stirn: hoch.
8. Augenbrauen: grau meliert.
9. Augen: blau, grau.
10. Nase: mittel, gerade.
11. Mund: energisch, mittel.
12. Bart: glatt rasiert, kurz geschnittener Schnurrbart.
13. Zähne: vollständig.
14. Kinn: rund.
15. Gesichtsbildung: oval.
16. Gesichtsfarbe: frisch.
17. Gestalt: aufrechter, militärischer Gang und Haltung.
18. Sprache: hochdeutsch.
19. Bekleidung bei der Flucht: graues Jackett, graue Hose, keine Weste, Oberhemd mit Einsatz, weichen Kragen, braune Halbschuhe, ohne Kopfbedeckung.

Leipzig, den 14. Juli 1923.   1052/19

Der Oberreichsanwalt.

The poster read: *The accused Captain Hermann Ehrhardt, against whom the investigation for high treason, perjury, and suborning perjury is now in progress, slipped out of prison at about 5 o'clock in the afternoon on July 13 in Leipzig.*

In order to end his flight and place him in the nearest prison, I request information by telegraph and the national administration has offered a reward of 25 million marks for anyone who gives factual information which leads to the apprehension of the accused fugitive.

Personal description of the accused fugitive Ehrhardt:
1. Family Name: Ehrhardt
2. Forenames: Georg Hellmut Hermann
3. Birthplace: Diersburg in Baden
4. Age: born 29.11.81
*5. Size: 1 meter 70 to 72 centmeters (*about five-foot-nine to five-foot-ten*)*
6. Hair: dark blond, grey mixture, cut short, slightly thin on top.
7. Brow: high
8. Eyebrows: mixed with grey
9. Eyes: blue, grey
10. Nose: middle length, straight
11. Mouth: energetic, middle size
12. Beard: shaved off smooth, close-trimmed moustache
13. Teeth: full set intact
14. Chin: round
15. Shape of the face: oval
16. Color of the face: fresh
17. Bearing: upright, military walk and posture
18. Language: High German
19. Clothing at the time of the escape; grey jacket, grey pants, no vest, overshirt with a simple, soft collar, brown ankle-length shoes, no head covering.

Leipzig, the 14th of July, 1923 1052/10
The senior government attorney

Chapter Twenty-Six
The Beer Hall Putsch

Hermann Ehrhardt, still wanted by the police, enjoyed a stay with his friend, Prince Carl Eduard at the castle of Callenberg, and a longer stay in Switzerland before her re-surfaced in Munich on September 29, 1923. He found that those members of the Ehrhardt Brigade who hadn't left for full-time jobs or fortunate marriages were still overwhelmingly loyal to him. Dr. Otto Pittinger, temporary leader of the moderate Right, a war-time medical officer, a Bavarian monarchist and a staunch anti-Hitler man, had urged the Ehrhardt Brigade men to sign up with his own *Neudeutsche Bund* (New German League). Hermann Ehrhardt, however, was a *Prussian* monarchist for practical and personal reasons: he owed the Kaiser his commission and he understood that the smaller German states like Bavaria and Hannover could not withstand France or Poland any more than the shifty Weimar politicians could. Ehrhardt was no more interested in supporting the secession of Bavaria than he had been in supporting the secession of Hannover. He shared Dr. Pittinger's suspicious dislike of Hitler but that was their only point of agreement. Ehrhardt's executive officer, *Kapitänleutnant* Eberhardt Kautter, and his contact man inside the SA, Hans Ulrich Klintzsch, had kept in touch with Ehrhardt while he was in prison. The Ehrhardt Brigade had re-formed under Kautter's command as the *Wiking Bund* [Viking League], a sports group that as actually a small private army loyal only to Ehrhardt. The group may have had as many as 10,000 members, included those who resided at home. The *Wiking Bund* was somewhat more loosely affiliated than the 2nd Marine Brigade, which had been a government-supported military unit before the government disbanded it after the Kapp Putsch. The men of the former Ehrhardt Brigade and their affiliates were now wearing an embroidered blazon with a Viking *Drakkar* [Dragon Ship] on one sleeve by the time Ehrhardt himself quietly showed up in Munich. Kautter and Klintzsch had followed Ehrhardt's orders to the letter.

Ehrhardt did some work on his appearance to avoid resembling the *Steckbrief* photograph. He let his beard and mustache grow rather shaggy instead of keeping them neatly trimmed. He adopted an M-16 steel helmet—minus the white swastika—as his parade headgear instead of his officer's cap. He kept his service ribbons but staggered their order. His Iron Cross First Class, worn large, and the ribbon for his Iron Cross Second Class were left in place. The sum total of decorations and medals was impressive: Ehrhardt also had an honor cross for front-line fighters, a naval medal for the combat-wounded, a special buckle issued to staff officers of the 2nd Marine Brigade in 1920 and a couple of house orders from Prussia and—oddly enough—from Saxony.

Hanfstaengl noted that The Ehrhardt loyalists maintained a considerable distance from the Nazi SA. "There were not so many swastikas and the processions were always headed by the German war standard. The S.A. usually marched together with the Viking Bund, who were Ehrhardt's militarized formations. The music was Bavarian but the colors were the the black, white, and red of the Kaiser's Reich."

"The Viking Bund wants to knock out the NSDAP," the former flying ace Hermann Göring warned on July 4, 1923, even before Ehrhardt escaped from prison. "That is not good. The Viking Bund is in no way connected with the SA or the NDSAP. It is true that it was for a time related to the SA, but it has now declared war against the Party and against the SA."

The situation in Munich was moving toward another putsch—but the goals of this one were so enigmatic as to make the slapdash Kapp Putsch of 1920 look like General Maercker's meticulous envelopment of Braunschweig in the spring of 1919. Hanfstaengl said that some Bavarian nationalists wanted to restore the zany but beloved Wittelsbach Dynasty, some wanted Bavaria to become independent, some wanted to install the Wittelsbachs as the new German Imperial family, and some wanted to leave the mostly Protestant Prussians and Hanoverians and the Free Cities of North Germany behind and federate with Catholic Austria. Hitler, in company with Ludendorff, wanted to march on Berlin, overthrow the Weimar Republic, and repudiate the Treaty of Versailles—with himself as the new ruler or political power behind Ludendorff.

Hermann Ehrhardt wanted none of these things. Friedrich Wilhelm Heinz, one of Ehrhardt's inside men, was a decorated war veteran who had volunteered for the Front out of school and finished the war in a hospital bed with both iron crosses and multiple injuries. Heinz had picked up another unofficial combat wound in Silesia and was also a veteran of the Kapp Putsch. Literate, intelligent, a heavy drinker and fiercely patriotic, Heinz told Ehrhardt what was brewing almost as soon as Ehrhardt arrived in late September.

"Would it not be possible to bring Ehrhardt, Buchrucker, Hitler and Kahr together, unite them, and direct them toward a common goal?" Heinz asked.

"Impossible!" Ehrhardt said bluntly. "There can be no reconciliation with Ludendorff! I reject Hitler completely. He failed miserably on the first of May (*during a scuffle*) and he will always fail. You yourself know what we think of Buchrucker.... I alone am the strongest." Within two months of his arrival in Munich after the jailbreak in Leipzig, Ehrhardt had Klintzsch and his other contact man, SA Chief of Staff Alfred Hoffmann, slip out of the Nazi movement.

Ehrhardt's opinions seemed accurate. When Major Ernst Buchrucker attempted to take over an arsenal in what was grandly styled "the Küstrin Putsch" in Prussia he was over-awed by government forces. One soldier was killed when he resisted arrest. Major Buchrucker—a serving officer, war veteran, and Rightist—got 10 years in prison, contrary to Professor Gumbel's take-out on short sentences for the Right.

Official Munich was governed by Gustav Ritter von Kahr, who had succeeded the unlucky Johannes Hoffmann as prime minister-president after the *Freikorps* and the *Reichswehr* had destroyed the Soviet Republic of Bavaria. Kahr was a Protestant and his title of

Friedrich Wilhelm Heinz

knighthood [*Ritter*], based on a political decoration, was not hereditary. Kahr's stronghold was the Bavarian Peoples' Party, political Catholics who corresponded to the national Catholic Center in Prussia and Hannover. General Otto von Lossow had been named the commander of the VII military district—Bavaria. The Bavarian State Police were headed by Colonel Hans Ritter von Seisser. All three of these Bavarian monarchists had been plotting a putsch of their own to demand that the National Assembly organize an army large enough to protect Germany from the French and Belgians or the Poles. The Ludendorff-Hitler faction

wanted to take over Berlin and throw out the "Reds" but they had no desire to establish Bavarian independence or to federate with Austria.

Hanfstaengl, still hypnotized by Hitler's oratory if not his strange personal life and unsavory friends, was asked to round up support of the march first on the *Bürgerbräukeller*, where Kahr, Lossow and Seisser were expected to announce their own putsch, and later on Berlin. Ehrhardt has earlier reacted to requests from Hitler through Ernst Röhm to join with the SA: "*Herrgott*, what does that idiot really want?" Hans Ulrich Klintzsch, a former naval officer, had kept *Korvettenkapitän* Ehrhardt informed about the strange personalities around Hitler. To Ehrhardt, anti-Semites of Jewish ancestry and homosexual he-men were two very bad bets for the sort of emotional stability needed in take-overs and confrontations with the National Assembly.

Ehrhardt had instead assembled his troops near the Thuringian border for a separate putsch at "Red" Saxony, a state which was a liability in desperate times. But General Hans von Seeckt struck first and removed the Red officials with *Reichswehr* troops before Kahr had a chance to send in the officially disbanded Ehrhardt Brigade for a little revenge and some political house-cleaning. At the end of October, Ehrhardt had most of his own private army back at his disposal—but not at Hitler's disposal.

A day or two before the planned putsch of November 8, Ernst Hanfstaengl called Ehrhardt's headquarters on Hitler's behalf to urge him to hurry with preparations: "I had rung up there... [I] got a crossed line and thus was witness to a conversation which made it clear that Ehrhardt's men were disposing of some of the mutual stocks of arms in a distinctly suspicious manner.

"Nor was Ehrhardt the only one. Captain [*Eberhard*] Kautter, another of Göring's *aides,* also switched allegiances and defended Kahr's ministry when the time for decision came. Pöhner, who had been superseded as police president, but still wielded great influence, was another doubtful ally."

Hanfstaengl blamed the whole thing on the way that some of Hitler's faction had offended Catholic sensibilities by dabbling in atheism or the occult. He was on the way to the truth but his awe of Hitler's hypnotic rhetoric kept him from an even more logical conclusion: Hitler was a maniac and his entourage contained a large inventory of degenerates. Captain Ehrhardt, Captain Kautter, and Lieutenant Klinzsch wanted no part of people like Hitler, Röhm, Rosenberg and Streicher ruling Germany. The weapons Hanfstaengl had heard Ehrhardt's men diverting were headed for Upper Franconia. So was most of the Ehrhardt Brigade. Captain Ehrhardt, having taken Hitler's measure, was planning to kill Hitler and destroy the SA if Hitler actually captured Munich. Hitler would have far less local support in the countryside outside Nuremberg than he would in Munich, and Bavarian civilians were less likely to be killed by the artillery manned by Ehrhardt's experienced naval gun crews. Hitler saw battle as a political statement. Ehrhardt knew the real meaning of war: to kill the enemy with as few losses on your own side as possible.

Hitler was in a state of delusion as he blundered ahead with the putsch. He put out the word for his loyal forces to turn out in field-grey windbreakers and field caps, with pistols. Rifles were handed out, and at 8:30 on November 8 Hitler pushed his way into the 3,000-seat *Bürgerbräukeller* with Hermann Göring, a 22-victory fighter ace, and Ulrich Graf, a bulky butcher, backing him up on either side. He was wearing a trench coat and holding a pistol.

"Not even a man equipped with the greatest executive powers can rescue the *Volk* [the German people] without the nationalist spirit and energetic help of the people themselves...." President Gustav Kahr was saying.

Hitler climbed onto a chair and fired his pistol twice into the ceiling. Kahr stopped speaking.

Adolf Hitler, Hermann Ehrhardt, Trebitsch-Lincoln and Julius Streicher in Munich, 1923.

"The national revolution has broken out," Hitler shouted. "This hall is occupied by six hundred heavily armed men, and no one may leave it."

Some of the blue [*civil*] police put on swastika armbands and a police lieutenant helped some SA men in steel helmets lift a heavy machine-gun onto a table. Some of the men in the audience shouted out "*Mexico!*" as a metaphor for political banditry. Some women fainted.

"The Bavarian government and the Reich government are deposed," Hitler lied. "The barracks of the Reichswehr and the State Police have been occupied. Reichswehr and State Police are advancing under the banner of the swastika."

Hitler, who had a flabby build, was wearing a cutaway coat and white starched shirt under his trenchcoat. Some of the aristocrats and bankers in the audience thought he looked like a waiter.

Hitler ordered Kahr, Lossow and Seisser into a back room at gun-point. "I have four bullets in my pistol," Hitler told the three Bavarians. "Three are for you, my colleagues, if you fail me, the last one for myself." Ludendorff showed up, wearing a tweed hunting coat instead of his Imperial uniform. He was grumpy after a high-speed automobile ride through the streets of Munich. Hitler explained to Kahr, Lossow and Seisser that Ludendorff would command the restored German Army from now on and that he himself would take over the role of political leader. Kahr, Seisser and Lossow would have important roles in the new Hitler government.

"Play out the comedy," Kahr whispered to Seisser and Lossow.

While Hitler was negotiating with Kahr, Seisser, and Lossow in the back room, Hermann Göring and Ernst Hanfstaengl were trying to keep the 3,000 people in the main room calm. Hanfstaengl later claimed that he talked the SA men out of shooting a couple of the delegates who seemed unimpressed with Hitler. He undoubtedly used his own money to buy mugs of beer for some of the cabinet ministers who had been locked out of the back-room conference. The Weimar inflation had raised the price of a mug of beer to a billion marks and Hanfstaengl ran out of cash and had to stay thirsty himself. Max Erwin von Scheubner-Richter, a White Russian with a fake German title, pointed out some of the delegates he wanted arrested. The Storm Troopers marched them off at gunpoint.

The Storm Troopers also broke windows, roughed up the staff and smashed furniture at the offices of the *Münchener Post*, the Socialist newspaper they denounced as "that poison kitchen." The *Münchener Post* was, in fact, a conduit for inflated death tolls by Rightist murderers and atrocities that could never be documented, such as the murder of the 10 or 11 Red nurses in 1919 just before the destruction of the Soviet Republic of Bavaria—the fictional murders which had sparked the factual Red murders at the Luitpold Gymnasium. The *Münchener Post* also printed the bogus charter of Organization Consul which declared war on 'Judah' some months after *Die Rote Fahne* listed Herman Ehrhardt as a rescuer of Jews and suspect in the World Jewish Conspiracy.

With less justification, some of the Storm Troopers rounded up a number of Jews as hostages. Those Jews who were anarchists or Bolsheviks had been killed or driven into exile in 1919, so the Nazi SA thugs leafed through telephone books and arrested people with Jewish-sounding names and no political profiles. Kahr, Seisser and Lossow all pledged their support. As soon as they were released, they took off for the barracks of the 19th Infantry, where the *Reichswehr* troops had warded off a take-over and were now in a state of siege. Hitler released the Jewish hostages: "We want no martyrs." Trebitsch-Lincoln was also backing Hitler in Munich in November 1923—dressed up to resemble Julius Streicher—and may have used his influence.

The top success of the putsch had come when the Nazi Captain Ernst Röhm, a wounded combat veteran and an avowed homosexual invert, led the 400 men of his own private army in the capture of the *Reichswehr* military headquarters building in Munich. Gerhard Rossbach, another purported homosexual and a *Balti-kumer*, was able to convince one *Reichswehr* unit to go over to the putschists. This triggered storm trooper warnings in Berlin.

General Hans von Seeckt, "The Sphinx with the Monocle" and commander of the *Reichswehr*, had expressed considerable respect for Hermann Ehrhardt in 1920. He expressed no respect whatsoever for Adolf Hitler in 1923. Seeckt ordered General Lossow to put down the putsch or step aside. The National Assembly declared the Nazi Party illegal. Prince Rupprecht of Bavaria indignantly refused to have anything to do with Hitler. Endorsed by Michael Cardinal Faulhaber, the Archbishop of Munich, Prince Rupprecht sent a message: "Crush this movement at any cost. Use troops if necessary."

A message sent out on all German radio stations: "State Commissioner General v. Kahr, Col. v. Seisser and Gen. v. Lossow repudiate the Hitler putsch. Expressions of support extracted by gunpoint [*are*] invalid. Caution is used in misuse of the above names.

v. Lossow"

In Berlin, President Friedrich Ebert asked General Hans von Seeckt: "Tell us, please, whom does the *Reichswehr* obey? Does it obey the government or the mutineers?"

"Herr Reich President, the Army obeys *me,*" Seeckt said.

Seeckt sent out orders that anyone who supported the Munich putsch would be guilty of high treason. He bluntly told Lossow: "All necessary measures for the crushing of this attempt and the restoration of order have been taken...."

The Putschists flew into rage or panic. No one knew what to do now that the local Bavarian monarchists and the *Reichswehr* commander, who enjoyed far more respect than the National Assembly, had both declared the putsch an act of treason.

The only man who seemed to be capable of making a decision was Erich von Ludendorff: "We march!"

Hitler, with his usual slavish respect for authority, promptly agreed with Ludendorff. Scheubner-Richter, with his mystical White Russian fatalism, told Hitler that this would be their last walk together. Hitler then tried to talk Ludendorff out of it. Ludendorff said bluntly: "We march."

The commanders rushed to present an impressive front by clustering men with similar field jackets and caps in the same groups. The lead elements wore steel helmets and carried K-98 Mauser rifles, some on slings and some with fixed bayonets. The band struck up the Badenweiler March, Hitler's favorite. About 2,000 putschists marched through the center of the city to relieve Captain Ernst Röhm, still holding the military headquarters building behind barbed wire, with machine-guns jutting from the windows. At the Ludwig Bridge, the Nazi lead element pushed through the thin contingent of "green [*national*] police" after the police failed to fire. The Storm Troopers beat some of them up. Orators ran ahead shouting Nazi slogans. The marchers ripped down the black-red-gold national flag at the Rathaus and pushed through another thin cordon of leery "green" police.

This photo would have been taken during the plans to kill Hitler if the Beer Hall Putsch got outside Munich

At the Feldherrnhalle, the Hall of Field Marshals, a platoon of "green police" appeared at a run and formed a line that covered the road. The military-style police unit was armed with rifles and pistols as well as rubber truncheons. Their commander, *Oberleutnant* Michael *Freiherr* von Godin, was a baron, a war veteran and a second-generation professional soldier from a devout family of Bavarian Catholic patriots.

"No shooting! General Ludendorff is coming!" Ulrich Graf shouted at the sight of the steady police cordon. Hitler, trembling with a bad case of nerves, was arm-in-arm with Scheubner-Richter, who had predicted his own death a few hours before. Ludendorff walked bolt upright.

The Nazi line crashed into the police cordon and blows with rifle stocks and thrusts with bayonets were exchanged. *Oberleutnant* Godin saw the policeman next to him take a bullet. The "green police" fired a ragged volley into the Nazi formation. Scheubner-Richter was struck dead and fell backward, dragging Hitler down with him. Some of the Nazi troops panicked and ran away. Others fired back. While Hitler was running frantically in the opposite direction, Erich von Ludendorff marched straight ahead and through the police cordon. A lone Storm Trooper named Hans Streck followed him. Hermann Göring stood up and shot it out until he was bit in the groin. Two Jewish sisters named Ballin pulled Hermann Göring off the street into their house and called a doctor. Everyone else who wasn't dead or serious wounded ran for it.

Ludendorff walked up to a "green" policeman and stood stiffly in front of him without visible fear.

"Excellency, I must take you into protective custody," the "green" policeman said.

"You have your orders. I'll follow you." Ludendorff said.

Ludendorff was an eccentric crank and an occultist anti-Semite, but John H. Wheeler-Bennett, the Oxford historian of the German military called Ludendorff's march "the one almost redeeming feature of an otherwise sordid and disreputable affair. It was the last gesture of the Old Imperial Army."

Hermann Ehrhardt had temporarily won by default. Hitler, he believed, was finished as a threat to Germany. Ehrhardt was a conscious Christian and not an atheist or an occultist anti-Semite of the Thule Society variety like Ludendorff. But Ludendorff and Ehrhardt both understood the same rigid code of military honor: no self-respecting German officer would ever support a coward like Hitler after he ran out on his own 16 dead and more than 100 wounded in a battle he had provoked by his own blundering and then lost without a serious fight. Or so it seemed at the end of 1923.

Chapter Twenty-Seven
Losing by Winning

Hermann Ehrhardt's relief after the Beer Hall Putsch seemed to have destroyed Hitler was shared by a great many people, but it was premature. To Ehrhardt, the fact that 2,000 Nazi storm troopers lost a street gun-fight with a few dozen "green" policemen indicated that the storm troopers were as degenerate as Hitler himself. Ehrhardt's *Wiking Bund* included hundreds of veterans of the 2nd Marine Brigade, the unit that had helped take Munich back from thousands of Communists and dissident Bavarian renegades armed with howitzers and machine guns in 1919, for the lost of a handful of men. Ehrhardt and Waldemar Pabst had failed to hold Berlin in the Kapp Putsch of 1920 only because the generals who originally urged them on had betrayed them with apathy or outright cowardice. The absurd upshot of Kapp's *matzoh* flour confiscation had cost Ehrhardt some journalistic and financial support. But Ehrhardt could have smashed Hitler—and undoubtedly would have killed him—if Hitler had ventured into Upper Franconia after the march through Munich in November of 1923.

But Hitler had a secret weapon—his hypnotic powers of oratory and the fact that the German people understandably felt threatened by the French and Belgians in the Ruhr, Polish disruptions in Upper Silesia and eastern Prussia, and above all by Russian Communism. Lower-class anti-Semitism focusing on economic envy or the visible number of Russian Jews among the Bolshevik leaders was a secondary issue. So was the absurd Weimar inflation. The danger of a French-Belgian-Polish or Soviet Russian take-over was a palpable threat all over Germany.

The trial of the Beer Hall Putsch leaders began on February 26 of 1924 as Hitler, Ludendorff, and eight other defendants were hailed into the Munich People's Court at the former Infantry School on Blutenburgstraße. County Court Director Georg Neithardt presided over the tribunal with a legally-trained assistant judge and three lay judges who were ordinary citizens: Christian Zimmermann, an insurance inspector, Phillipp Hermann, an insurance clerk, and Leonhard Beck, a merchant. The alternative lay judge was a tobacconist, Max Brauneis. Ludendorff had walked through the armed police like a bullet-proof automaton but he made a fool of himself in court rambling on about "international Jewry" and also against the Catholic Church, which was not a clever tactic in Munich. But Ludendorff had intentionally or inadvertently convinced the court that he had mental health problems. The madly eccentric former hero was set free.

Hitler had characteristically run away and hidden himself behind a woman's skirts after the putsch failed. The woman was Helene Niemeyer Hanfstaengl. When Hitler theatrically tried to shoot himself in her presence, Helene wrested his revolver away from him and tossed it onto a 200-pound barrel of wheat flour that the Hanfstaengls had set aside due to economic uncertainty. Hitler couldn't grope for the revolver without covering his tuxedo with flour. Eventually the police appeared and Hitler gave himself up.

But once he had an audience, the cringing fugitive from the battle of the *Feldherrnhalle* managed to turn himself into a defiant hero. Hitler accepted the credit—rather than the blame—for the putsch where 20 Germans had killed one another and said that his movement would continue to fight for the restoration of German military power against France: "...the scattered bands will become battalions, the battalions, regiments, the regiments, divisions... the old cockade will be rescued from the filth; and the old banners will be unfurled again."

Hitler and three other defendants—former Munich Police Chief Ernest Pöhner and two military leaders—were sentenced to five years of fortress arrest, the most respectable of prison sentences for political dissidents. They would be eligible for parole in six months. The other four were all military leaders, including Ludendorff's stepson, Lieutenant Heinz Pernet, and Captain Ernst Röhm. They each got a year and a half. Professor Gumbel's charges of extreme leniency for Rightists actually found some instances of confirmation at the Hitler trial. But the Nazi Party was officially abolished and Hitler's ultimate release after nine months was wisely stipulated on a gag rule that forebade him to speak in public.

The ultimate victim of the verdict was Hermann Ehrhardt. Hitler had made himself a hero to the extreme right, and to some extent of the Bavarian people in general, by turning what was essentially a failed power play into the defense of the German nation against the predatory French, the vengeful Belgians, and the revolutionary Russian Jewish Communists. Ehrhardt, though he never acted against law-abiding Jews, had a far better claim to these honors than Hitler did—but Hitler had the center of the stage. Ehrhardt, by refusing to back Hitler in the ludicrous Beer Hall Putsch, was characterized as a traitor to the extreme right—and Ehrhardt was still a fugitive from justice who could expect local protection by Rightists in Munich but nowhere else in Germany. His own men and the Prussian monarchist faction of the Right still supported him, but the Bavarian monarchists and the defeated but numerous Nazis saw him as The Enemy. Hermann Ehrhardt was now fair game for any informer who wanted to collect an inflation-adjusted reward without being murdered by his own right-wing friends.

The verdict on Hitler came down on March 27, 1923. Ehrhardt was out of Munich by April and headed for Austria where Waldemar Pabst, who shared his low opinion of Hitler, was managing the *Heimwehr*, a national defense militia to defend isolated Austria from either the left-leaning Weimar Germans or Mussolini's Italian fascists. Pabst held Ehrhardt in considerable respect after the Kapp Putsch and was only to glad to quietly help him out.

The French occupation of the Ruhr gradually dissipated after the American Charles Dawes, a Midwestern banker and former short-term war-time general, reported that the Ruhr occupation and the defiant German strikes and disruptions made it impossible to expect reparations. The Dawes Plan was far more lenient than most Germans expected. But the Reichstag had to dissolve because the delegates split over whether to accept it or not. The moderate parties took a political beating from the Communists in the May 4 Reichstag Elections: the Independent Social Democrats finally gave the deed a name and voted as Communists, which gave the Reds 62 seats. The Nazi Party candidates took 32 seats even though the Nazi Party was illegal. Voters knew which candidates were Nazis under other names. The Social Democrats lost 71 seats and the moderate liberal Democrats lost 11.

In August, the effect of the Dawes Plan was already leading to an economic resurgence in Germany. When the third Reichstag election was convoked on December 7, 1924, the extremist parties took a beating: the Nazis lost 17 Reichstag seats, and now had only 14. The Communists lost 18 seats and now had 43. The Social Democrats gained 13 seats for 131 seats in the Reichstag, while the German Nationalists—the conservative Christian monarchists with no use for Hitler—took 103 seats and became the second largest party. The moderate Catholic Center Party came in third with 69 seats and the Catholic Bavarian Peoples Party had 51 seats. The moderates now outnumbered the conservatives and the liberals—the two Catholic parties, taken together, were within 11 seats of the once-overwhelming Social Democrats—but even the conservatives outnumbered the extremists of both the Left and the Right. The results of the election were so encouraging to moderates and liberals that the Weimar National Assembly foolishly let Adolf Hitler out of prison more than four years ahead of schedule and eventually granted him a full pardon so that he could speak in public again. Like Hermann Ehrhardt a year before, the Weimar government assumed that Adolf Hitler was a loud-mouthed buffoon who was washed up in politics.

On February 28, 1925, President Friedrich Ebert died in office. Ebert's own Social Democratic party's left wing had turned against him when he called out troops during the attempted Communist take-overs in Berlin. Hermann Ehrhardt and Waldemar Pabst had chased President Ebert out of Berlin during the Kapp Putsch, though Ebert returned to office with quiet satisfaction when the Kapp Putsch collapsed. Ebert had weathered the Hitler Beer Hall Putsch with far less trouble. But Ebert had been engulfed in a number of subsequent scandals: the one that probably killed him came when he was charged with favoritism and accepting bribes from of Julius Barmat, a Russian Jewish food contractor who had reportedly obtained unfair advantages due to friends in Ebert's Social Democratic Party. Nazis and Communists alike defamed him—the Nazis as a tool of their favorite villain, the Jewish World Conspiracy and the Communists as a tool of their own favorite villain, the International Capitalist Conspiracy. Ebert, a liberal patriot who had lost two sons in the war, insisted on testifying in his own defense and neglected his internal health problems until it was too late. He died at the age of 54 after his appendix ruptured.

Hugo Preuß died later that year on October 9—of natural causes unrelated to Operation Consul. The half-Jewish Nazi Alfred Rosenburg slandered Preuß as the Jewish influence behind the Republic, but the Christian monarchists like Hermann Ehrhardt knew that while Preuß was a liberal, he had never betrayed the Kaiser or conspired with Bolsheviks—and he was never a target for OC. Two of his sons later fled Germany when Hitler took over some years later and one son, a combat veteran of the trenches, committed suicide.

President Friedrich Ebert's death meant holding the Weimar Republic's first election for president. Friedrich Ebert himself had been handed the power by the National Assembly when the Kaiser abdicated and Ebert bumped Philipp Scheidemann. The run-off election featured the ludicrous Ludendorff, not Hitler, as the National Socialist candidate. Ludendorff's final moment of bravery at the Beer Hall Putsch had been eclipsed by his paranoid performance at the trial that followed. Ludendorff received only 200,000 votes. The top vote-getter in the first round of voting was Dr. Karl Jarres of the Nationalists with 10.7 million votes, followed by the Social Democratic candidate, Prussia's prime minister, Otto Braun, with 7.8 million votes, Wilhelm Marx of the Catholic Center with 4 million votes, and Ernst Thälmann of the Communists with 1.8 million votes.

The Nationalists reshuffled their deck. The nationalist candidate for the run-off election was Field Marshal Paul von Hindenburg, the victor over the Tsarist Russians at Tannenberg in 1914 and the commander of the Kaiser's army for the last two years of the war. The Democrats and some of the Social Democrats agreed to throw their support behind Wilhelm Marx—a moderate Catholic despite his insidious last name. Ernst Thälmann ran for the Communists, ignoring an appeal from more moderate leftists to throw his votes behind Marx—Wilhelm, not Karl—to stop Hindenburg at all costs. The run-off brought Paul von Hindenburg 14,655,766 million votes and the presidency of the Weimar Republic. Catholic liberal Wilhelm Marx received 13,751,615 votes and Communist Ernst Thälmann received 1,931,151 votes. The contemporary journalist Stefan Lorant, who did not admire either Hindenburg or Thälmann, noted glumly that a coalition of Left and Center could have stopped Hindenburg dead in his tracks and given the presidency to a moderate-to-liberal Catholic. The Soviets, however, had bluntly forbidden their German Communist Party tools to make any compromise with either the Social Democrats or the political Catholics.

Paul von Hindenburg was a linear descendant of Martin Luther and of Polish royalty, as he often mentioned. His mother was from a respectable middle-class family without a title, which he did not often mention. A product of cadet schools since late childhood, Hindenburg admitted that he had never read a novel or work of poetry or philosophy in his life. He had fought against Austria-Hungary (and Bavaria and Hannover) in the Seven Weeks' War, Bismarck's North German consolidation in 1866, and had served in the sweeping victory over France in the Franco-Prussian War of 1870 – 1871. Hindenburg had been a witness to the actual founding of the German Empire at Versailles on January 18, 1871. He was intensely patriotic—and a quiet monarchist who hoped to see a Hohenzollern back on the throne of Prussia.

Hindenburg's election was a windfall for Hermann Ehrhardt. The new president proclaimed an amnesty in the spring of 1926 and in October, Hermann Ehrhardt was able to return to Germany without fear of arrest. The *Wiking Bund* had largely fallen apart during Ehrhardt's exile so he affiliated with the *Stahlhelm* [Steel Helmet], a right-wing veterans group independent from the Nazi movement and not dominated by drastic anti-Semites.

The Dawes Plan and the Soviet shift away from violent revolution had reduced the strength of the Communists and the left wing of the Social Democrats so much that in 1926, an initiative by Communists and Social Democrats to break up the inherited estates of the German nobility reached the Reichstag—which refused to adopt an expropriation law with a vote of 236 to 141. The Leftist parties demanded a national referendum to take away the land of the landed gentry, notably in Prussia, the largest of the German states. They needed 50 percent of the popular vote. They didn't get it. The 14,455,184 votes that the Communists and Social Democrats polled on June 20, 1926, turned out to be 4.5 million votes short of victory. The nobility kept their castles, their manor houses, and their tenant farmers, who themselves generally voted conservative. In another confrontation, President Hindenberg ordered German embassies abroad to display both the black-white-red flag of the Empire and the black-red-gold flag of the Republic.

Matters at home soon required Ehrhardt's attention. Ehrhardt's wife had gone through immense turmoil due to his political misadventures and her nerves were frayed. Ehrhardt had always treated his step-daughter as if she were his own child, but she had had a different father, his widowed wife's deceased first husband, and had been born with the baronial von Gilsa title. She was coming of an age for courtship and marriage. His presence in the household, even though it was now legalized, must have been something of a handicap. Princess Margrethe had also been compromised in the all-important royal marriage sweepstakes and was past 30. Intensely religious, and also caste-conscious, she was—through a relationship Ehrhardt had essentially tricked her into—a female ex-convict and the suspected paramour of a man with 354 reported murders to his credit. She had no hope of the sort of marriage that her royal status and religious good character entitled her to. Hermann Ehrhardt had unintentionally but effectively ruined her life. Ehrhardt dealt with this in the way he felt fate required: he obtained an amicable divorce from his first wife, whose family was still independently solvent and could support her and the children in comfort if not in luxury. As the son of a clergyman, Ehrhardt was immune to any caste distinction and his proposal of marriage could be accepted without disgrace—even by a princess. The fact that the princess now controlled a sizable personal fortune was not a deterrent, but fortune-hunting by officers who had no private fortune was a respected part of the tradition of the Prussian officer corps. Hermann Ehrhardt offered his hand in marriage to the woman he had compromised, and she accepted. On August 13, 1927, Margrethe Viktoria Princessin zu Hohenlohe-Öhringen married Hermann Ehrhardt at Neuruppin, Germany. Any compulsion for marriage was purely legal and not clinical. Their first child, Marie Elisabeth Ehrhardt, was born on May 6, 1929, almost two years after the wedding, and their son, Hermann Georg Ehrhardt, was born on November 15, 1930.

Ehrhardt's attempt to meld the *Wiking Bund* with the *Stahlhelm* shattered, based perhaps on a perceived lack of importance. The *Wiking Bund* was a fighting group disguised as a social group, while the *Stahlhelm* was a political group whose members once knew something about fighting. A stable government and the restored economy reduced extremism on both extremes—Communist and Nazi. The presentable and educated young men who had started out in the 2nd Marine Brigade began go back to the universities, to marry, and to enter the professions or family businesses or take over estates. With two million Germans dead and a million crippled in the war, the adage was that any man who couldn't marry a blonde heiress wasn't trying very hard. Ehrhardt disbanded what was left of the *Wiking Bund* on April 27, 1928.

The Nazi SA, unlike the *Wiking Bund*, was full of young men who were not presentable or marriage material and had come to see violent politics as a form of full-time employment. The idea that the Nazi movement in its early years was secretly supported by industrialists has been widely disputed: Hitler got most of his own pocket money from rich older women and from the sale of his best-seller, *Mein Kampf.* The rank and file Nazi storm troopers lived with their parents and collected street donations in tin canisters. The Nazis were fighting for their own economic survival and self-importance, and often to ward of hostility to their unconventional lifestyles. Walther Rathenau's personally fatal alliance with the Soviet Union, followed by the American-sponsored Dawes Plan, had all but eliminated the climate of fear and despair that had driven German politics a for the first three years after the war ended.

The Nazi movement used some ruthless tactics to recruit respectable Germans who might improve their thuggish image. One of the candidates was Alexander von Falkenhausen, winner of the *Pour le Mérite*, was a remote but recognized royal—he was descended from an early elector of Brandenberg from the days before Prussia had kings, through a recognized mistress, which ws not at all unusual. Falkenhausen had thwarted a potentially lethal Turkish round-up of Jews in Palestine toward the end of World War I and was not an anti-Semite. He was the staff officer who had helped Hermann Ehrhardt disband the 2nd Marine Brigade at Münster without bloodshed. In 1927, Falkenhausen had been named commander of the military district around Dresden in "Red" Saxony.

"It was clear to me that the younger people sought a political idea and that they declined interest in the major parties," Falkenhausen wrote in 1950. "In 1929 I spoke with the former *Reichswehr* Minister [*General Wilhelm*] Groener when he came to Dresden and I had many hours of conversation with him, in which we discussed the splintering of the major parties and the growth of National Socialism: this made it necessary for the young people to have ideals before them of fulfillment of duty, love of the Fatherland, and obedience. Only through a movement like this could one keep the younger people away from National Socialism."

In 1930 Falkenhausen was dismissed from the Army. He soon learned that the Nazi press had leaked a rumor that he had joined the Nazi Party: this rumor had caused his dismissal from the Army, which was still anti-Nazi. The insidious Nazi movement then offered the unemployed Falkenhausen a salaried position with the SA and a seat in the Reichstag. Falkenhausen was quietly appalled and declined the Nazi 'honor' as a disgrace. He joined the *Stahlhelm* instead. Falkenhausen also joined the German National Party of Alfred Hugenberg, the Rightist group which was opposed to Nazi policies including the extreme anti-Semitism, the rejection of royals, and the quasi-Socialist economics. Hugenberg was nicknamed "the hamster" both for his girth and for his lack of ferocity. Both the Nationalists and the *Stahlhelm* moved closer to the Nazi position as the World Depression after the Wall Street Crash of 1929 revitalized the Communists. "Many people in the middle-class parties granted themselves the delusion that they could use Hitler as a 'drummer.'" Falkenhausen wrote. "To me however, it was clear that a bonding between the German Nationalists and the National Socialists could only be compared to the friendship of a defenseless lamb and a hungry wolf. So I left the German Nationalist Party. I stayed with the *Stahlhelm* until [*Stahlhelm co-commander Franz*] Seldte made a deal with [*Ernst*] Röhm and then I left the *Stahlhelm* as well."

Hitler had marginalized the *Stahlhelm* from the inside. He had backed Theodor Duesterberg, the more anti-Semitic of the two *Stahlhelm* co-commanders against Franz Seldte until the SA had absorbed the legitimate veterans' group. Then the Nazis professed to have discovered—at just the right moment—that Theodor Duesterberg, himself an anti-Semite, had had a Jewish grandfather. The *Stahlhelm*, once bigger and certainly less perverse and more respectable than the SA, ceased to play a prominent role in German politics.

Falkenhausen, who spoke fluent Japanese he had learned at staff colleges, was shortly invited to help reorganize the Nationalist Chinese armies of Chiang Kai-shek, who also spoke fluent Japanese, but no English or German. As an anti-Communist who liked the Chinese people, Falkenhausen accepted instantly and was able to restore some morale in the Nationalist Chinese Army and to organize defenses of Northern China that stopped the Communist Chinese encroachment. He taught the Chinese to build block-houses supplied by truck that left the Communists without supplies in the austere climate of Northern China. Falkenhausen and Chiang also provided a safe Chinese haven, except for some fighting against the Communists, for *Reichswehr* commander Hans von Seeckt, "The Sphinx With A Monocle," when Hitler later dismissed Seeckt, the acknowledged organizer of the post-war *Reichswehr*. The Nazi cover story was that Hans von Seeckt was dismissed because his wife, '*nee* Jacobsohn,' was Jewish. Seeckt's wife, in fact, was a highly Nordic Frisian and he and she were both mildly anti-Semitic, though they never came close to Nazi standards. The real reason Hitler got

Alexander von Falkenhausen

rid of Seeckt was that Seeckt was a furtive monarchist and had openly and effectively opposed the Beer Hall Putsch. The triggering event came when Seeckt allowed a Hohenzollern prince to take part in Army maneuvers. The Nazis hated Jews but they both hated and feared Prussian royals.

As the *Stahlhelm* lapsed into Nazism followed by extinction and as Falkenhausen headed for China, Hermann Ehrhardt saw one potential but dangerous alternative to Hitler: the Black Reichswehr. The Black *Reichswehr,* also known as the

Black Front—black meaning secret, with overtones of a suicide pact—had been hastily formed when the French and Belgians occupied the Ruhr. Recent soldiers and surplus recruits were formed into labor battalions where they undertook drainage and construction projects—with military training on the side. The *Reichswehr* with its 4,000 officers and its grand total of 100,000 men was gradually augmented by as many as 80,000 "labor battalion" troops who were trained with close-order drill and weapons and more than willing to defend Germany in case of foreign invasion. Younger reservists, regional and municipal militia units, "blue" police armed with pistols and rifles and "green" police armed with rifles and machine guns brought the Weimar Republic's military manpower close to a million men. But the whole thing was secret—and secrecy in uncertain times led to informers.

Paul Schulz was a part of the Black Reichswehr. Born in Stettin in eastern Prussia in 1898, he left school without his *Abitur* as a 16-year-old war-time volunteer and fought on the Eastern Front. He spoke some Russian and was expert at taking and interrogating Russian prisoners. Schulz was transferred to the epic slaughter battle at Verdun in France, then back to Russia before the Tsar's army collapsed, and finally to the Western Front as an elite shock trooper in the Kaiser's army. "It was especially *Vizefeldwebel* Schulz who was responsible for the success of just about every patrol he took part in," the regimental diary of Infantry Regiment 130 recorded. Schulz was promoted to reserve officer. He was shortly one of only 216 enlisted men promoted to the rank of *Leutnant* in the regular officer corps. After the German defeat, Schulz took part in the *Freikorps* movement, though he took no part in the Kapp Putsch because he had no orders to take part. Schulz was rigidly dedicated to following orders. His excellent combat record on both fronts and his penchant for working long hours convinced his superiors to keep him on in the *Reichswehr* even after the officer corps was reduced to 4,000 men under pressure from the Treaty of Versailles, when some members of the aristocracy were let go.

Schulz, however, became a secret but subsequently acknowledged member of Hermann Ehrhard's notorious Organization Consul. Some of Schulz's off-duty hours were reportedly spent dispatching sergeants to murder informers who revealed arms caches to the Allied Control Commission. *Oberleutnant* Schulz was hauled into court early in 1925 and faced with a possible death sentence. He was offered leniency if he informed on Captain Ehrhardt or members of OC. Paul Schulz held to his honor and never informed. The only plausible evidence showed that Schulz had served as a driver for two *Reichswehr* men before their lethal interviews with two arms-cache betrayers. The informer, Carl Mertens, was basically a grown-up messenger boy turned pacifist who had no real inside knowledge of OC and the court let the case slide. But Paul Schulz was arrested again in March of 1925 for ordering the murder of *Reichswehr* Sergeant Wilhelm Wilms, another accused arms-cache informer.

"At the defendant's bench sat a well-brought-up young man, with a smooth face and a short neck," the liberal *Berliner Tageblatt* reported. "He favored a small beard. His eyes were small and penetrating. He bowed to the judge's bench as well or as clumsily as he would to a comrade at the Front. All and all, he gave the impression of a non-commissioned officer type from the old Army." Mertens, the uninformed informer, portrayed Schulz as a howling homicidal maniac. But General von Ludendorff appeared for the defense and testified that Schulz was a good soldier doing his duty. With the moderate President Friedrich Ebert now dead and the Social Democrats listing heavily to the Left, Schulz's trial was dragged out until on

March 26, 1927, when he was convicted of murder and sentenced to death. The National Socialists and the non-Nazi German People's Party both protested the death sentence and proclaimed that Paul Schulz was a martyred war hero. Schulz's sentence was reduced to imprisonment and in early 1930 he was released and eventually pardoned.

Paul Schulz

The Black Reichswehr, as Falkenhausen had observed, had gradually been gobbled up by the Nazi SA and its supporters while Schulz was in prison. Schulz accepted Hitler's leadership of the German Right as Ehrhardt gradually faded from politics. But Schulz never actually subscribed to the principles of the SA—he was not a hard-core anti-Semite and he was quietly a monarchist. There were other obstacles. After Schulz had worked with the SA for a few months, and helped put down a purported internal mutiny by Hitler's bosom buddy, Captain Ernst Röhm, Schulz wrote Hitler a letter of June 2, 1931. Otto Strasser, one of Hitler's Party rivals, leaked a copy to the Social Democratic *Münchener Post*.

Schulz—who had recently married and eventually became the father of three children—attempted to draw Hitler's attention to "the dangers ...necessarily entailed, in my opinion, by the employment of morally objectionable persons in positions of authority...." Schulz offered a list of names including Ernst Röhm, Berlin SA Chief of Staff Karl Ernst, a former bouncer in gay bars and his former male paramour ex-Captain Paul Roehrbein, and a number of Röhm's personal aides. Schulz said that male prostitutes all over Berlin knew about Röhm and Roehrbein and were bragging about their own influence in the Nazi Party. "Captain Röhm makes absolutely no secret of his disposition; on the contrary he prides himself on his aversion to the female sex and proclaims it in public.... Things have now reached the stage where rumors are being spread in Marxist quarters that you yourself, my most esteemed *Führer*, are also homosexual. Widespread incomprehension reigns, among the intelligentsia as well, that there are far more homosexually inclined officers in the *Braunes Haus*."

Hermann Ehrhardt

Hermann Ehrhardt also read the newspapers, and Friedrich Wilhelm Heinz, an Ehrhardt loyalist, somerimes wrote for them. On February 1, 1932, Wilhelm Heinz, scored a hit in the *Montagsblatt*, [Monday Pages] sponsored by Otto Strasser and by Hermann Ehrhardt. The Rightist but anti-Hitler paper was the first to point out that the Austrian citizen Hitler had taken up formal residence in a German state and was about to obtain his German citizenship. The goal may have been to stop this legal ploy in its tracks. If so, it failed. The *Stahlhelm* veterans group was also fading fast as a non-Nazi option for anti-Communists. But Ehrhardt shared Paul Schulz's distaste for homosexual bully-boys posing as he-men and for anti-Semitism taken beyond the realm of logic. He was not alone in this. Otto and Gregor Strasser, two brothers who were leaders of the Black Reichswehr and later of the so-called "Left Wing of the Nazi Party," had developed their own Jewish policy: the 80,000 Russian Jewish immigrants, seen without sentimentality as an economic liability in hard times and a bewildered matrix for Marxism, would be asked to leave Germany. The German-born Jews would individually vote on their own future. Those German-born Jews who wanted to retain German citizenship would retain it. Those who requested membership in a protected minority group would be shut out of politics but protected in their persons and property. There were, in fact, several protected minorities in the Kaiser's Germany: the Wends, a Slavic tribal people, had lived in the Spreewald section of Berlin since the high Middle Ages; the Nordic Frisians, acquired when Prussia defeated Denmark in 1864, lived at the base of the Jutland Peninsula and on the North Sea Frisian Islands and also enjoyed minority status. Hamburg, the second largest city in Germany and the largest seaport, had a venerable and exotic Chinatown with 2,000 to 3,000 Chinese residents from the 18th Century until the Nazi era, but this neighborhood shortly failed to survive a brutal combination of Nazi racism and the Allied air raids and vanished in the 1940s. Ernst Hanfstaengl, still enthralled by Hitler's hypnotic oratory, countered the relative humanity of the Strasser brothers by pointing out that an artist like he himself could tell that the Strassers appeared to be of Jewish ancestry.

Hermann Ehrhardt had seen the Nazi movement dominating the Right and began to form an opposition. He and Hartmut Plaas, a veteran officer of the 2nd Marine Brigade and Organization Consul, founded the "*Gefolgschaft*"—The Followers—with about 2,000 members, including Nazi Party drop-outs and German-born Communist drop-outs. The leader they were following was no longer Adolf Hitler but Hermann Ehrhardt, whose opposition to Hitler was based primarily on Hitler's twisted psychology, as influenced by his own brief contact in Munich and by the opinion of Paul Schulz. Ehrhardt saw that Hitler was a dangerous demagogue and that no movement so obviously opposed to Christianity could be permitted to rule in Germany. Hartmut Plaas, the son of a senior forester who had himself been a Nazi and an anti-Semitic roughneck until he grew disillusioned with Hitler, understood the Followers to support opposition to international capitalism and opposition to class warfare. Plaas himself was a strong advocate of the rights of peasant landowners.

When Paul von Hindenburg's seven-year term as president expired in April of 1932, the chancellor, Heinrich Brüning, tried to extend the 84-year-old war hero's presidential tenure by a constitutional amendment. Adolf Hitler and Alfred Hugenberg, the Nationalist leader in the Reichstag, both refused to approve, so a new election was scheduled. The election posters told Germans who to vote *against*: The Social Democrats pointed out that a vote for Ernst Thälmann, the Communist candidate, was really a vote for Hitler—the implied idea being that the Social Democrats could beat Hitler but Thälmann couldn't hope to win, since most Germans outside Berlin and Leipzig were staunchly anti-Communist. Hindenburg's posters said *"Vote for a man, not a party"*—perhaps a comment on Paul Schulz's letter about the Nazi version of "back-room politics." Hitler's posters showed burly workingmen rather than his actual suporters, German women and the lower-middle class, because the Social Democrats and even some uneasy home-grown Communists who didn't trust Stalin and his tool Ernst Thälmann might be the swing vote. The first round of voting failed to give any candidate a majority but eliminated Hugenberg the Hamster's non-Nazi Nationalists and the once-mighty Social Democrats. Hindenburg won the run-off election with 53 percent of the vote. Hitler had needed special legislation as a councilor of the conservative city of Braunschweig to be declared a German citizen so he could run for president, as Heinz and Ehrhardt had already reported. Hitler received 36 percent. Thälmann, whose posters had called for "Socialism"—Communism sounded Russian and brought back memories of 1919—received 10.2 percent. Bad as things were, most Germans didn't want Russian-backed Communism in any way, shape, or form. But Hitler's followers took 237 seats in the Reichstag and the Nazi Party was now the single most heavily represented, with 37.27 percent of the vote. The Social Democrats took 133 seats and the Communists took 89 seats.

Ehrhardt had campaigned "vehemently" for Hindenburg. Ehrhardt's conscience left him no choice, since Hindenburg's amnesty had gotten him back into Germany legally, just as Kaiser Wilhelm had kept him in the Navy after his poaching antics in the Danish islands. Loyalty, as Ehrhardt said at the end of "Adventure and Fate," was his most respected virtue. "Hamster" Hugenberg's Nationalists with Duesterberg as their presidential candidate clearly had no chance of stopping Hitler, and stopping Hitler was Ehrhardt's ultimate goal, short of bringing back the monarchy by free election rather than an armed counter-revolution.

Another Hitler opponent was Dr. Helmuth Mylius, an industrial entrepreneur and trained economist and leader of the Party of the Radical Middle Class. Dr. Mylius was the publisher of a weekly newspaper, *Die Parole*, which had begun denouncing Hitler as a mountebank in 1930 and proclaimed that the Nazi Party was dominated by bogus economic theories and by crooked politicians. Mylius also disapproved of Hitler's extreme anti-Semitism. He understood that massive inheritance taxes motivated by the envy of the loser class and concepts like substituting "work" for "gold" as support for paper money were nonsense. How does one transfer 'work' from bank to bank? Dr. Mylius was "radical" in the sense that he recognized the need for drastic economic reforms, particularly because one German worker in three was unemployed during the early years of the World Depression that had started on Wall Street. He was also "middle class" in the sense that he was anti-Communist and a nationalist rather than an internationalist. Dr. Mylius was better educated than Hitler and had a far better grip on economics—but Mylius lacked any real voter appeal. He was fat, he always wore glasses, and he had a heavy face and combed his back hair sidewise over his pate in a vain attempt to hide the fact that he was completely bald on top. His speeches may have asked people to think too much. Hitler never made that mistake.

In the *Reichstag* election of November 1932 the German Middle Class Radical Party received 60,246 votes—about 0.17 percent of the tally—and no seats in the Reichstag. Hitler himself actually experienced a drop in votes and came nowhere near a majority. But Hitler's Nazis were still the top vote-getters with 11,737,021 votes, 33.09 percent. With 196 seats in the Reichstag, they had 65 more seats than the Social Democrats and 96 more than the Communists. The upper classes, moderately rightist and alarmed by the 11-vote Communist vote increase, began to see Hitler as the lesser of two evils. As Hitler became more and more inevitable, Hermann Ehrhardt and his Followers prepared for the worst.

Chapter Twenty-Eight
Stalking Hitler

Franz von Papen was one of those people like Napoleon III whom Queen Victoria once described as "too clever by half." He always seemed to trip over his own perceived genius for intrigue and to take others down with him.

Papen was a Catholic scion of the minor nobility from Westphalia, not a Lutheran Junker from east of the Elbe like Hindenburg or most of the senior German generals. And he was a cavalry officer in the Kaiser's army, which essentially meant that his family could pay for a string of horses—Prussian cavalry officers had to bring their own mounts. Cavalrymen in most armies had a reputation for being brave but somewhat brainless—Winston Churchill, a superb writer and patriotic historian, had been ultimately granted a cavalry commission through the Royal Military Academy at Sandhurst because he couldn't cope with algebra or learn French: he was thus disqualified as an infantry officer because the infantry had higher mental standards.

Franz von Papen, however, learned excellent French and English. After his regimental service Papen attended training courses for the German General Staff, where intelligence was a prerequisite. Waldemar Pabst was one of his classmates.

Captain von Papen was military attaché at the Imperial German Embassy in Washington DC during the early days of World War I, when the United States was officially neutral but was shipped huge amounts of food and ammunition to Britain and France. Papen felt Americans were primitives and declared "that he was unable to understand how they had made the transition from walking on four legs to two... probably someone had indicated there was more money to be made by walking that way." Americans, Papen said, displayed "a national ability to fall into decadence without the customary intervening and uplifting period of civilization."

Papen developed his own technique to reduce the flow of ammunition to Britain. He bought up future orders of shells and explosives intended for Britain and kept them warehoused in the United States because they could not be shipped through the British blockade to Germany. Other German agents had more spectacular methods. On July 30, 1916, the Lehigh Valley Railroad terminal on Black Tom Island—actually a man-made peninsula in Jersey City just across the Hudson River from New York City—was hosting two million tons of ammunition intended for Britain and France and stored in freight cars and on barges. A series of small fires broke out. Sequential ammunition explosions broke windows 25 miles away and the earth shook as far away as Philadelphia and Maryland. Seven people were thought to have been killed. Some bodies simply disappeared. A ten-week-old infant who reportedly fell from a chest of drawers may have been a casualty of convenience to collect damages. Extensive investigation traced the bombs that started the Black Tom explosions to German agents—two of the guards at Black Tom were said to have been German agents—as well as to Irish patriots and Asian Indian freedom fighters opposed to Britain, and to Communists. The Kingsland plant in nearby Lyndhurst, New Jersey, caught fire and exploded on January 17, 1917, touching off a panic at the insane asylum at Snake Hill, but causing no deaths. Britain had a lot of enemies in the heyday of Empire and a large percentage of New York and New Jersey's policemen and firemen were of Irish ancestry—some of them the same people whom, when the United States declared war on Germany, rushed to enlist to serve the country they loved even as an ally of the country they hated far more than they hated Germany. According to urban legend, when Franz von Papen was repatriated after the American declaration of war, his meticulous personal checkbook with the list of recipients was confiscated by U.S. Customs and 125 New York and New Jersey public employees were quietly fired as bad security risks. Papen spent the rest of the war in Turkey in military and diplomatic roles.

Papen, while he considered himself an adept schemer, was exploited by an even more adept schemer when he ventured into German politics. One of Papen's skills was flattery that went just far enough and never too far. He became a friend of Paul von Hindenburg and a closer friend of Hindenburg's son and confidante, Oskar. Under the Weimar constitution, the president was democratically elected. The president then had the power to appoint at will the chancellor who ran the country on a day-to-day basis. The system was based on the German Empire of the Middle Ages when the *Kaiser*, usually one of the five dukes ruling the larger Germanic tribes, was elected by other dukes and bishops. The elected Kaiser appointed a chancellor, usually a churchman, as the executive and financial leader.

Kurt von Schleicher, a war-time general in charge of supplies who was described as brusque and opinionated, had slipped into Hindenburg's good graces by influencing Hindenburg to replace the capable Chancellor Heinrich Brüning with Franz von Papen. Schleicher then bumped Papen, who has no power base in politics. Even Papen's own Catholic Center party preferred Hitler as a bulwark against Communism. Schleicher, whose name literally means "the sneak," took over as chancellor. Papen was driven into a fury of envy and indignation and plotted a terrible revenge. He worked overtime to make Adolf Hitler acceptable to Paul von Hindenburg, who disliked and distrusted Hitler almost as much as he hated the Communists. Hindenburg especially despised the homosexual component of the SA, notable "that breech-loader" Ernst Röhm. Hindenburg yearned for the days of the Kaiser when the homosexual Captain Röhm would have been handed a loaded pistol and sent to a quiet room....

Kurt von Schleicher, fighting for his political life, tried to use Gregor Strasser and the "Left Wing" of the Nazi Party to undermine Hitler. Schleicher also made a drastic effort to restore the troubled German economy by breaking with the hide-bound conservative tradition: he obtained a half-billion revalued marks for quick-fix working-class and entry-level jobs to stimulate the economy and help unemployed workers who had exhausted their national unemployment benefits to feed their families and start paying taxes again. The municipal benefits offered by the cities as a fall-back measure were so low that workers often had to choose between food and rent payments. They either lost their flats or ate in soup kitchens and breadlines. At a private dinner, Schleicher even mentioned a deal he had covertly worked out with the foreign powers to bring back a conscripted German national militia and eventually to bring back conscription of regular soldiers, shelving the Treaty of Versailles limitation to a 100,000-man army. The world depression had raised fears of Communism all over Europe and German troops would be a welcome addition to the garrison of the Western democracies like Switzerland and France, and constitutional monarchies in the Netherlands and Scandinavia.

Finally, afraid that Hindenburg might re-appoint his friend Papen, Schleicher cut his own deal with Hitler: Schleicher agreed to support Hitler as chancellor if Schleicher, the former general, could become defense minister in direct control of the army.

Waldemar Pabst, the consummate string-puller running the Austrian *Heimwehr*, attempted to reach President Paul von Hindenburg and warn him that letting Hitler into the government would be like "riding a tiger."

"The people want Hitler, it's either him or Thälmann," Hindenburg told Pabst. "We're better off with Hitler and he arranged it for us."

Pabst was at a loss. Hitler's last-ditch electoral effort in the small Protestant state of Lippe was a Nazi disappointment. The Nazi orators extolled the Teutonic victory of Arminius, the tribal leader and the original Hermann, over the Romans in AD 9. But some people in Lippe said that the militaristic, authoritarian Nazis reminded them of the invading Romans: according to the Roman historian Tacitus the Germanic warrior tribes let themselves be ruled only by consensus and drowned homosexuals in swamps. The Nazi Party was also running out of money. Members were quitting in alarming numbers or stopping their dues as Brüning's and Schleicher's stimulus packages had gradually re-started the economy. Liberal newspapers sighed with relief that Hitler was done for. The Nazis had still never won a majority in a fair election. Hindenburg's fear, Ernst Thälmann, was a war veteran with the Iron Cross, but Thälmann was also a war-time deserter and a Stalinist tool. Yet Hindenburg was a grand old man out of touch with reality. An OC stick hand grenade tossed through Thälmann's apartment window in June of 1922 had already served notice on him. His wife and daughter were in another room and he himself was not at home when the library blew up. Several hundred Organization Consul veterans would have been available to shoot Ernst Thälmann on his way to the podium if he were to show up at his own inauguration.

Hindenburg may have been deluded but he was still a German officer with a strong sense of honor. Franz von Papen, who wanted to ride Hitler's coat-tails into office, worked out a codicil to Hitler's acceptance: all Jews who were veterans of the Kaiser's army would be allowed to retain their civil service jobs, which constituted a large and safe portion of the German national economy. Hitler, who was subservient to President Hindenburg almost to the point of cringing, pledged his word.

"I never knew that anybody that could speak like that could lie like that," Franz von Papen told Waldemar Pabst years later with intense remorse. Abandoned by almost everybody, Kurt von Schleicher resigned after 57 days as chancellor and took over as defense minister. A deal had been made behind closed door. Franz von Papen became Hindenburg's vice chancellor under Hitler as chancellor.

On January 30, 1933, Adolf Hitler became chancellor of the Weimar Republic. Less than a month later, on February 27, the Reichstag building caught fire in the middle of the night. A Dutch-born Communist named Marinus van der Lubbe was caught and charged with the spectacular arson in the middle of Berlin. Hitler suspended civil liberties. Enemies of the regime were rounded up in large numbers and imprisoned or sent to concentration camps. As the 81 Communist *Reichstag* delegates and two dozen left-wing Social Democrats were locked up, Hitler demanded the adoption of an "Enabling Act" that gave him virtually absolute power for four years. The Nazis controlled 288 votes and their Nationalist allies had another 52. The remnants of the Catholic Center—Otto von Bismarck's most consistent opponents—voted with Hitler. Some right-wing Social Democrats joined them. The vote came in at 444 for the Enabling Act and 94 opposed. On March 23, 1933, the Weimar Republic essentially voted itself out of existence.

Attempts on Adolf Hitler's life began even before he was appointed chancellor. His car had been fired on in Thuringia and in Leipzig in 1923. In 1929 a renegade SS guard reportedly left a bomb under the speaker's platform at the Berlin *Sportpalast* but ducked out to use the men's room before Hitler appeared to speak.

The SS renegade somehow found himself locked in the men's room and missed his chance to trigger the bomb. On March 15, 1932, a fusillade of shots greeted Hitler's train on the tracks between Munich and Weimar, but missed him. Other ambushes were planned but Hitler failed to show up. A dissident SA man was captured near Hitler with a loaded handgun in 1933 near Obersalzberg. Shots missed Hitler's car on the road between Obersalzberg and Rosenheim.

Hermann Ehrhardt had yet to be heard from. Early in 1933, undoubtedly with the help of Princess Margrethe's money, he bought the estate of the Counts of Bredow in Klessen, Westhavelland, a low-lying marsh and thicket region of the central Prussian state of Brandenberg so isolated that it was a favorite for star-gazing. One June 28, the *"Westhavelländische Tageszeitung,"* the daily newspaper, reported that *"Kapitän* Ehrhardt has joined the NSDAP," and that Ehrhardt had accepted the authority of the SS over the Ehrhardt Brigade.

Hermann Göring protested against the acceptance of Ehrhardt and his men into the SS, since they had openly campaigned for Hindenburg and against Hitler in the presidential election. Ehrhardt was a hero to the anti-Communist forces in Germany but he was an avowed monarchist and had no profile as a Nazi-style anti-Semite. Göring was suspicious—and he was right. Ehrhardt's agreement to serve as a SS auxiliary after being actively recruited as a name-brand war hero was part of what became known as the mysterious "Mylius-Ehrhardt Plot." Dr. Helmut Mylius and Hermann Ehrhardt understood that simply killing Hitler could lead to another attempt at a Communist take-over: some governmental structure under Nationalist auspices had to be organized. Ehrhardt also understood that the bungling amateurs who fired shots at Hitler's train were useless assassins. Hitler—true to the Prussian stereotype of a rural Austrian—sometimes skipped appointments entirely and often showed up late. Ehrhardt's SS affiliation was actually an attempt to infiltrate the Nazi SS with as many as 160 of his own gunmen and to work out Hitler's schedules to arrange Hitler's rendezvous with certain death. Ehrhardt's own feelings about the role he had to play is reflected in photographs taken on him on parade with the SS runes on his sleeve below his trademark dragon ship blazon: his face reflects a mixture of disgust and defiance.

But a chance for deception arrived on July 17 of 1933 when Hermann Ehrhardt and his loyalists from Organization Consult and the old Ehrhardt Brigade turned out in force at a memorial service for Erwin Kern and Hermann Fischer, the assassins of Walther Rathenau 11 years before at Burg Saaleck. Kern and Fisher had been reviled and repeatedly betrayed after the Rathenau murder in 1922. Hitler made them into national heroes—perhaps hoping that some of their doomed defiance would rub off on his own lackluster performance in the battles against the Bolsheviks that had made Hermann Ehrhardt a national hero even to rank-and-file Nazi supporters. Ernst Werner Teschow, the driver in the assassination, was a guest of honor and members of the SA and the SS also turned out in force. Teschow had spoken to Rathenau's grieving mother in 1922 while he was in jail awaiting trial and had been appalled to learn that Rathenau had donated the equivalent of a million dollars to anti-Communist groups. He and Ehrhardt came to honor their dead comrades—who were almost palpable to German military veterans—but they had other reasons for being there.

Dr. Mylius reached out to conservative economists and friends in the German Army and they set a tentative date of May 1935 for the Rightist putsch against Hitler. But just before Christmas of 1933, Heinrich Himmler, national chief of the SS, received word of the plot from infiltrators of his own, that "the knife being sharpened was not our own knife, and the heads and necks at risk were that of the Führer and our own." On February 1, the reconstituted Ehrhardt Brigade was disbanded again under Himmler's orders with no stated reason. Hermann Ehrhardt, however, remained a member of the SS along with Hartmut Plaas.

On June 30, 1934, Hermann Ehrhardt was at home on his farm in Westhavelland. His wife was not at home. A SS detachment pulled up in a car and Ehrhardt spotted them, knew what they wanted, grabbed two hunting rifles, and vanished into the woods behind the house. Ehrhardt was a renowned marksman and the SS may have been afraid of a confrontation. Austrian and French sources suggest that Ehrhardt still had infiltrators or admirers in the SS and they let him escape. When the SS left the farm, Ehrhardt slipped back into the house, notified his wife, and then slipped away to Switzerland. He had no doubts as to why they were looking for him.

Herbert von Bose, a count and a former General Staff officer and conservative opponent of Hitler, was an old friend of Hermann Ehrhardt's. Bose's wife came from Alsace. Her maiden name was Eschwege, which may have been the source of Ehrhardt's fictional name in November of 1922. Bose had supported the Kapp Putsch and was photographed standing just behind Ehrhardt in the signature photo of the take-over in 1920. In 1934 Bose had already give away his seal ring, suspecting that something was about to happen. On June 30 between 10 a.m. and 11 a.m., when three SS men walked into his office at the Palais Borsig in Berlin, Bose was officially working as Vice Chancellor Franz von Papen's press secretary and secretly conspiring against Hitler. He was arrested and was taken to a side room "to be interrogated." As *Graf* von Bose sat down to talk, he was shot in the back 10 times. Albert Speer, who decided to become a "good German" after Hitler lost the war, remembered seeing Bose's bloodstains on the carpet when he took over the office, but tried not to think too much about it. Bose's friend Edgar Julius Jung was also murdered.

Oberleutnant Paul Schulz, former teenaged war-time shock trooper and terror of arms-cache informers in the 1920s, was also picked up at his office and taken to Gestapo headquarters. Three Gestapo agents in plain clothes took Schulz for a ride to a forest near Potsdam in a civilian car. The SS men parked and told him to get out of the car. They shot him in the back. The three Gestapo killers went back to the car for a blanket to wrap around the body. Schulz, severely wounded but not dead, jumped up and ran into the woods. He was able to elude the Gestapo men and found his way to the home of a right-wing admiral, who smuggled him to Switzerland.

The wave of murders that engulfed Germany came to be known as "The Night of the Long Knives" after a revolt against the British rule of India almost a century before. The official crowd-pleaser story was that homosexuals within the SA had been plotting to betray Hitler and that he had struck first. In fact, the known homosexuals in the SA were loyal to Hitler, but they had become a public embarrassment and a personal threat. Now that Hitler had political control of the German Army he no longer needed his old buddies. Ernst Röhm, Edmund Heines, Karl Ernst, Paul Röhrbein and several other unrepentant homosexuals were among the victims of June 20—but they were not typical victims. The primary targets were monarchist officers like Ehrhardt, Schulz, Bose, and other conservatives who actually *were* plotting to get rid of Hitler. They were all married, most with children, many of them believing Christians. All were adherents to the code of honor of the Kaiser's officer corps where exposed homosexuals were expected to commit suicide or vanish into obscurity, not play at power politics.

Gregor Strasser, the Nazi Party drop-out and leader of the "Black Front" which promoted more socialism and less anti-Semitism, was murdered. So was Father Bernhard Stemple, the defrocked priest who had helped Hitler and Rudolf Hess write "Mein Kampf." Several Catholic politicians in good standing with their church were also murdered because they had stood their ground when most members of the Catholic parties stampeded to support Hitler as a last-ditch anti-Communist.

Secondary targets included mainstream political figures who had dared to oppose Hitler. Gustav Ritter von Kahr, who 'betrayed' the Beer Hall Putsch, was hacked to death with pickaxes and dumped in a swamp. Former Chancellor Kurt von Schleicher had actually helped revive the German economy with two million make-work jobs that helped restore cash flow and confidence. When Schleicher's young wife Elisabeth answered the doorbell on June 30, she was shot dead because she happened to be at home. Schleicher was also shot dead. Ferdinand von Bredow, a close friend of Schleicher's and an admirer of Hermann Ehrhardt, and whose family was intermarried with the Bismarcks, was also murdered—reportedly shot in the face when he answered his door at his home on Berlin-Lichterfeld rather than fleeing when he learned that Schleicher and his wife were murdered and knew what to expect. Bredow, a holder of the Iron Cross and Hohenzollern house orders, whose family had a splendid military history even before Frierich von Bredow's legendary Death Ride in the Franco-Prussia War of 1870, was a long-time opponent of Hitler and a major general in the *Abwehr*, the intelligence section of the German Army and a present and future nest of covert anti-Nazi activities.

Hermann Ehrhardt (x) at the Kapp Putsch in Berlin in 1920, second from the right is Herbert von Bose

Sheer envy probably caused the murder of Hans Peter von Heydebreck, whose *Wehrwolf Freikorps* had smashed an attempt by Polish irregulars to ignore the 1921 plebiscite and capture Upper Silesia. Heydebreck was a World War I amputee who led his troops in France and Italy while his mangled arm gradually disintegrated into a stump. He was also a *Freikorps* leader and later an SA senior officer, a minor Prussian aristocrat and a minor anti-Semite. One of his last letters began with the phrase "I live for my Führer." Hitler supposedly questioned Heydebreck personally after his arrest at Munich and then ordered him shot at Stadelheim Prison along with the arrested SA homosexuals. Heydebreck was a national hero and an avid Nazi. He appears to have been sexually normal, but his mangled arm was an obstacle in pursuit of women. But Heydebreck's war record and *Freikorps* record must have aroused Hitler's envy or fear of rivalry.

Sheer stupidity led to the death of Willi Schmid, music critic of the *Münchner Neueste*. The Nazi death squad thought he was Ludwig Schmitt, an SA group leader, who was murdered separately.

No one knows how many people were murdered during the Night of the Long Knives and the aftermath—official figures ranged from 75 to 84 but the actual number may have been in the hundreds. Dr. Helmut Mylius somehow escaped notice. He later served in the *Wehrmacht* during World War II as a reserve major in the quartermaster corps.

The Mylius-Ehrhardt Plot actually hit Hitler with a ricochet. Hitler's massacre of the SA leadership had left him with some very violent enemies among his former followers: they wanted revenge. According to a story told by Otto Strasser, the murdered Gregor Strasser's surviving brother, an SA man named Heinrich Grunow knew the road approaches around Hitler's residence at Berchtesgaden and set up a personal ambush. Grunow knew that on one sharp turn Hitler's chauffeur and sometime look-alike double, Julius Schreck, would slow the car down to 15 miles an hour and make Hitler a plausible target. Grunow, nurturing his grief for his fallen comrades and possible lovers, was waiting with a loaded pistol when the car came up the road. He shot the man in the back seat three times. The he shot himself. But the man in the back seat was Julius Schreck, the look-alike chauffeur. Hitler, who had been afraid to drive before he was 30, had impulsively asked Schreck to let him drive while Schreck sat in the back seat. Schreck was shot three times, once in the chest, one in the jaw, and once in the temple. He died instantly. Hitler panicked at the wheel and stopped the car instead of speeding up as Grunow committed suicide on the spot. Hitler escaped with shattered nerves but without a scratch.

Germany was too hot for Hermann Ehrhardt. He made his way from Switzerland to France where he lived for two years. In 1936, he resettled in independent Austria and purchased the baronial estate of Brunn am Wald in the region of Krems on the Donau. The castle had been built in the 11th Century and the manor house Ehrhardt and Princess Margrethe bought dated to 1584. They sold the

Prussian estate in 1937 and apparently had no intention of venturing back to Germany.

In 1938, Germany came to Hermann Ehrhardt. Hitler annexed Austria with the agreement of the Austrian majority after a brave resistance by a cluster of conservative Austrian patriots. Nazi officials assured Hermann Ehrhardt—still a hero to those German monarchists and anti-Communists who had any doubts left about Hitler—that he had nothing to worry about as long as he kept away from politics. Ehrhardt gave his word, but as Franz Maria Liedig had told him in Munich in 1922, an oath under duress has no standing in law.

Ehrhardt's home at Schloss Brunn

The Night of the Long Knives had put Germany on warning: most people in what had once been a legalistic nation no longer had serious objections to executions without court orders or death warrants as long as the victims were Communists, outspoken homosexuals, or renegade Nazis. Field Marshal August von Mackensen, a World War front commander, was a spry anachronism who turned out at royal funerals in a white eagle moustache and tight-fitting hussar uniform as if he were living in the 1830s and not the 1930s. Mackensen, who was 85, actually had the courage to demand that Schleicher and Bredow be exonerated from Nazi charges of treason to preserve the honor of the German officer corps. General von

Schleicher and General von Bredow were quietly exonerated—but since they were already dead only Mackensen achieved much satisfaction. Hindenburg himself was glad to be rid of Röhm and the SA perverts. So were a lot of other people. SA active and honorary membership plunged from 2.9 million at the time of the Röhm Purge to 1.2 milion in April of 1938. The officially homophobic SS in their sharp black uniforms with silver trim took over as the intimidation experts of German society.

Hitler, having betrayed the SA who had helped bring him to power, now betrayed the Jews whose charity kitchens and hostels he had used to avoid hard work in Vienna. In 1935, intermarriage between Christians and Jews became illegal. Before Hitler, more German Jews had married Christians than had married other Jews. Commercial sex between Aryans and Jews was also criminalized. Jews were forbidden to hire Christian women below the age of 60 as household help. The Jews were restricted or banned from a number of professions. The American Jewish community was livid and after a brawl outside a movie theatre, boycotts of German shops and stores took place in New York City. Reciprocal boycotts in German cities led to broken windows and at least some broken heads. Americans who wanted to do business with Germany rationalized that Jews were no worse off in Germany than African-Americans were in the South: many U.S. states had "miscegenation" laws forbidding whites to intermarry not only with blacks but with Chinese-Americans and Japanese-Americans. Hitler's make-work programs—which actually started as Schleicher's make-work programs before he was forced out as chancellor and later murdered—were widely copied, minus most of the militarism, in Franklin Delano Roosevelt's New Deal programs such as the Civilian Conservation Corps, a direct knock-off of the Black Reichswehr.

But Hitler had some internal problems—the "Old Fighters," those SA members who had helped bring him to power, wanted to take over the civil service jobs staffed by a great many Jewish veterans. The German civil service was notably efficient and notably self-protective, and even after Hindenburg died at 86 on August 2, 1934, his word and his will remained law to large elements of the German Army. One exemplary Jewish veteran in the diplomatic corps was Georg Rosen. A Rhodes Scholar from a notable family of Orientalists who had left a safe posting in Portugal when Portugal declared war in the Kaiser, Rosen volunteered for the Western Front and served with honor. Rosen later assisted the Hamburg businessman John Rabe in rescuing Chinese survivors during the Nanking Massacre in December of 1937. As far as the German military advisers around Alexander von Falkenhausen and the Chinese Nationalists were concerned, Georg Rosen was a German combat veteran and his job was bullet-proof—but capable and honorable people like Georg Rosen stood between the leftover SA types and soft office jobs.

Events back in Germany took a drastic turn early in 1938. A number of groups who disliked and distrusted Hitler began to coalesce in the summer 1938. Conscription was revived in 1935 to the intense relief of Germany's anti-Communist majority as well as those who remembered the seizure of the Ruhr by France and Belgium. Hitler, having capitalized on the revival of a full-sized German Army —made possible by international agreements with anti-Communist Britain, France, Italy and Poland during Schleicher's brief chancellorship—re-occupied the Rhineland in 1936 with kettledrummers on horseback and marching troops. The British and the French did nothing. Everybody in Europe knew that the Allied seizure of the Rhineland had been an act of bullying and nobody wanted to fight to defend it.

"All Quiet On The Western Front," as a book and as a movie, had made the idea of killing Germans unpopular—even though its author, Erich Maria Remarque, was an early and consistant anti-Nazi. Some Hollywood movies like *"Music In The Air"* (1934), with music by Jerome Kern, who was Jewish and lyrics by Oscar Hammerstein, whose father was Jewish, portrayed Bavarians as wholesome and lovable rustics. American 'Empire' movies like *"The Charge of the Light Brigade"* (1936) —the murderous Russian advisor is a dead ringer for Stalin—depicted the Russians as the world's great villains. As later as 1937, in *"Lancer Spy"* with George Sanders, the Kaiser's German officers were seen as rigid men of honor. The bad guy was seen as a simpering homosexual like Count Harry Kessler. In Percival Christopher Wren's somewhat racist novel *"Beau Geste,"* published

Field Marshal August von Mackensen in hussar uniform

in 1924, the sadistic French Foreign Legion sergeant is French and his pet squealer is Italian. In the Hollywood movie with Gary Cooper made in 1938—but released in 1939—the sadistic sergeant and the squealer are both Russians. The Germans were depicted as brusque but defiantly brave. *"You know what I think of the French Army...."*

But Hitler overstepped it. The Rhineland was authentically German and has been since Roman times. The Sudetenland, on the Czechoslovakian border, included ethnic Germans among its residents but had never been an integral part of the Prussian Kaiser's empire. Hitler used the pretext of the abuse of Germanic settlers to demand annexation of the Sudetenland and its belt of Czech fortresses, The Czechs refused. The German officers knew that their newly restored army was not

ready for war with the patriotic Czechs. The latest Czech tanks were far better than German tanks. The Czech border fortifications were extremely strong. The massive munitions industry left behind by Austria-Hungary at Skoda could supply the Czechs endlessly. The German generals believed that Britain and France would help defend Czechoslovakia, which was a power in itself.

Colonel Hans Oster of the *Abwehr*, the son of a Lutheran pastor and a deeply religious man, saw the Nazi movement as anti-Christian, just as Hermann Ehrhardt had. Colonel Oster organized a group of like-minded conservatives into a new plot: if Britain and France balked Hitler in his demand for the Sudetenland—or threatened to declare war on Germany—Hitler's popularity would dip even after the Rhineland triumph: he could be arrested and put on trial at a sort of sanity hearing and locked away for life. Most of Oster's fellow officers wanted to avoid a point-blank assassination if at all possible because they believed that assassinating a head of state was treasonous.

Common sense intervened with two Ehrhardt loyalists. Friedrich Wilhelm Heinz and Franz-Maria Liedig were both teenaged volunteers for World War I. Both had joined the 2nd Marine Brigade, both had served with Organization Consul, and both—Heinz in particular—had had personal experiences with Hitler, and was serving in his bodyguard in 1938.

"Hitler alive has more weight than all the troops at our disposal," Heinz said bluntly. He and Liedig, trained as a lawyer, convinced Colonel Oster that Hitler would have to be killed in a gun battle with their own heavily armed 28-man assault company, a unit right out of the 2nd Marine Brigade. Heinz, Liedig, and Colonel Oster concealed the plans for the fatal gunfight—but not the action to eliminate Hitler—from the dozen senior officers involved. Military officers and senior police officers who were part of the "Oster Plot" were also alerted. Safe houses in Berlin were set up for the designated hitters.

On September 14—as the Heinz-Liedig assault group was already poised for action—British Prime Minister Neville Chamberlain unexpectedly announced that he would fly to Munich to meet with Hitler. Chamberlain essentially accepted some of Hitler's worthless assurances. France more reluctantly joined Britain in abandoning the Czechs, and the Poles bit off a piece of Czech territory. Mussolini and Stalin both stood pat. Hitler's popularity soared, this time not due to re-possessing territory that was rightfully German but by annexing part of another country. Hitler became so popular—more because war was averted than because the Sudetenland was annexed—that the Oster Group realized they would be seen as traitors if they struck him down. The Oster Plot collapsed without firing a shot.

Hitler blundered again a few months later. A resentful 17-year-old male prostitute named Herschel Grynszpan, evicted from Germany to France because he was a Polish-born Jew who had never become a German citizen, murdered his male lover, a German diplomat named Ernst vom Rath. Everbody in the Paris gay community knew that Grynszpan and Rath were lovers. But political types on both sides tried to pretty the story up. The Nazi propaganda mill claimed that Rath was a clean-living humanitarian who had just resigned from the Nazi Party when he was murdered by a Jewish fanatic. In fact, Ernst vom Rath was a member of both the SA and (illegally, since he was homosexual) of the SS. Rath and his brother had both been treated for anal gonorrhea, a signature homosexual disease. The anti-Nazi propaganda mill said that Grynszpan had avenged his mother and father, who froze to death on the Polish border when neither the Germans nor the Poles would grant them food and shelter. In fact, the Poles had somewhat grudgingly granted them food and shelter and Grynszpan's parents were still alive in Israel 20 years after the Rath shooting. The homosexual homicide—not revealed as such in Germany or in the Jewish community—was the catalyst for the outburst of *Kristallnacht*—a massive Russian-style *pogrom* in which Nazi thugs and other riffraff burst into Jewish shops and homes to steal, and sometimes to kill.

In one case, a Nazi broke into a fifth-floor apartment and a brave little lapdog ran up and barked at him. The Nazi threw the dog out the window. When the old lady who owned the dog screamed, he shot her. About 91 Jewish Germans were murdered, and 25,000 others were beaten up or sent to concentration camps for more abuse, awful food, and continuous threats. Jewish Germans who had once hoped that Hitler would fall from power began a frantic scramble to get out of Germany and Austria. The terrified Jewish refugees encountered the ghosts of Rosa Luxemburg and Eugen Leviné-Nissen at every turn. Many European countries were still haunted by the German Spartacists of 1919. They were reluctant to accept Jewish refugees because of the fear of Marxist or Bolshevik revolutionaries concealed among them. Cordell Hull, U.S. Secretary of State—whose wife's father was Jewish—refused to allow a German ship loaded with 936 Jewish refugees to dock in the United States while agitators warned of Communism. The ship took the refugees back to Europe, where several hundred of them died fighting Hitler or in the concentration camps. The British in particular were reluctant to accept Jewish refugees. As a compassionate move, the British began a program called Kindertransport, which accepted only *unaccompanied* Jewish children under the age of 17. Kindertransport saved about 10,000 children's lives. The United States eventually allowed German Jews to use Germany's 1924 immigration quota, which permitted 27,000 people per year to come to the United States. For most desperate Jews, refuge included the Dominican Republic, Mexico, and the Japanese Empire, which sheltered about 40,000 Jews in occupied Shanghai or Manchuria but usually not in the Japanese home islands. Millions of other Jews died because of two sets of maniacs: the minority of Russian Jews who tried to take over Socialist Germany as if they were still in Tsarist Russia, and the minority of incredibly perverse and twisted German Nazis who saw hatred and murder as solutions to their own inadequacies as human beings.

Kristallnacht got the Jews out of the German civil service, but it also ruined German's international standing as a civilized country as long as Hitler was alive.

Even Charles Lindbergh, whose Darwinian beliefs at that time could accommodate the non-violent aspects of Nazi Aryan racism—Lindbergh was actually more worried about "Asiatics" than Jews—cancelled serious plans to move his family to Germany. Hermann Ehrhardt apparently cancelled any plans he may have had to move back to his home town. But he had one more act to play in German history after his first brush with fame under another name in a Hollywood movie.

Chapter Twenty-Nine
Half-Lives in Hollywood

For most Americans who have never heard of Hermann Ehrhardt, the definitive account of Germany in World War I was "All Quiet On The Western Front" by Erich Maria Remarque. Yet people who have read Remarque's books beyond his first novel, or those who have seen Hollywood movies based on his novels, have encountered Hermann Ehrhardt not once but twice—shortly to be followed by a movie portrayal of "Captain *Ehrlich*" by John Wayne.

The Ehrhardt figure first turns up simply, or not so simply, as "Lieutenant Heel"—no first name—in "The Road Back," Remarque's second novel and in some ways a sequel to "All Quiet On The Western Front." The name "Heel" doesn't mean anything in German—*Ferse* is the German word for the back end of the foot—but Remarque, a World War I veteran born in 1898, came from Osnabrück in Hannover, British territory until 1837. Like most Hanoverians, he understood some English. Remarque probably knew what a "heel" was in Anglo-Saxon slang, especially after Jack London's 1908 novel, "The Iron Heel," made the term a synonym for a bully, a tyrant or a dictator. The character in "The Road Back" lives up to his name. Heel keeps his weary men fighting off a pointless British attack until the last minute before the Armistice.

Heel gets his last chance to fight when the British make a wanton grenade attack on the last night of the war.

Wessling, a farmer with a family and a four-year veteran of the Front, is fatally wounded by a British bullet in the last fight. The other men, most of them teenagers, haul him to the aid station. Heel covers their retreat and kills four British soldiers with a grenade and two with a pistol.

Wessling dies—through no fault of Heel's, who emerges as a front-line officer long on courage if somewhat short on compassion for the enemy. But Heel has some bad moments ahead of him. The news of the Revolution at home threatens to destroy him. Max Weil—the Kurt Eisner figure—suggested that Heel give up his epaulets. Heel bluntly refuses—but when he learned that the Kaiser has fled and officers have been deposed, he is mortified.

Max Weil—a probable stand-in for the anarchist-socialist Kurt Eisner of Munich, as Heel is for Hermann Ehrhardt—tries to get the younger men to set up a Soldiers' Council but most of them aren't interested. Finally they elect Weil, Ludwig Breyer, the company's junior lieutenant, and Adolf Bethke, a steady soldier with a wife and kids at home. Max wants Ludwig and Heel to give up their epaulets. The enlisted men decline to confront Heel—a real fighter who has their respect —but he gives his epaulets up voluntarily.

Later that night some of the soldiers hear a noise and quietly sneak up to peer into the company office.

"Heel is seated at the table. His blue officer's jacket, the litefka, *is lying before him. The shoulder straps have gone. He is wearing a private's tunic, His head is in his hands and—but no, that cannot be, I go a step nearer—Heel, Heel is crying.... Next morning we hear that a major in one of the neighboring regiments shot himself when he heard of the flight of the Emperor....* [Next morning] *Heel is grey and worn with sleeplessness. Quietly he gives the necessary instructions. Then he goes again. And we all feel just terrible. The last thing that was left to us has been taken away—the very ground cut from under our feet."*

Remarque's birth name was Erich Paul Remarck—originally Remarque for French ancestors who had fled to Germany, as Pabst's wife's family had, during the anti-religious phase of the French Revolution. He was a wavering Catholic, a pacifist and humanitarian, and a *bon vivant*, and perhaps for these reasons he was not a Communist supporter. In "The Road Back" Remarque sets up Lieutenant Heel— the Hermann Ehrhardt figure—as the spokesman for the monarchists and militarists, but not for the Nazi movement, which Remarque always detested. Max Weil—the Kurt Eisner figure—becomes the spokesman for the Leftists and Social Democrats, perhaps even the Independent Social Democrats, but not for the Russian-led Bolsheviks. Each adversary stands on the opposite political perimeters of Right and Left without verging into mania. The last civil conversation between them as the company is demobilized captures this.

"Heel tries to say a few words in farewell. But nothing will come. He has to give up. No words in the world can take the field against the lonely, empty barracks square and these sorry ranks of the survivors...."

"*Heel passes from one to another and shakes hands with each man. When he comes to Max Weil, with thin lips, he says 'Now your time begins, Weil—'*

"*'It will be less bloody,' answers Weil quietly.*

"*'And less heroic,' Heel retorts.*

"*'That's not the only thing in life,' says Weil.*

"*'But the best,' Heel replies. 'What else is there?'*

"*'Weil pauses a moment. Then he says, 'Things that sound feeble today, Herr Lieutenant—kindliness and love. These also have their heroisms.'*

"*'No,' answers Heel swiftly, as though he had already long thought upon it, and his brow is clouded, 'They offer only martyrdom. That is quite another thing. Heroism begins where reason leaves off; when life is set at a discount. It has to do with folly, with exaltation, with risk—and you know it. But little or nothing with purpose. Purpose, that is your word. Why? Wherefore? To what end?—Who asks these questions, knows nothing of it.'*"

Max Weil looks at Heel in his new private's tunic and thinks his side has won. But a decade had passed between the publication of "The Road Back" and the Berlin riots and murderous second Soviet Republic of Bavaria. Remarque knew, without taking sides, who the winner was in those showdowns: the idealistic anarchists let in the murderous Bolsheviks and were then murdered themselves by outraged citizen soldiers or policemen—not, as the myth would have it, by Organization Consul, which did not yet exist. The final show-down between Heel (Ehrhardt) and Weil (Eisner) takes place during what could be the Berlin Transportation Strike just after the signing of the Treaty of Versailles.

"*Rumors are flying. It is said that the military have already fired on a procession of demonstrating workers.... The Reichswehr has taken up position in front of the Town Hall. The steel helmets glean palely.... It would be madness to go farther. The machine gun is covering the square.*

"*But one man is going out all alone. Behind him, the seething crowd surges down the conduits of the streets, it boils out about the houses and and gathers together in black clots....*

"*'Back!' calls a clear sharp voice... that was Heel's voice... a warning volley crackles out upon the air. Suddenly, the man wrenches himself free. But no, he is not saving himself. He is running toward the machine gun!*

"*'Don't shoot, Comrades!.'*

"*Still nothing has happened. But when the mob sees the unarmed man run forward, it advances.... The next instant a command resounds over the square. Thundering, the tick-tack of the machine gun shatters into a thousand echoes from the houses, and the bullets, whistling and splintering, strike on the pavement.*

"*We see an officer come down the steps. Without knowing quite how, we are suddenly all standing there beside Ludwig awaiting the coming figure that for a weapon carries only a walking stick. He does not hesitate an instant....*

"'I congratulate you, Lieutenant Heel, the man is dead... Do you know who it is?...'

"Heel looks at him and shakes his head.

"Max Weil.'

"I wanted to let him get away,' Heel says pensively....

"Ludwig looks at him coldly, 'A nice piece of work....'

"Then Heel stirs. 'That does not enter into it.' he says calmly. 'Only the purpose—law and order.... My men stay where they are! If they withdraw, they would be attacked tomorrow by a mob ten times as big. You know that yourself. In five minutes I occupy all the road heads. I give you till them to take off this dead man.'"

Remarque then describes the agony of spectators hit by stray bullets and the rage of the crowd, but his treatment is oddly—perhaps professionally—even-handed. Max Weil is shown as naïve and deluded, the crowd is a genuine threat, and while Heel is shown as arrogant he neither fires without warning nor continues shooting after the crowd breaks up. Above all, Heel is no coward. He walks into the group around Weil's body, armed only with a stick, with no fear for his own life—a contrast to Hitler's cowardly flight from the firing at the Beer Hall Putsch, where Hermann Ehrhardt had refused to support Hitler and was plotting to ambush him on the road to Nuremberg, and where Ludendorff strode bolt upright through the police firing line protected only by his fame and reputation. Ehrhardt had already shrugged off Ludendorff after the Kapp Putsch. He saw Hitler as perverse and vicious—and so did Remarque. Lieutenant Heel was Hermann Ehrhardt as of Remarque's reading in 1931.

Remarque's novel of refugees, "Flotsam," published ten years later in April of 1941, became the fulcrum for the Hollywood feature film "So Ends Our Night," where an Ehrhardt character played by Fredric March becomes the undisputed hero of the movie.

The character of Josef Steiner, an anti-Nazi and a decorated World War I veteran, becomes the older friend of two young Jewish refugees, and the casual mentor of other anti-Nazi or anti-Communist fugitives. The original novel "Flotsam's" Josef Steiner bears some circumstantial resemblance to Ehrhardt but the resemblance is inconclusive: Remarque may not have been ready to give a hard-core militarist and monarchist a clean bill of health—even one like Ehrhardt who sometimes rescued Jews and always hated Hitler. The novel's mature hero, Josef Steiner, respects Jews and Judaism, as Ehrhardt often did—but Steiner also befriends Russians, and Ehrhardt was no fan of Russians, though Remarque doted on them if they were not Bolsheviks. Steiner doesn't much like Poles. Ehrhardt liked Poles as much as politics permitted, though he may have considered the Poles as potential German reinforcements against the menacing Russians. Remarque's Josef Steiner cracks some blasphemous jokes. Hermann Ehrhardt was reverential in his speech and advocated Christian principles to anyone who would listen. Remarque's Josef

Steiner, holder of the Iron Cross after a patrol in 1915, clearly served as an Army enlisted man, not a naval officer.

But the fim version, "So Ends Our Night" as directed by John Cromwell and written by Talbot Jennings, also the son of a clergyman, is actually a tribute to Hermann Ehrhardt with everything but his name on the cast of credits. In the film version, the *Gestapo* agent played by Erich von Stroheim shows Steiner (the Ehrhardt character) his paperwork file with the word *HAUPTMANN* (Army Captain) visibly stamped on them and congratulates him on his escape from Dachau, said to be more impressive than any of his wartime exploits, which were numerous. The Gestapo agent is called "Steinbrenner" in the novel and "Brenner" in the movie. In the movie, Brenner (Stroheim) admires Steiner (Ehrhardt) as a sort of heroic role model despite political differences. In Remarque's novel, Steinbrenner of the *Gestapo* appears to be a homosexual sadist maliciously coming on to a straight guy who isn't interested. The movie Ehrhardt character, at one point, appears to let himself be picked up by a rather eager woman but later resents not so much the woman as himself, for his betrayal of his own politically estranged wife. He later passes up a chance at another attractive woman who dotes on him. In the movie, an Austrian policeman slaps Steiner (Ehrhardt) in the face and Steiner takes the policeman's name and tells him his rudeness is out of line. The movie Steiner later stalks the Austrian policeman, taps his shoulder, and knocks him out cold in an alley with a single punch, like an American he-man. In Remarque's novel, Steiner (Ehrhardt) knocks the policeman's teeth out and stomps him, perhaps fatally. Then he feels guilty about it. The real Ehrhardt knew enough *jujitsu* to kill without weapons or any theatrical brutality. Hitler was shrewdly afraid to be in the same room with him. Ehrhardt, however, is never known to have murdered either Judge Metz or Prosecutor Schneider: had either of them turned up dead, he would have been on the short list of suspects.

Some scenes in the novel and the film are very much the same. Joseph Steiner slips into Germany and urges the estranged wife he still loves to divorce him so that she, at least, will be safe from Nazi vengeance. When she later becomes fatally ill, Steiner slips back into Germany again to console her on her death-bed. This alludes to two real Ehrhardt incidents covered in "Adventure and Fate": Ehrhardt's risky ride from Münster to Hamburg to spend time with his dying mother—followed by the arrest warrant served at her grave—and his heroic or foolhardy attempt to convince Judge Metz in Munich that Princess Margrethe never knew her house guest, Baron Eschwege, was really Herman Ehrhardt. As the movie Josef Steiner (Ehrhardt) gets on the train for Germany to re-join his dying wife, the award-winning background music plays Ludwig Uhland's "*Ich hatt' einen Kameraden,*" written in 1809, set to music in 1825, and known around the world as the official funeral song of the German soldier. As Josef Steiner (Ehrhardt) crosses into Nazi Germany, customs officials check his passport and ask him for his address. Steiner says: "*Krems on the Donau.*" Krems on the Donau [*Danube*] was Hermann

Ehrhardt's Austrian town address after his escape from the Gestapo following the Mylius-Ehrhardt Plot to kill Hitler in 1934: Josef Steiner's address in Remarque's novel FLOTSAM is simply given as "*Graz.*" Steiner's wife dies at peace after his loving vigil. Steiner's final interview with the Gestapo reveals no names of his friends in the resistance. Steiner's friends in the anti-Nazi resistance back in Paris slip the two Jewish kids an envelope he sent them with 2,340 *francs* and two tickets for Mexico—which accepted large numbers of Jewish refugees—on a Portuguese freighter.

"He wasn't afraid to hate evil—to live with self-respect—to die to keep it," his younger friend Ludwig *Kern* says with admiration. Ehrhardt also had a friend named Kern—Erwin Kern, whose unwarranted murder of Walther Rathenau landed him in prison.

The real Hermann Ehrhardt had 30 years of life left to him when "So Ends Our Night" was released in American movie theaters in 1941. The cinematic treatment might very well have made him blush had he ever seen it. He would have known instantly that the movie's Steiner character was supposed to be him, but he also might have thought the Steiner character was a little too good to be real. But the Nazi censors obviously didn't authorize this sort of movie for release in Occupied Europe, let alone Germany.

In 1943 a cinematic German anti-Nazi named Franz *Erhart*—close to the actually spelling, if not the actual Ehrhardt personality—also cropped up in "Hitler's Children" played by the versatile Lloyd Corrigan. The "Hitler's Children" Franz Erhart, a minor figure in a B movie, helps re-unite the two tragic lovers without any drastic heroics on his own part. Lloyd Corrigan's Erhart role is not especially prominent but the film, which actually featured cowboy star Tim Holt and Nancy Drew prototype Bonita Granville as the two starred-crossed and doomed Aryan Hitler protestors, reached Number Four at the U.S. box office once America was at war with Germany.

The films, however, suggests how Hermann Ehrhardt was able to survive his final clash with Hitler—the one that followed the July 20 Bomb Plot explosion. He was already a legend and the known execution of a heroic anti-Bolshevik leader would have been unacceptable to the Army and to the German public.

The last previous plot against Hitler had lapsed when the British and the French failed to show any fight on behalf of the Czechs at Munich and Hitler's seize of the Sudetenland in 1938.

Hitler's subsequent seizure of Czechoslovakia, a genuine European democracy, had made him a hero to the dimmer or more racist sections of the German public. Ehrhardt's agents, Friedrich Wilhelm Heinz and Franz Maria Liedig, the inside men on the 1938 arrest-and-execute plot with Ehrhardt's advice and support, knew that Hitler had suddenly become so popular that any armed resistance to his rule would have been heavily outnumbered. Hitler, however, was deluded by his own success and reasoned that once he was allied with Stalin in 1939, the British and the French would never fight for the Poles, just as the British and French had decided not to fight for the Czechs. Hitler miscalculated. The German troops who marched off to World War II marched off to dead silence. The American journalist William L. Shirer, on hand in Berlin, saw that even the Nazi leaders themselves were stunned and depressed. Hitler, of course, had already predictably blamed the war he himself had launched on the Jews in a speech at the Berlin *Sportpalast* on January 30, 1939. Long before the invasion of Poland: "If the world of international financial Jewry, both inside and outside of Europe, should succeed in plunging the nations into another world war, the result will not be the Bolshevization of the world and thus a victory for Judaism. The result will be the extermination of the Jewish race in Europe." Hitler's logic was splattered: those Jews who played a role in international banking were hoping for his death by assassination or apoplexy but had no more interest in fostering Bolshevism than Albert Ballin, Walther Rathenau, or Hugo Preuß.

Prince Oskar, the oldest son of Prince Oskar von Hohenzollern and a lieutenant in the German Army, was killed serving with his regiment in Poland on September 7, 1939. His valor won approving mention from fellow officers and in the German press. The stunning German victories in Belgium, the Netherlands and France starting on May 10 in 1940 and the British evacuation at Dunkirk—billed as some sort of moral victory to the intense disgust of the French, who were still fighting two weeks after the British left—brought a second wave of German euphoria, but a warning as well. Princess Viktoria Luise reported that hundreds of German officers stopped off to visit her father, the former Kaiser, at his mansion of exile in Doorn in the Netherlands and insisted on shaking hands with him. When the Kaiser's grandson, Prince Wilhelm von Hohenzollern, son of the priapic Prussian Crown Prince Wilhelm, was fatally wounded courageously leading German troops against the French on May 23, 1940, his military funeral at Potsdam brought out 50,000 mourners despite Nazi disapproval of Prussian royals. Most of them were generals, colonels, or decorated combat officers of the *Wehrmacht*. The huge turnout was seen as sparking renewed interest in restoring the monarchy, at least by many senior officers and combat leaders. Hitler shortly ordered that no more members of German royal houses would take part in combat. The Battle of Britain was a second warning: the Royal Navy and the Royal Air Force, aided by Canadian, Czech and Polish pilots and a handful of American volunteers, handed Hitler his first serious defeat. Hitler and his generals now knew that if either Russia or the United States came to Britain's aid, the war was as good as lost.

German officers stationed in Belgium did some talking over their drinks and Belgian Communist waiters passed along the word that Russia was next. Stalin apparently didn't believe that Hitler would betray their alliance. But the NKVD believed it and dispatched Vitaly Pavlov, a 27-year-old secret agent, to reactivate Harry Dexter White, an intense Communist sympathizer in the U.S. Treasury Department who had dropped out of the espionage network a few years before due to fear of exposure by his Communist courier, Whittaker Chambers. The Soviet NKVD desperately wanted a war between the United States and Japan so that Russia would not have to fight on two fronts at the same time—Germany in Poland and Japan in Siberia and Mongolia. White was an expert in Chinese and Japanese economics and the principal brain transplant for U.S. Treasury Secretary Henry Morgenthau Jr., one of Franklin Delano Roosevelt's neighbors and best friends. Henry Morgenthau Sr., as Woodrow Wilson's ambassador to Turkey, had tried to foster the myth that the Kaiser's Germans had planned and supported the Turkish deportation of Armenians during World War I based on their genetic penchant for homicide. The British had detained the German military commander in Turkey, Otto Liman von Sanders—who was himself of mixed Jewish ancestry—and after two years of questioning found no evidence of German complicity in the Turkish mass murders of Armenians. Franz von Papen and Alexander von Falkenhausen had both served in Turkey—but so had Rudolf Höss, future commandant of Auschwitz. The name 'Morgenthau' was familiar to the Germans of the Hitler era.

On June 22, Hitler invaded the Soviet Union. Red Army resistance, after years of Soviet purges of generals and colonels, seemed to be collapsing. The Nazi armies were followed by *Einsatzgruppen*, Special Action Groups, which murdered Jewish men and boys in cold blood as a possible threat to security and left the women and girls to the non-existent mercies of local anti-Semites who blamed all Jews for Communism. Henry Morgenthau Jr. was no Communist, and he was no genius either, but as an observant Jew he had logical reasons to hate Nazism and Hitler. With the Soviet agent Harry Dexter White pulling his strings, and the Anglo-Saxon State Department lawyer Dean Acheson and Asian expert Stanley Hornbeck, descended from a New England Tory family, backing a tough policy with Japan, "Morgenthau"—actually White—drafted the background policy memorandum for the Hull Note of November 26, 1941—an ultimatum to Japan so drastic that Japan, both insulted and threatened, finalized plans to attack Pearl Harbor long before the United States or Britain were ready for war in the Pacific.

Hitler now had the scapegoats he needed. He had two important American Jews supposedly helping to start a war with the United States—even though Hitler himself and the Japanese had both declared war on the United States before FDR had the chance to declare war on Japan and Germany. Those Jews overrun during the Nazi invasion and not murdered outright were drafted into a slave labor program: On June 20, 1942, the Jewish slave labor program was officially turned into a mass murder program as Hitler had promised in 1939.

"With regard to the Jewish question the Führer has decided to make a clean sweep of the table," a broadcast reported on December 12, the day after Hitler declared war on the United States. "He prophesied to the Jews that if they again brought about a world war, they would live to see their annihilation in it. That wasn't just an idle phrase. The world war is here, and the extinction of the Jews must be the necessary consequence."

The Wannsee Conference of June 20, 1942, finished up the details. Jews would be used up on construction projects under brutal conditions and those who actually survived the overwork and starvation rations would be eliminated to make sure the Jewish population did not experience "a natural selection" of tough survivors breeding tougher Jews. The head instigator was Reinhard Heydrich, a hawk-nosed avid violinist who had been dogged all of his life by rumors of mixed Jewish ancestry. Admiral Wilhelm Canaris reportedly had photo-copies of documents that proved Heydrich's mixed Jewish ancestry and used them to blackmail Heydrich for his own protection. Neither of them would outlive the war. Heydrich was assassinated by Czech partisans in 1942 and Canaris—an old friend of Hermann Ehrhardt and Waldemar Pabst from the 1920s—was executed for his role in the last of the plots against Hitler.

Most of the 34 to 42 documented plots to kill Hitler were the work of individuals—a Christian theological student, idealistic Jewish kids, befuddled home-grown Communists, or dissident lower-ranking Nazis. As the word got out that huge numbers of Jews were being murdered and that the German Army would have to explain these murders after the inevitable defeat, middle-ranking German officers serving mostly on the Russian Front became a matrix for plots that involved large numbers of professional soldiers who knew something about killing, at least in battle. But the officers didn't know much about assassinations. Some of them appear to have tapped Hermann Ehrhardt, whose legendary career as the head of Organization Consul and whose long-standing hatred of Hitler made him the ideal prompter. Most of these men, like Ehrhardt himself, were opposed to extreme anti-Semitism. More to the point, most of them were also hoping that once Hitler and his crew of criminals were dead or under arrest, the Western allies would either agree to an armistice or to a separate peace which would allow them to defend Germany against Russia. At the very end, some of them, like Henning von Tresckow, once the youngest lieutenant in the Kaiser's army, and Baron Claus von Stauffenberg, who had suffered multiple battle wounds in North Africa, saw their deaths as a spiritual expiation for tolerating Hitler's murderers and the Nazi movement for as long as they had.

When Stauffenberg's briefcase bomb detonated in Hitler's headquarters on July 20, 1944—coincidentally, Princess Margrethe's 50th birthday—a brief organized revolt broke out in Berlin. But Hitler was alive and he called out the SS and the police for a crackdown on conspirators and suspects. Most of the first-string would-be Hitler-killers were executed as quickly as possible to shut them up nd hide the fact that they were war heroes.

Hermann Ehrhardt was arrested immediately. He knew about the plot and had failed to inform, and he probably had helped organize some of his old followers and admirers to take on the SS—but with Hitler alive, the indecision if not the cowardice of the many senior officers who sat on the fence doomed the courageous if belated take-over bid, just as the indecisive senior officers had doomed the Kapp Putsch.

Ehrhardt himself remained in custody at various prisons for the remainder of the war. A surprising number of non-violent suspects were not executed. One was Alexander von Falkenhausen, who had briefly served as Ehrhardt's chief of staff during the demobilization of the 2nd Marine Brigade. Falkenhausen, a consistent and stubborn anti-Nazi, after having served as Chiang Kai-shek's top German military adviser until 1937, later uneasily took over as military governor of Belgium and Northern France in 1940. Falkenhausen had quietly offered to support the plotters with his garrison troops if they succeeded in killing Hitler. He had pointedly failed to inform on them. Falkenhausen had dragged his feet about deporting Belgian Jews for slave labor but had no such qualms about deporting Communists. He had no qualms about supporting the removal of Hitler either. Falkenhausen spent a year in Dachau and other concentration camps.

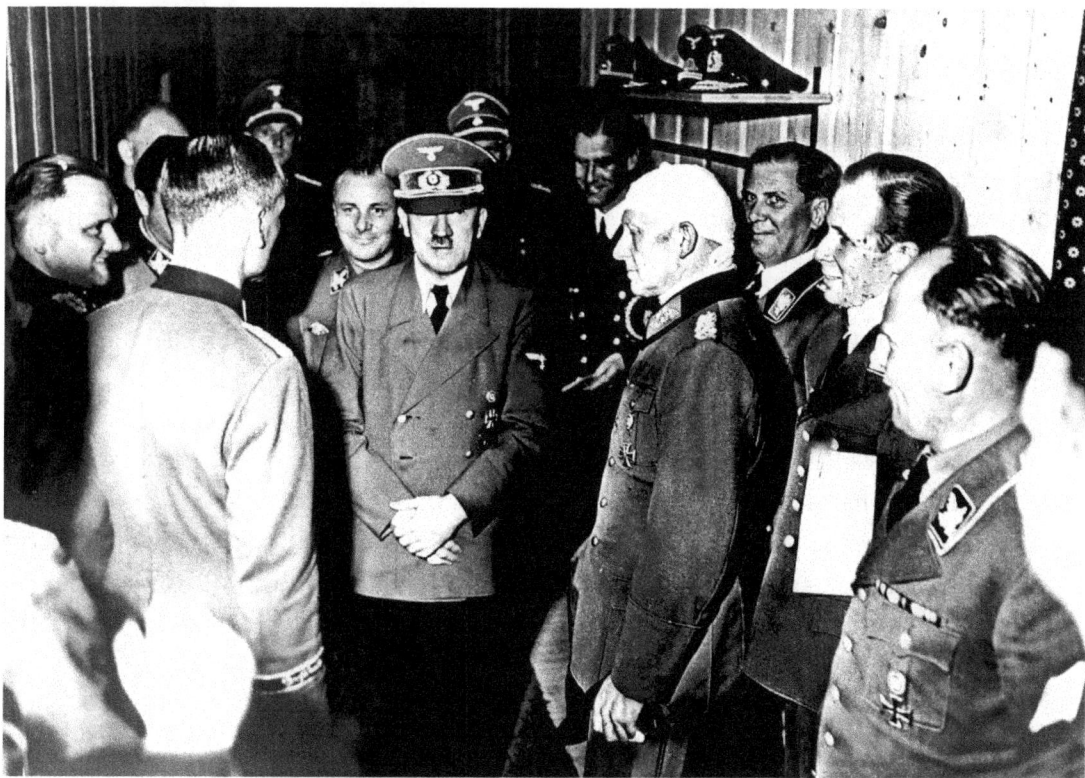

A seemingly unshaken Hitler

Franz-Maria Liedig was also picked up and spent a year in concentrations camps once paperwork disclosed his role in the Oster Plot of 1938 and his failure to inform in 1944. Perhaps the Nazi authorities were saving Falkenhausen, Liedig, and Ehrhardt for a hecatomb at what some fanatics still hoped would be the eventual Nazi victory due to the V-2 ballistic missiles raining on London and the new late-war German jet fighters and high-speed Type XXI U-boats. Perhaps nobody wanted to execute Alexander von Falkenhausen, a remote Prussian royal and holder of the *Pour le Mérite,* or Hermann Ehrhardt—the single greatest leader against the Bolsheviks in 1919 and 1920 and a man whose friends might avenge him at all costs. Hartmut Plaas, who had escaped arrest previously due to his adolescent anti-Semitism and his agitation for oppressed small farmers, a favorite Nazi cause, was caught in 1944 and shot by a firing squad at Buchenwald. Friedrich Wilhelm Heinz, once trusted as Hitler's bodyguard leader, deftly went underground and survived the war. He later worked as a journalist for *Time* and *Life.* Waldemar Pabst had served in the German Army—all the better to get at Hitler—but he was tipped off in time and beat it to Switzerland before the Gestapo closed in. Falkenhausen and Liedig were located and released from Nazi custody in South Tyrol with a group of mostly foreign political prisoners after Hitler killed himself.

Hermann Ehrhardt was set free at the end of the war in May of 1945 and returned to Princess Margrethe and his two teenaged children.

Ehrhardt was done with politics once and for all. A monarchist at heart, he appeared to have no desire to attempt a regeneration of the generation that had abandoned the Kaiser and tolerated Hitler. The Germans of the generation that had supported Hitler filled him with dismay. The feeling may have been mutual. Many late-in-the game Nazi backers like Franz von Papen used the threat of Communism as an excuse for Nazism: Hermann Ehrhardt, Germany's leading anti-Communist in 1919 along with Waldemar Pabst, had tried to block Hitler's rise to power through electoral politics, and planned or backed multiple Hitler assassination attempts long before the Tresckow-Stauffenberg July 20 Bomb Plot. Ehrhardt did what more Germans should have done. Some Germans never forgave him for doing things they themselves never dared to do.

On May 4, 1945, as the Third Reich collapsed, Heinrich Tillessen was arrested and was charged with the murder of Matthias Erzberger in 1921. Tillessen, once a deck officer with Hermann Ehrhardt's 17th Torpedo Boat Division, later an officer of the Ehrhardt Brigade and a gunman for Organization Consul, was let off the first time because he had received an unconditional amnesty from President Paul von Hindenburg – and Chancellor Adolf Hitler – in March of 1933. Tillessen, however, had served Hitler's Kriegsmarine as a Naval desk officer during World War II. The West German court under Allied supervision shrugged off the amnesty and tried Tillessen for murder but trial ended in acquittal. Tillessen was then tried a second time by a West German court under French auspices. The proposed death sentence was rejected, but in 1947 Heinrich Tillessen landed in prison for 15 years. Ehrhardt had been delighted with Erzberger's murder, had said so in print in Adventure and Fate, and had also stated his admiration for Tillessen as a patriotic idealist. Erzberger's demand for a "Jew Census" had helped foster Hitler's rabid anti-Semitism and his political demand for a "Peace Resolution" in the Reichstag hand undermined the attempts by both Pope Benedict XV and Wilhelm II's Jewish advisors to broke an equitable armistice. Herman Ehrhardt thought in straight lines and must have reasons that Erzberger was a traitor to Germany just as Hitler was a traitor to humanity. Both belonged dead. Ehrhardt can only have decided that he had had his day in court in Allied-occupied Germany with Judge Metz in 1922 and had no desire for a replay.

Waldemar Pabst, Franz-Maria Liedig, and Paul Schulz ventured back to West Germany to help rebuild the nation, or what was left of it once the Morgenthau Plan for massive destruction was scrapped and the Marshal Plan took over. Waldemar Pabst became extremely wealthy in the munitions industry. Paul Schulz had pulled strings during the war to get several Jews he personally knew to be patriotic Germans excused from deportation—which is to say, near-certain death. While supplying information to the *Abwehr* to defend Germany from the Allies, Schulz was able to divert at least one transport of Jews being sent to Auschwitz to safety in Switzerland. After 1945, Schulz worked in his deceased father-in-law's company providing Germans with jobs in a tough economy. Franz Maria Liedig, despite his damaged health after months in concentration camps, became a force in Catholic politics and was one of the founders of the present Christian Socialist Party. Georg Rosen, the German Jewish rescuer at Nanking, pulled strings at Oxford so that Germans could once again qualify for Rhodes Scholarships as he himself once had. Emil Ludwig, the German Jewish biographer of Napoleon who had Hitler's number in World Book Encyclopedia, did what he could to revive memories of German culture from the era of Goethe and Schiller. John Weitz, whose father had been decorated in World War I and who had himself served in the American O.S.S., forerunner of the CIA, attempted to revive German and American business contacts and rebuild the German economy.

But Hermann Ehrhardt didn't seem very interested in Germany after the war. Perhaps he had expected too much from the average German. He had noted several times that the Communists seemed to thrive on the timidity of the average *Bürger* and that Nazis also relied on citizen apathy induced by fear. The estate at Brunn am Wald and its water-moated castle became a little kingdom in itself, a place out of time where the tenant farmers took his reasonable orders. In 1948, Herman Ehrhardt became an Austrian citizen. The estate grew gradually under Ehrhardt's management and he renewed his boyhood fondness for pot hunting, hands-on agriculture and property improvements. By 1961 the estate had enlarged from the historic 36 households in 1840 to 42 houses with 205 residents, and a few other houses were constructed in the next decade. Many of the new houses were vacation rental units rather than full-time residences for tenant farmers. Hermann and Margrethe, tolerant Lutherans in a Catholic neighborhood, built a chapel in 1966. They must have had a lot to meditate on there. The house that Ehrhardt's father-in-law had built for Hermann and his first wife in Wilhelmshaven had been completely destroyed in an Allied air raid in 1944. One of his first wife's relatives, General Werner von Gilsa, had kept quiet about the 1944 plot to kill Hitler at the risk of his own life. General von Gilsa was later surrounded by the Russians during the capture of Dresden. He opened up the military food stores to the civilian population and then committed suicide on May 8, 1945 rather than surrender even after the war was over. Werner von Gilsa was Hermann Ehrhardt's kind of German. He lived and died a nobleman along with an estimated 223 German generals killed in combat or in suicide. Many other German generals had not done so. One of Ehrhardt's own former men, Baron Manfred von Killinger, escaped the 1934 purge but later had become a Nazi murderer of Jews in Romania. Killinger also committed suicide to avoid capture for less creditable reasons. Most 2nd Marine Brigade veterans served more honorably, but they served Germany under Hitler all the same. Ehrhardt never did.

But Ehrhardt had one more Hollywood appearance. "The Sea Chase," starring John Wayne as "Captain Ehr*lich*"—whose name means "honest"—was released in 1955. David Farrar, as a rather stiff British naval officer, plays a friend and admirer of "Captain Ehrlich," a former Imperial German Navy captain who is now a merchant skipper in charge of a rust-bucket freighter about to be interned in Australia in August of 1939.

"I see you still carry the old Imperial flag," the British captain tells Captain Ehrlich. "Within a week, we'll be fighting Nazism and you've been fighting it since its inception.... Have you ever in your life made a compromise with a conviction?"

The original novel by Andrew Geer (1905 – 1957) was adapted by John Farrow, the producer and director and himself a former naval officer, with the help of James Warner Bellah, a sometime screenwriter for John Ford, and by John Twist. They made some notable changes to the plot. In Andrew Geer's original novel, the ship's captain is a villainous fanatic and the sensible first officer is the actual protagonist. In the movie, the first officer is a creepish Nazi murderer played by Lyle Bettger, the female lead is Lana Turner playing a blonde *femme fatale* whom the captain reforms through true love, and the captain is obviously Hermann Ehrhardt. "Captain Ehrlich" mentions prayer with respect, quotes the Greek classics, and behaves with a tough but genuine nobility and great maritime skill. His rough-neck crew is on the verge of mutiny when the voyage starts but by the end they respect him almost to awe. "Captain Ehrlich" actually slips though the British in the old freighter but is betrayed by the Nazi authorities off the storm-bound coast of Norway for political reasons. He puts the crew safely over the side in the lifeboat, throws the swastika out of the flag locker and onto the deck, and tramples it on his way to the halyard.

"That's not the swastika!" a younger British officer asks David Farrar. "What flag is *that?*"

"The Imperial Battle Flag!" Farrar says. "And I think he's going to try to ram us."

Bosley Crowther of the New York Times probably never heard of *Korvettenkapitän* Hermann Ehrhardt, of his anti-Communism in 1919, or his four plots to kill Hitler between 1923 and 1944. Crowther's review dismissed the premise of "a fine, firm Imperial German with an odd desire to get back to the 'homeland' so he can spit in Hitler's eye... a ridiculous motivation for a fellow to run the gauntlet he does." The fact that the most organized Hitler resisters were members of the officer caste rather than liberals or Leftists was one of the best-kept secrets of the FDR propaganda mill. FDR, lost in time, drink, and toward the end on regular does of medical cocaine, blamed the whole war on the Kaiser's Prussian officer corps. His leftist advisors, including columnist Dorothy Thompson, who claimed to be of German ancestry but wasn't, chortled with him over how many German officers they would be able to execute or castrate once they won the war. U.S. Treasury Secretary Henry Morgenthau and his brain transplant, the Soviet agent Harry Dexter White, schemed to break defeated Germany into five de-industrialized zones where the peasants could live on potatoes: Morgenthau and White considered 20 million civilian Germans dead by starvation and disease to be an acceptable figure—an echo of Clémenceau at Versailles. Once Roosevelt was dead, Stalin was turning defiant, and Harry Truman was apprized of how many Soviet agents were concentrated in Morgenthau's U.S. Treasury Department, Henry Morgenthau and Harry Dexter White were both run out of the government. Postwar German civilian casualties including refugees from eastern Germany and late-war soldiers who surrendered voluntarily were closer to 4 million than the Roosevelt administration's gleefully anticipated 20 million.

At the end of "The Sea Chase," the narrator raised the faint hope that "Captain Ehrlich" and his beloved found a tiny island where they lived peacefully after the sinking of the old freighter with the Imperial Battle Flag still flying from the mast. Brunn am Wald was the tiny island and while Princess Margrethe was hardly a *femme fatale* before her path fatefully crossed Hermann Ehrhardt's, they were linked for good and remained together at their fiefdom near Krems an der Donau.

Bernd Rottenecker, a local historian in Germany, reported that sometime in the 1960s Ehrhardt made a sentimental journey back to Diersburg in Baden. Rottenecker, a teacher at the Erich Kästner-Realschule there, reported that by 2013 there were few members of the Ehrhardt family left in Diersburg. Members of the family, in fact, report in from time to time from Austria, Britain, Brazil and the Midwestern United States where they generally have respectable positions in their communities but seem understandably confused as to what Hermann Ehrhardt's life was all about. Large segments of the liberal press find his career as a Hitler-stalker while Hitler was still alive to be rather embarrassing. Hermann Ehrhardt had started trying to kill Hitler while most law-abiding Jews and anti-Nazi patriots could still be saved, while the Anglo-Saxon nations and France did little or nothing or invested money in Nazi Germany, and while Stalin became Hitler's ally with the support of American Communist sympathizers like Harry Dexter White and a large part of the Treasury Department, and Hollywood screenwriters like Dalton Trumbo.

During his last years, Ehrhardt continued to avoid the limelight and to make improvements on the estate. He presided over the construction of a new fire-house, a lodge for social gatherings, and some road improvements to make life more enjoyable for the residential tenants and the guests, and for his family. He appeared to have found peace.

Chapter Thirty

The Epitaphs

On September 27, 1971, Hermann Ehrhardt's death was recorded, without any fanfare, by Eva Schwarz, an Austrian town clerk from Brunn am Wald. The death certificate mentioned no cause of death: Ehrhardt was two months and two days short of the age of 90. His daughter Marie Elizabeth Ehrhardt had married into the British peerage and discretion was advisable. Princess Margrethe Viktoria Luise von Hohenlohe-Öhringen, 14 years Ehrhardt's junior, lived on until 1975 and then passed quietly at 81 without much outside notice. Hermann and Margrethe are buried together in the cemetery at Lichtenau with a tall stone cross to mark the grave.

Ehrhardt's quiet death came as such a surprise that *DER SPIEGEL*, perhaps West Germany's most important news magazine, ran the obituary several weeks after the fact, on October 11, 1971.

"Hermann Ehrhardt, 89. The *Freikorps* leader fought against the black-red-gold Weimar Republic as long as it existed. In 1920 he and his 'Ehrhardt Brigade'—password 'black-red-gold—unbelievable'—marched with assault packs and hand grenades toward Berlin. Together with the East Prussian General Agricultural Director Kapp and General Lüttwitz he staged a putsch against the national administration. After the shattering of the Kapp Putsch he dived for cover and later founded 'Organization Consul,' to whose account are laid the murders of Foreign

Minister Rathenau (1922) and Finance Minister Erzberger (1921). As the Republic broke into pieces and Hitler came, Ehrhardt went out of the country. He died—as first became known this past week—the Monday before last in the lower Austrian town of Brunn am Wald."

The German legal code provides fines and prison time for 'Holocaust Denial' and German reference texts today sometimes describe Hermann Ehrhardt as an "anti-Semite." The source for this poisonous description is the bogus charter of Organization Consul cobbled together from reports from informers and printed in a Social Democratic newspaper in 1922 while Ehrhardt himself was in prison in Leipzig. The Organization Consul charter article, augmented by the Stalinist Professor Gumbel's fake casualty list, appeared a few months after the Communist newspaper *Die Rote Fahne* mocked Ehrhardt as "Ahasuerus in an armored car," a tool of the International Jewish Conspiracy, at a time when lower-class Leftist movements, Nazi and Communist alike, used the imaginary Jewish World Conspiracy to recruit useful thugs. Wolfgang Kapp, not Hermann Ehrhardt, had ordered the stupid *matzoh* flour confiscation during the Kapp Putsch. Ehrhardt and Waldemar Pabst welcomed Kapp's speedy departure after this publicity disaster. The Kapp Putsch, in fact, had aimed to maintain the active strength of the German Army at a level to be able to withstand Poland or Belgium, not to topple the Weimar Republic and bring back the Kaiser. The fact that Ehrhardt himself sent men from the 2nd Marine Brigade to rescue law-abiding Jews from proto-Nazi thugs has never been disputed—simply ignored as incongruous. Members of the Rathenau murder team told Otto Friedrich that Ehrhardt never ordered the death of Walther Rathenau, though no such disclaimer can safely be attached to the death of Matthias Erzberger, who was not Jewish and whose own career had some anti-Semitic overtones. Ehrhardt seemed to savor Erzberger's murder and the warning attack on the Prussian Phillip Scheidemann with prussic acid—Ehrhardt's kind of pun.

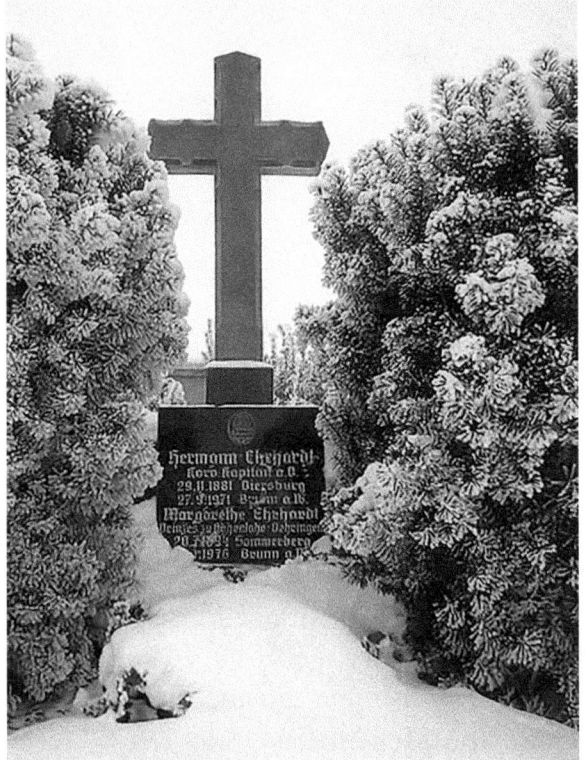

Hermann Ehrhardt's Grave

Waldemar Pabst, who was less averse to public life than Hermann Ehrhardt, might have the last word. Pabst had Jewish in-laws and a French wife and had helped his half-Jewish half-sister Martha to escape from the Third Reich during the early Hitler years. But in 1962, when critical documents were revealed, Pabst finally admitted that he had ordered the executions of Karl Liebknecht and Rosa Luxemburg. He told *Der Spiegel*:

"It was not only in the best interests of Germany that we protected ourselves from the fate that Herr [*Walter*] Ulbricht and his wire-stringers were preparing for us. The victory of Communism in Germany in 1919 would have brought about the collapse of the entire Christian West. The termination of this danger outweighed the removal of two political seducers."

Hugo Preuß, who was Jewish, a Democrat, and the father of thee German soldiers, concurred independently of Pabst, a Christian and a monarchist, that Germany's internal defeat of Bolshevism had saved the West. Count Harry Kessler, an ultra-liberal known as "the red count" and a homosexual, agreed that Luxemburg and Liebknecht had been asking for it. The Liebknecht-Luxemburg official status as martyrs for freedom came only as a reaction to the horrors of the Hitler era.

Pabst was speaking a year after the completion of the Berlin Wall. As he spoke, the Berlin Wall was already stained with the blood of youngsters who had no personal memories of Hitler but had gambled their lives for freedom. Berlin during the Weimar Republic had voted 60 percent Socialist or Communist. The labor unions of Berlin claimed credit for the defeat of Hermann Ehrhardt and Waldemar Pabst with a general strike during the Kapp Putsch but nobody serving in the Ehrhardt Brigade missed a meal. Ehrhardt and Pabst blamed their defeat on the cowardice of the conservative senior officers who had refused to support them. At the climax of World War II in April of 1945, the Soviet Red Army broke into once-Leftist Berlin, killed 50,000 civilians or recent conscripts, raped 50,000 German women—4,000 women committed suicide to avoid rape and hundreds of rape victims killed themselves afterwards. The Soviets also stripped the city of machine tools and industrial gear, not to mention most of the liquor and good clothing and all available wristwatches. By June of 1953, dissident Leftist workers in Berlin and other East German cities like "Red" Leipzig were throwing stones at Soviet T-34 tanks and cursing their Russian "liberators" as they once cursed Hermann Ehrhardt and Waldemar Pabst. Soviet Russian and East German puppet troops killed at least 55 dissident East Germans, perhaps as many as 125, in June of 1953: the Soviets and East German Communists blamed the German working-class riots on the influence of American movies like "The Wild One" starring Marlon Brando and "Rebel Without A Cause" starring James Dean. East Germans began to escape from East Berlin and East Germany by fence-climbing, by tunnel, by home-made balloon, and by long-distance swimming at the Baltic seashore. East German athletes and musicians regularly defected any time they toured a Western nation.

Sophie Liebknecht, Karl Liebknecht's widow, was living in venerable comfort in East Berlin in 1962 when Pabst's confession to her husband's 1919 murder broke into print. Sophie Liebknecht urged the government of the German

Democratic Republic—the same people who built the Berlin Wall and killed the teenagers and young men trying to escape—to indict Waldemar Pabst for the murder of her husband and of Rosa Luxemburg. She apparently shrugged off the 1,000-plus people killed in Berlin when Liebknecht had declared war on the Weimar Republic.

When West German and American reporters heard about Sophie Liebknecht's tearful request and confronted Waldemar Pabst on a stroll through Berlin, they asked him for a quote. He didn't let them down. He glanced at the Berlin Wall.

"If you like what you see over there, you can call me a murderer."

Hermann Ehrhardt is an inconvenient figure in history. Most Anglo-Saxon biographers of Hitler—Robert L.G. Waite and William Shirer in particular—portray Ehrhardt with broad strokes as a proto-Nazi. Waite, a state-side soldier who spent World War II guarding a bridge between the United States and Canada, assembled some important facts and quotes in the 1940s that suggested that Ehrhardt had actually despised Hitler. Yet Waite's decent horror at Auschwitz shortly gave him the Darwinian perspective that Ariovistus, the Germanic opponent of Julius Caesar, and Arminius, the Germanic opponent of Augustus Caesar, were progenitors of the Third Reich before there was a First Reich and had ostensibly ventured into Gaul looking for Jews to murder rather than more plausibly looking for better farmland and for wine instead of home-brewed beer. Shirer "revealed" the horrors of medieval German history without any reference to the African Slave Trade—no Germans took part—the destruction of the American Indians—few Germans took part—or the Anglo-Saxon encroachments India, China, Mexico and the Philippines. Shirer never enlarged on why so many of the Kaiser's stalwart Germans, included some he knew personally, had French or Polish surnames, or why the Kaiser had so many Jewish supporters, some of them downright heroic. Frank Capra took the same perspective in the "Why We Fight" film series, seen by 12 million American service personnel. Capra portrayed Otto von Bismarck and Kaiser Wilhelm II as proto-Nazi types plotting to take over the world. U.S. Ambassador to Turkey Henry Morgenthau Senior told the world that the evil in German genes had led to the murder of the Armenians—an action that not only German Christian missionaries but German generals and politicians had opposed while it was still in progress. The Amish and the Mennonites of eastern Pennsylvania are full-blooded Germans and speak and pray in a clear German regional language. They don't seem to spend a lot of time planning mass murders and no tanks appeared to be concealed in their haystacks.

The facts don't stand up to the rhetoric. The Nazi horrors—most of them absolutely real—were circumstantial and not genetic in origin. The very plausible threat of Communism and the events of 1919, inflated and exploited by Hitler and his psychopaths, not the heritage of Caesar's Gallic Wars or the music of Richard Wagner's operas, were the prelude to the Holocaust. Hermann Ehrhardt was an intuitive Christian. He never persecuted Jews and sometimes rescued them, as his Communist detractors charged in the Communist newspaper *Die Rote Fahne*. His hatred of Hitler was abundantly evident in the four Hitler assassination attempts he plotted or supported. Ehrhardt's role in breaking up the Bolshevik attempts to annex Germany may have staved off a worse catastrophe than World War II, since Stalin demonstrably murdered twice as many innocent people as Hitler did.

Hermann Ehrhardt may have become intolerable in the post-war era because his existence offended people who had something to hide—people like the covertly Stalinist Professor Gumbel and his absurd figure of 354 Rightist murders that went unpunished, a lurid fiction still quoted in responsible histories to this very day. The heirs of the Communists portray their heroes Rosa Luxemburg and Karl Liebknecht as martyred rebels for freedom and try to obscure the fact that the Spartacists were not trying to overthrow the Nazi movement, nor ever the Kaiser, but the Social Democrats who were themselves liberal Marxists who merely tolerated Christianity and private property. Had Hermann Ehrhardt and Waldemar Pabst and their supporters not taken extreme measures, Europe might well have become part of the Stalinist Soviet Union—and a German-Russian coalition would probably have conquered Western Europe with ease. Ehrhardt and Pabst were not responsible for the Holocaust. The shifty cowardice of both liberal and conservative politicians left the assassination of Adolf Hitler as a decent man's only option. Herman Ehrhardt tried it or supported it four times. He failed, but so did everybody else.

Waldemar Pabst himself may have felt some envy for Hermann Ehrhardt, though they never betrayed each other and often functioned together as a team, whether fighting the Bolsheviks or plotting against Hitler. Ehrhardt was taller and stronger than Pabst, which can be important among military men. Ehrhardt also had two sets of children and Pabst married twice but never had any children at all. Pabst, with his affinity for the French language and French civilization, was nicknamed "the little Napoleon." Ehrhardt was simply nicknamed *Der Chef*—"the chief," or in German usage, "the boss." Ehrhardt had a greater capacity to inspire loyalty, though Pabst, the former General Staff officer, was better at cool-headed analysis and negotiation. Ehrhardt also remained quietly uneasy about Pabst's murder of a woman—even a Rosa Luxemburg. When Pabst and Ehrhardt were both past 80, Pabst blamed Ehrhardt for only one mistake: Pabst said that Ehrhardt should have arrested the Ebert-Noske administration on the first day of the Kapp Putsch in 1920 but had honorably but foolishly delayed because he had given his promise to his superior officers when he held them at gunpoint after they broke in on his nap. Ehrhardt said he had telephoned ahead to have the Ebert-Noske team arrested or pursued by automobile, or even by aircraft, but that the generals had failed him. But Pabst and Ehrhardt both saw one another as friends and comrades betrayed by cowardly old men more worried about the future of their pensions than the future of their nation.

"*Korvettenkapitän* Ehrhardt was a man whose fate beckoned him to a very influential role in politics, not to seek political power for itself, but because he believed he understood how to get Germany out of its perilous situation better than any of the administrations after 1918," Pabst wrote. "Based on this opinion, he led the way for Germany, which he loved with a glowing, yes even a fanatical love, to break the shackles of the Treaty of Versailles, so that a fully armed people could once again be taken seriously in world politics and would have a voice among the nations. No other German statesman dealt with this insight with the understanding that for Germany, it was a matter of life and death.

"Ehrhardt had an unusual capacity for the handling of his troops, and his people backed him with near-idolatry and the highest feelings of honor. Many saw him as the man selected by fate to save Germany. All in all, *Korvettenkapitän* Ehrhardt was his own man."

Bibliography

BOOKS

Allen, Frederick Lewis. *Only Yesterday: An Informal History of the 1920s.* New York: Harper & Rowe, 1931.

Ballard, Robert D. *Exploring the Lusitania: Probing the Mysteries of the Sinking That Changed History.* New York: Warner Books, 1995.

(Ballard deplored the loss of life on the ship but noted (Page 143) that "the Germans were being blockaded by the Royal Navy; most believed (correctly) that the British had no compunction about starving them to death—men, women, and children—if that was what it would take to win. The majority of Germans saw the Lusitania affair as a heroic attack on an 'armed auxiliary cruiser,' a brilliant blow against a merciless enemy who should be heartily condemned for his conduct.")

Beamish, North Ludlow. *History of the King's German Legion.* London: Thomas and William Boone, 1837.

Berg, A. Scott. *Lindbergh.* New York: G.P. Putnam's Sons, 1998

Braunschweig, Herzogin zu und Lüneburg Viktoria Luise. *The Kaiser's daughter: Memoirs of H. R. H. Viktoria Luise, Duchess of Brunswick and Lüneburg, Princess of Prussia.* Upper Saddle River, NJ: Prentice-Hall, 1977 (Offers an inside look at the Prussian-Hanoverian royalty conflict. She married the heir to the non-existent Hanoverian throne after she saw him in a dream and they lived happily ever after, at least after Hitler killed himself.... Most British historians won't touch this topic with a fork....)

Brett-James, Anthony. *The Hundred Days: Napoleon's Last Campaign from Eyewitness Accounts.* New York: St. Martin's Press, 1964 (Account of Major George Baring.)

Brune, Werner. *Wilhelmshavener Heimatlexikon.* Wilhelmshaven: Brune-Mettcker Druck- und Verlagsgesellschaft, 1986. (Brief but detailed military and political biography of Hermann Ehrhardt)

Bullock, Alan. *Hitler: a Study in Tyranny.* New York: Harper Torchbooks, 1962. (The notorious Captain Ehrhardt is portrayed as an early supporter of Hitler, not as a subsequent drastic Hitler opponent and near-miss murder victim.)

Butler, Daniel Allen. *The Burden of Guilt: How Germany Shattered the Last Days of Peace, Summer 1914.* Philadelphia: Casement Publishers, 2010.

Butler, Smedley. *War is a Racket.* New York: Round Table Press Inc., 1935

Charlot, Monica. *Victoria: the Young Queen.* Oxford: Blackwell, 1991.

Charny, Israel W. *An Encyclopedia of Genocide.* Santa Barbara: ABC-CLIO, 1999.

Chernow, Ron. *The Warburgs: The Twentieth-Century Odyssey of a Remarkable Jewish Family.* New York: Random House, 1993

Collectif. *Grand Dictionnaire: Français-Anglais, Anglais-Français.* Paris: Larousse, 2010.

Connaughton, Richard. *Rising Sun and Tumbling Bear: Russia's War with Japan.* London: Cassell, 1988.

Crankshaw, Edward. *Bismarck.* New York: The Viking Press, 1981.

Diamond, Sander A. *The Nazi Movement in the United States 1924–1941.* Ithaca: Cornell University Press, 1974.

Dietrich, John. *The Morgenthau Plan: Soviet Influence on American Postwar Policy.* New York: Algora Publishing, 2013.

Dietrich, Otto. *The Hitler I Knew: Memoirs of the Third Reich's Press Chief.* New York: Skyhorse Publishing, 2010.

Doyle, Arthur Conan. *The Crime of the Congo.* London: Hutchinson & Company, 1909. (Frontispiece illustrations. Thomas Fleming reports in "*The Illusion of Victory*" that radical lawyer Clarence Darrow offered a bounty of $10,000 to any young Belgian who could show the stump of a lopped-off hand. He spent not one penny. Had Darrow made the offer in the Belgian Congo, he might have gone broke.)

Doyle, Arthur Conan and Leslie S. Klinger. *The New Annotated Sherlock Holmes*. New York: W. W. Norton, 2005.

Drury, Ian and Gerry Embleton. *German Stormtrooper 1914–1918*. Oxford: Osprey Publishing, 1995.

Ehrhardt, Hermann. *Kapitän Ehrhardt: Abenteuer und Schicksale—retold and edited by Friedrich Freska*. Berlin: August Scherl, 1924.

(Hermann Ehrhardt's autobiography from his childhood through the year 1924. The book has never been translated into English and apparently never reprinted in German after the 1920s, though it went through at least eight printings. All conversations in Chapter 2 are quotes or responsible paraphrases from *"Kapitän Ehrhardt: Abenteuer Und Schicksale"* unless otherwise noted. None are imaginary or based on conjecture. The meeting with the Princess is from Pages 2 – 2. A bean-flipper—Hermann grandly referred to his as a catapult—is a forked hard-wood stick with a rubber inner tube from a bicycle tire tied to both tines of the wooden fork, the middle of the tubing threaded through a leather pocket. The word Hermann used for Signor Farina—*Schuft*—describes people so low-down that, while not Jewish themselves, they hired out to clean up for Jews on the Sabbath when Jews were forbidden to do work. Among Germans serving in the U.S. Army during the Civil War, the use of the term by officers from Baden to describe officers from Prussia, where Jews were more widely accepted than in Baden, was considered grounds for a court-martial. In Germany it was grounds for a duel. CF: the papers of Colonel Joseph Spangenberg, 41st New York Volunteer Infantry, in the U.S. National Archives, quoted in "Scapegoats of Chancellorsville," *America's Civil War*, May 2008. *"Abgedankt"* literally means "thanked off" and has slightly contemptuous overtones, as when an officer is released from service for an inadequate performances of his duties or moral problems. The U.S. equivalent, "discharged for the convenience of the service" usually meant that the officer discharged was a known homosexual or habitual adulterer with married women or had serious drinking problems. "Convenience of the service" discharges were used to avoid a formal dishonorable discharge, which some officers saw as an incentive for suicide or terminal alcoholism. What Ehrhardt actually said was "I answered nothing further than what Götz von Berlichingen answered from the window." Erich Maria Remarque used the same euphemism for the same quote a few years later in *"All Quiet On The Western Front."* Ehrhardt's description of himself as a *Landsknecht* leader harkens back to the German and Swiss mercenary detachments of the late Middle Ages, though he was—at least at this point—an official officer of the Weimar Republic, very much to his own distaste. His description of German as a *Reich*— realm, but semantically, empire—shows what he really would have liked. Tactically he was an effective anti-Communist for the Social Democrats but at heart he still was a monarchist. '*O Deutschland hoch in Ehre*'—Oh Germany, High In Honor' was traditionally sung to celebrate victories. Ehrhardt, who knew his Roman history despite his defective Latin grammar, used the archaic German tribal word *welsche* for the French. Semantically, this refers to people who never keep their word—as in, "to welsh on a promise." No disrespect for the Cymric people who live in the southwest of the isle of Britain was intended. Archaic German refers to the Slavic people as *Wende*—wanderers. The Wends of the Spreewald neighborhood in Berlin descended from the original tribe of that name. Titled Prussians and Huguenot fugitives in the days of the Kingdom of Prussia enjoyed hiring Berlin's Wendish girls as maids and nannies. Ehrhardt used high German for the phrase: "*Lieber tot, als Sklave!*" In Frisian, the phrase is "*Lever duad as Slav.*" In Richard Wagner's opera, Lohengrin, the semi-villain, Friedrich Graf von Telramund, paraphrased the Frisian battle cry as "*Viel lieber Tod als Feig!*"—I'd rather be a corpse than a coward." English translators never seem to get this translation right. *The Internationale* was, of course, the Communist anthem of the proposed one-world government and featured the denial of God and rejection of all monarchies and other governments. The original lyrics in French dated from shortly after the Paris Commune of 1871, also an anti-religious revolt of ultra-leftists against a democratic republic. After the French *Communarde*s murdered the kindly old Archbishop of Paris, Monsignor Georges Darboy and six other priests along with a number of politicians and policemen, French Republican soldiers executed an estimated 10,000 *Communardes*, some of them by the use of prototypical machine guns. The *Gulaschkanone* [Goulash Cannon] was a horse-drawn cook wagon that served the soldiers hot meat-and-vegetable soup or beans as often as possible. Enlisted men otherwise carried black bread and butter or cheese along with an 'iron ration' of canned meat to be eaten only in emergencies. Company horses were coach or carriage horses used to pull cook wagons, baggage wagons and artillery pieces, not the thoroughbreds or hunting horses generally used in jumping or hunting. Ehrhardt had learned casual riding while growing up around the farmers in Baden and had no trouble staying on a horse, but was no polished equestrian. The horsemanship of the other naval officers appears to have been even more questionable. The Haller troops not only had French uniforms but some French FT-17 Renault tanks, which were busy elsewhere. The Polish leaders Jozef Haller and Józef Piłsudski saw the irksome German troops in Upper Silesia and East Prussia as a counter-weight to their avowed enemies, the Russians, who were threatening Warsaw and Polish independence itself. As Ehrhardt says, the pot shots by both sides were probably unauthorized. Ehrhardt, who was notably straight, doesn't mention this specifically, but some of the other *Friekorps* units, notably *Freikorps Roßbach* and Ernst Röhm's later *Sturmabteilung* (SA) seemed to attract a large number of he-man homosexual types, aggressive perverts who were happiest when proving their manhood far away from women. The Reds, who condemned prostitution and endorsed Free Love while they doled out pretty mistresses to ugly men, officially considered homosexuality a disease nurtured by Capitalism. They took full propaganda advantage of the scandals in some other units that Ehrhardt alludes to. The 'Green' Police, organized with the advice of Waldemar Pabst, functioned rather like U.S. Marshals: they dealt primary with crowd control and martial-law situations rather than criminal investigation. Like most policemen in Germany and everywhere else outside the Communist bloc, they tended to be socially conservative. A light machine gun was a gun fired from a bipod that could be carried by one strong man, and a heavy machine gun was a gun fired from a tripod or cradle mounting with a three-man or four-man crew. Both fired standard rifle ammunition.   The Seven Years' War is referred to as [the last] French and Indian War in American. In Europe, Britain, Hannover and Prussia fought Austria, French and Russia to a standstill. The Thirty Years War [1618 – 1648] killed six to eight million Germans through battle, murder, or disease. The tribal migrations restructured late Roman Europe as Germanic and Swedish [Gothic] tribes overwhelmed and blended with Roman and Celtic settlers in the Rhineland, France, Britain, Spain, Northern Italy, North Africa, Hungary and Rumania —allied with the Asiatic Huns. Ehrhardt here engaged in a pun with *Landsknechte* [Renaissance mercenaries] stuck on the native soil confuted with *Landschnecke* [land snails] stuck with their shells. His ability to counterfeit various regional dialects came from the same fascinating with word-play. This allusion is perhaps as close as Ehrhardt ever came to make an anti-Semitic remark: He otherwise refers to the people he was fighting as "Russians" with no religious reference, and he believed that native-born Germans Jews, veterans in particular, should have the same rights as other Germans. He also personally endorsed Munich Police Chief Ernst Pöhner as a good man. Pöhner—but not Ehrhardt—said that Russian-born Jews should all be deported back to Russia during the economic and political emergency. The *Großdeutsche Partei* was based in Austria and advocated unification of Austria with Germany under Christian Socialist principles. The party would be banned in Germany by the Nazi administration in 1934. Ernst Moritz Arndt was perhaps the most verbally bloodthirsty of the German patriotic poets of the Napoleonic era. He was also an opponent of Prussian serfdom which was abolished during the reforms that followed Prussia's defeat by France in 1806. Contemporary German historians assume that Ehrhardt was the concealed mastermind behind the Rathenau murder but admit that there is not a shred of documentary evidence or any confession to prove this. His guilt is simply assumed. His previous and subsequent attempts to kill Hitler are seldom mentioned. The Saxon dialect Ehrhardt refers to and imitates refers to the old Saxon kingdom of the Wettin Dynasty that was sometimes bonded with Poland, a nation that almost habitually selected its kings from foreign candidates. The ethnic term 'Saxon' as in "Niedersachsen," or Lower Saxony, refers to Hannover and the vicinity where the people speak *Platt*, a different language very close to simple English or to Dutch. *"Wat is dat?"* instead of *"Was ist das?"* means *"What is that?"* and *"he was"* in Platt covers *"er war"* in High German. Public education based on the language of the Luther Bible since at least the early 1800s had made High German the national standard for a hundred years by Ehrhardt's time but the regional languages were still widely spoken and understood. When the Ernst Jünger, a soldier and officer from the 73rd Infantry, garrisoned in Hannover and the youngest man ever to win the *Pour le Mérite*, spoke the *Platt* of Hannover to the middle-aged Flemish couple he roomed with while he was stationed in Belgium, they understood him perfectly and virtually adopted him. Ehrhardt elegantly used the French word *necessaire* for the manicure set. See *Unabridged Grand Dictionnaire Français-Anglais / Anglais-Français*. Ehrhardt incidentally gets a far better press in France and Spain than he does in the United States—or in many German publications. A *Schlag* in German means a blow, but can also refer to whipped cream used to top off gourmet coffee or cakes. Max Graetz, who lived from 1861 to 1937, was an electrical light and appliance designer and manufacturer with outlets in Berlin, Paris, and Bridgeport, Connecticut. As the widow of a nobleman who died in military service and the wife of a clergyman's son, both of them officers, Frau Ehrhardt was also entitled to some class or caste respect. Like all Imperial German Navy officers, Ehrhardt read and spoke English with some fluency. Franz-Maria Liebig, one of Ehrhardt's most loyal followers and, like Ehrhardt, a life-long anti-Nazi and Hitler assassination conspirator, had never seen Ehrhardt before Ehrhardt steered the rust-bucket freighter back to Wilhelmshaven after the mutinous Red crew jarred him out of his personal depression shortly after the Armistice. The similar names suggest that Frix and Frax were Karl Tillessen [1891 – 1979] and Heinrich Tillessen [1894 – 1984], brothers who served with Ehrhardt in torpedo boats during the war and later served in the 2nd Marine Brigade and in Organization Consul. Both re-joined the German Navy during the Hitler years but were not indicted as war criminals. Their eldest brother, Werner Tillessen [1880 – 1953] remained in the German Navy after World War I and also avoided extremist politics.)

Evans, Richard J. *The Coming of the Third Reich*. New York: The Penguin Press, 2004.

Evans, Richard J. *The Third Reich in Power*. New York: The Penguin Press, 2005.

Farago, Ladislas and Andrew Sinclair. *Royal Web: The Story of Princess Victoria and Frederick of Prussia*. New York: McGraw-Hill Book Company, 1982.

Felix, David. *Walther Rathenau and the Weimar Republic: The Politics of Reparations*. Baltimore: Johns Hopkins Press, 1971.

Fleming, Thomas. *The Illusion of Victory: America in World War I*. New York: Basic Books, 2002.

Fleming, Thomas. *The New Dealers' War: FDR and the War Within World War II*. New York: Basic Books, 2001.

Fluek, Toby Knobel. *Memories of My Life in a Polish Village 1930–1949*. New York: Alfred A. Knopf, 1990.

Friedrich, Otto. *Before the Deluge: A Portrait of Berlin in the 1920s*. New York: Harper Perennial, 1995.
(Says the boy laughed but doesn't affirm that he was killed. The youngest fatality listed in hospital or morgue records was 17 years old—hardly a child. All fatalities were adult males, mostly workers or graduate students in their 20s. Otto Friedrich quotes eyewitness Hans Staudinger, an admitted Socialist and later a teacher at the New School for Social Research in New York City: "I saw them shoot.... I ran over to watch, and then they were shooting. And the interesting thing is that the officers did not give the command to shoot. It was the young soldiers themselves who decided to open fire. Young people are very cruel." Incongruously, Otto Friedrich includes Hugo Haase in the list of murder victims: Haase was an Independent Social Democrat murdered by a demented leather worker name Johann Voss, apparently for being not far enough to the Left. Haase was shot on October 8, 1919, and died on November 7, 1919, at least six months before the founding of Organization Consul.)

Garraty, John A. and Mark C. Carnes. *American National Biography*. Oxford: Oxford University Press, 1999.
(Henry Steele Comager's collection of American treaties contains both documents. Nobody told the Koreans at the time and when a Korean scholar stumbled over the informal but binding Taft-Katsura Agreement in 1921, the Koreans, then ruled by Japan since 1910, screamed betrayal by America. Nobody paid attention. Entry on Rosika Schwimmer.)

Gilbert, Adrian. *Illustrated History of World War I*. New York: Brompton / Bison, 1988.
(British viewpoint with the "Rape of Belgium" brought into a somewhat responsible perspective.)

Goerlitz, Walter and Brian Battershaw. *History of the German General Staff 1647–1945*. New York: Frederick A. Praeger Publishers, 1953.

Hallahan, William H. *Misfire: The History of How America's Small Arms Have Failed Our Military*. New York: Scribner, 1994.

Hanfstaengl, Ernst. *Hitler: The Missing Years*. New York: Arcade Publishing, 1994.
(Karl Ernst had been a bouncer in a gay nightclub but was taken away from
his honeymoon to be murdered on June 30, 1934.)

Hanser, Richard. *Putsch! How Hitler Made Revolution*. New York: Peter H. Wyden,
1970.
(While highly readable and not at all in favor of Communism, offers a
sympathetic, rather sentimentalized look at Eisner leaving out some of the
rough spots. Spartacus-educational.com, which is pro-Eisner, and the
German language Wikipedia site—17 pages and neutral—offer details
which identify Eisner as a member of the Independent Social Democrats—
virtually Communists, despite statements to the contrary elsewhere. Erhard
Auer pulled through the Lindner shooting. Lindner was later arrested and
got 14 years in prison for murder. Hitler later claimed that three Reds came
to arrest him due to his neutrality speech but that he faced them down with
a carbine. Hanser and most other historians doubt this ever happened. The
fact that Hitler took no actual part in the Battle of Munich was notorious
even among Nazi Party members. Toller eventually lived in New York City,
saw his plays at first extolled and then largely ignored. He either committed
suicide after the Reds lost the Spanish Civil War or was secretly murdered
in 1939. The Protocols were conclusively described as a forgery by a Swiss
jury in 1934. They are in fact based on a completely fictional novel called
'Biarritz,' written in 1868 by an English-hating German posing as an
Englishman who had cribbed some of the plot from Alexandre Dumas' story
about an 18th Century Gypsy charlatan known as the Count of Cagliostro.
The German author was subsequently sent up for forgery in a separate case.
Henry Ford later printed the fictional Protocols as factual material. The
quote became one of Hitler's favorites. The Weimar Constitution was
republican and democratic, but that made no impact on Clémenceau. Since
Alsace is Germanic and Lorraine largely Germanic, Clémenceau was
probably making a political statement but the lethal implications were taken
very seriously in Germany. The "boy" "venture to hoot" and says he was
kicked to death with hob-nailed boots. Hanser also says an officer "snarled
an order." Ehrhardt and Mann both say the firing was spontaneous and the
only official order was to cease firing. Richard Hanser, though generally
accurate, states "Rossbach was, like Röhm, a homosexual, and, like
Ehrhardt, a veteran of battles in the Baltic." Hermann Ehrhardt never
served in the Baltic except as an officer in the Kaiser's navy during the war.
The charter of the 2nd Marine Brigade limited service to areas that had been
part of the pre-war German Empire. Rossbach's unit may have been
involved in the "scandals" that Ehrhardt mentioned when he cited the need
to rescue Jews being abused by *Baltikumer* units. Godin was sent to Dachau
when Hitler came to power and was later deprived of his citizenship. He

spent time in Austria, where Waldemar Pabst ran an anti-Nazi home defense unit, later to Switzerland where he advised Allen Dulles on how to counteract Nazi influence among the Swiss, who disliked Hitler and sheltered Jewish refugees but were also adamantly anti-Communist and hoped the Soviet Union would lose the war.)

Hetherington, Peter. *Unvanquished: Joseph Pilsudski, Resurrected Poland, and the Struggle for Eastern Europe.* Houston: Pingera Press, 2012.

Heydebreck, Peter von. *Wir Wehr-Wölfe Erinnerungen eines Freikorps-Führers.* Liepzig: K.F. Koehler, 1931.
Reprint: Toppenstedt: Die Uwe Berg-Verlag GbR, 2007.
(A one-armed hero of World War I, Peter von Heydebreck was credited with saving Silesia from Polish irregulars in the last *Freikorps* battles. Ultimately a Nazi supporter, he was later murdered by personal order of Hitler.)

Hickey, Des and Gus Smith. *Seven Days to Disaster: The Sinking of the Lusitania.* New York: G. P. Putnam's Sons, 1981

Hitler, Adolf. *Mein Kampf.* New York: Reynal & Hitchcock, 1940.
(Self-serving autobiography, revealing, more interesting as psychology than history.)

Hitler, Adolf and Raoul de Roussy de Sales. *My New Order.* New York: Reynal & Hitchcock, 1941.
(Hitler's speeches from 1918 to the invasion of Russia translated from German.)

Hoffmann, Peter. *The History of the German Resistance, 1933-1945.* Montreal: McGill-Queen's University Press, 1996.

Hohenzollern, H.R.H. Viktoria Luise and Robert Vacha. *The Kaiser's Daughter: Memoirs of H.R.H. Viktoria Luise, Princess of Prussia.* Englewood Cliffs, NJ: Prentice-Hall Inc., 1977.
(Since the Red rebels in Saxony had vowed to "slaughter the bourgeoisie without regard to sex or age" if 'Prussian' government troops invaded, the life of the Prussian princess was actually at substantial risk.)

Hohenzollern, Wilhelm II. *My Early Life.* New York: George H. Doran Company on Murray Hill, 1926.

Horn, Joseph. *Mark it With a Stone: A Moving Account of a Young Boy's Struggle to Survive the Nazi Death Camps.* New York: Barricade Books, 1996.

Horne, Alistair. *The Price of Glory: Verdun 1916.* New York: St. Martin's Press, 1962.

Howard, Michael. *The Franco-Prussian War.* New York: Dorset Press, 1990.

Jones, Nigel H. *Hitler's Heralds: The Story of the Freikorps 1918–1923.* New York: Dorset Press, 1992.

Keegan, John. *An Illustrated History of the First World War.* New York: Alfred A. Knopf, 2001.

Kershaw, Ian. *Hitler: A Biography.* New York: W. W. Norton & Company, 2010.

Kilduff, Peter. *Iron Man: Rudolf Berthold: Germany's Indomitable Fighter Ace of World War I.* London: Grub Street Publishers, 2012.

Knightley, Phillip. *The First Casualty: From the Crimea to Vietnam: The War Correspondent as Hero, Propagandist, and Myth Maker.* New York: Harcourt Brace Jovanovich, 1975.

(Knightley reviewed statements from all sides and said that about 10 percent of the Belgian casualties were murdered and the others were legally executed for fighting out of uniform.)

Konrad, Rüdiger. *Waldemar Pabst: Noskes "Bluthund" oder Patriot?* Schnellbach, Beltheim: Bublies Verlag, 2012.

(This recent book contains a substantial amount of Pabst's own writings about his role in the *Freikorps* movement and about Ehrhardt, whom he acknowledged as the greatest of the *Freikorps* leaders. Ehrhardt's disguise as demonstrated by two photographs show how Ehrhardt's extremely regular features, minus the recognition factor of his trademark Oriental-style mustache and goatee, and with an air of friendly affability rather that fierce dedication, could have fooled people who had not seen him for a few years. He apparently could change his appearance from menacing to downright innocent at will. Contains most of the biographical details of Pabst's life and extensive excerpts from his writings. An essential source. There was nothing especially unusual about a Christian-Jewish marriage in the Kaiser's Germany in 1895. Until the Hitler era, in most given years more German Jews married Christians than married other Jews. This obviously changed drastically when Hitler came to power and became illegal in 1935. Martha Steiner, stigmatized as a half-Jew—and perhaps also as the half-sister of the notable anti-Nazi Waldemar Pabst—fled from Hitler's Germany in the 1930s and died in obscurity in the United States some years later. The word Julius used to describe Waldemar, "*Bengel,*" defies exact translation but "dork," meaning a clumsy male adolescent, comes close. Julius Camillo Pabst lived to be 93 and audibly regretted that he didn't make 100. Pabst uses the term *Kriegsminister*, War Minister, but since the war was now temporarily over, Defense Minister seems more appropriate except in direct quotations. Labeled Marxist sources continue to cite 300 executions at the *Vorwärts* building. Neutral or official sources indicate that there were only a handful of casualties. County Harry Kessler was a sort of upscale gay Aryan version of Stefan Lorant. The son of an Anglo-Irish girl and [possibly] of a much older German banker—and dubiously said to be the natural son of Kaiser Wilhelm I. Like Lorant, who was straight, Kessler enjoyed Weimar, especially the sexual liberation, but distrusted and disliked Hitler. An officer and diplomat during World War I, Kessler died in exile in France in 1937. Gustav Noske used similar phrases when arguing for non-violent resistance, especially to the French. Noske shortly waffled. Waldemar Pabst disliked Reinhardt whom he described as "brave, indecent and foolish." Reinhardt's own independent Freikorps, *Freiwilligen-Regiment Reinhard*, had assisted in the anti-Communist actions in December 1918 and January 1919 with hard fighting and with what Pabst himself felt was excessive brutality. Pabst summed up General Georg Ludwig Maercker, another Freikorps leader, with one simple description: "*Schwein.*" Wilhelm Reinhardt later became a Hitler supporter and Waffen-SS field commander and was detained for extensive Allied interrogation after World War II. Pabst uses the word *Kerl*, which cognates with the Anglo-Norse word "churl" but is not at all pejorative. In old Anglo-Germanic-Norse societies, the churl was a free warrior and small land-owner who could be depended on to fight in battle for his tribe.)

Koster, John. *Operation Snow: How a Soviet Mole in FDR's White House Triggered Pearl Harbor.* Washington, DC: Regnery History, 2012.
(For confirmation see *The Battle of Bretton Woods* by Benn Steil, *Stalin's Secret Agents* by M. Stanton Evans and Herbert Romerstein, *The Venona Secrets* by Herbert Romerstein and Eric Breindel, and a brief mention of White in the documentary *World War II Behind Closed Doors* by Laurence Rees.)

Krafft-Ebing, Richard von. *Psychopathia Sexualis: The Classic Study of Deviant Sex.* New York: Arcade Publishing, 2011.
(When used by Germans born before 1900, the term "*psychopath*" often refers to a homosexual. Dr. Richard *Freiherr (Baron)* von Krafft-Ebing defined normality as any male-female copulation which, if unprotected, could produce offspring. Krafft-Ebing implied that anyone who deviated into opposite-sex non-genital copulation or any same-sex copulation with humans or animals was a psychopath. Krafft-Ebing was not unduly hostile to homosexuals as individuals: he simply saw humans as part of the natural world in which sex served purely reproductive purposes. Dr. Krafft-Ebing popularized the terms *heterosexual* and *homosexual* and *sadism* and *masochism*.)

Langer, Walter C. *The Mind of Adolf Hitler: The Secret Wartime Report.* New York: Basic Books, 1972.
(Based on O.S.S. study from 1943.)

London, Jack, King Hendricks and Irving Shepard. *Jack London Reports.* Garden City, NY: Doubleday & Company, 1970.

Lorant, Stefan. *Sieg Heil! An Illustrated History of Germany from Bismarck to Hitler.* New York: W. W. Norton, 1974.
(As a basic outline of events. Lorant covered much of what he described as a newsman. He was caustically anti-Nazi and strongly anti-militarist but admired many German artists and liberal politicians and many aspects of

German culture. Parenthetically, Lorant believed that the Weimar Republic would have succeeded except for the predatory economic demands of the Allies which ruined the German economy and paved the way for extremists of the Left and Right, and ultimately for Hitler. Lorant pioneered photo-journalism and admired the Weimar Republic. He credits Adolf Joffe, a Russian Jew and former Russian refugee with a long record as a Social Democrat and later a Communist, with doling out the guns, but Joffe had been banished from Germany just before the Armistice. Like his friend and mentor, Leon Trotsky, born Lev Bronstein, Adolf Joffe fell afoul of Stalin and committed suicide rather than go to a 1920s labor camp. Trotsky fled to Mexico and was murdered in 1940 by a Stalinist agent who hit him in the head with an ice axe. Lorant, who was Jewish, hated Hitler and the thugs around him but admired many things German and was not a Communist. He saw the moderate / conservative election results of 1919 as a tragic lost opportunity to rehabilitate Germany but "... *The 'big four' at Versailles made critical blunders. Their harsh demands, their greed, their moral righteousness, their determination to keep the Germans under their thumbs and 'make them pay' for their deeds in the war paid ghastly dividends.*" The irony expanded: British and French propagandists had blamed the Germans for so many fake atrocities in World War I that many Germans and neutrals scoffed at rumors of Hitler's genuine atrocities in World War II until it was too late to prevent most of them. Lorant lists 42 strikers dead in a confrontation with "the police." Ercih Maria Remarque, previously author of "*All Quiet On The Western Front*," depicted the massacre in his second novel, "*The Road Back*" where the drastic reactionary Lieutenant Heel is a sinister stand-in for Hermann Ehrhardt. After Ehrhardt's role in two attempts to kill Hitler, Remarque turned Ehrhardt into the near-perfect hero of "*Flotsam*." Lorant reproduces a grainy UFA newsreel frame photo showing the crowd fleeing from a unit of the troop column followed by what appears to be the Whippet tank. A few women and children are seen in the crowd but none appear to have been hit. His caption mentions a "senseless massacre" with "machine guns" but has no mention of the pistol shots that wounded soldiers and the grenade that wounded horses a few moments before, nor of the fact that working-class adult demonstrators tried to grapple rifles away from soldiers who were sometimes teenaged volunteers. The soldiers probably selected those hecklers they found most threatening or obnoxious. Some of the bullets may have passed through two or three men at close range. Mann published his account in 1921 and Ehrhardt published in 1924. Their versions, which confirm one another, were not contested at the time of publication. Germany had no tradition of separation of church and state. Many Jews, like Hugo Preuß, gravitated to the practice of law, but the Jews got along so well with the Lutheran majority and Calvinist and

Catholic minorities in Prussia, Hannover, and the Free Cities that the main threat to the Jewish population in pre-Hitler times was not massacre but assimilation through intermarriage or voluntary conversion. The attempt to take the Crucifixes out of the schools in Catholic Oldenburg was to touch off a near-revolt in early Nazi times. The Crucifixes stayed in place. Hitler is photographed locking a protective hug around Ernst Schmidt's shoulder. Schmidt looks somewhat embarrassed, as does Karl Lippert, the Jewish soldier who Hans Mends said sometimes volunteered for dangerous missions—unlike Hitler. Henry Ashby Turner Jr. also mentions the Soviet ban on supporting Social Democrats in *Hitler's Thirty Days to Power*. Marinus van der Lubbe was portrayed by the Left as a retarded boy who had been framed for the Reichstag Fire. Modern research suggests that he was of ordinary intelligence and actually set the fire to show opposition to Hitler.)

Lovell, Mary. *The Sisters: The Saga of the Mitford Family*. New York: W. W. Norton & Company, 2001.

(Unity Mitford had such a mad crush on Hitler—predictably never consummated—that she shot herself in the head with his pistol when war broke out between Britain and Germany in 1939. Hanfstaengl said that he once had to drag Diana and Unity Mitford into a washroom to scrub off their heavy English make-up so the *Hitler Mädchen* didn't beat them up at a Party Day rally in 1934. The Mitford girls said they didn't wear heavy make-up and he seemed to be enjoying himself a bit too much....)

Lukacs, John. *The Hitler of History*. New York: Alfred A. Knopf, 1997.

(Nothing on Ehrhardt.)

Luxemburg, Rosa and Georg Adler et. al. *The Letters of Rosa Luxemburg*. London: Verso, 2011.

Machtan, Lothar and John Brownjohn. *The Hidden Hitler*. New York: Basic Books, 2001.

(Machtan also reports that General Lossow kept a Munich police dossier of Hitler's homosexual affairs with young men in 1918 – 1919 for purposes of blackmail, but the dossier does not appear to be extant. The Schulz letter appeared in Leftist newspapers all over Germany and sparked at least one homosexual blackmail attempt. The blackmailer was arrested and "hanged himself" in prison.)

Manchester, William. *The Last Lion: Winston Spencer Churchill: Visions of Glory 1874–1932*. Boston: Little, Brown & Company, 1983.

Martin, David G. and Irene Brummerstedt. *Carl Bornemann's Regiment: The Forty-First New York Infantry (DeKalb Regt.) in the Civil War*. Hightstown, NJ: Longstreet House, 1987.

MacDonogh, Giles. *The Last Kaiser: The Life of Wilhelm II.* New York: St. Martin's
Press, 2000.

(Macdonough sums up what is the usual understanding of anybody who
mentions the Papal Peace Proposal at all: The Germans were willing to
accept it, the Austrians were almost desperate to accept it, and the Allies felt
they could get a better deal with the use of fresh American troops and a little
more starvation.)

Mann, Rudolf. *Mit Ehrhardt Durch Deutschland: Erinnerungen eines Mitkampfers
von der 2. Marinebrigade (Quellentexte zur Konservativen Revolution).*
Toppenstedt: Die Uwe Berg-Verlag GbR, 2004

(Mann was Ehrhardt's quartermaster officer and took part in the *Freikorps*
battles of Brunswick and Munich, the occupation of Silesia and the Kapp
Putsch. Mann uses the term "coolie"—*Kuli* in German—not as a racial
epithet but as a social context. The Kaiser's army was limited by the Social
Democratic control of the budget to 55 percent of available manpower and
tended to avoid drafting lower-class working men, preferring farmers, skilled
tradesman and graduates of the *Gymnasium* [classical high school] or the
Universities [as reserve officers after a year of service in the ranks] because
these men were more responsible and had more of a stake in national
interests. German military organization was intensely regional. Each
regiment was based in a city or district and every corps in one of the *Länder*
which had been independent countries before unification in 1871. The
German General Staff tapped regiments for especially talented officers and
the Prussian Guard was open to the tallest and sturdiest enlisted men. Saber
dueling for first blood, not for death, was a popular sporting among some
university groups, eventually with face protectors covering everything but the
forehead, cheeks and chin. The *Schmiss*, or dueling scar, was an upscale
status symbol. German adults into the 20th Century sometimes illegally
fought in lethal pistol duels over matters of honor, and were often pardoned
by the Kaiser. See Kevin McAleer's *Dueling* for a full description of the cult.
Dum dum bullets were cut on the tips lead tips of bullets to blow large holes
in opponents: their use was forbidden by the Hague Convention of 1907,
Article IV, 3, but the United States, which had re-introduced the .45-caliber
pistol of the Indian Wars to disable tribal Muslim bolo men in the
Philippine Islands, refused to ratify this article. The soft=lead slugs from
British .445 Webley revolver had the same essential impact. The Kalmuks
are an Altaic Asian tribe related to the Koreans, Manchus, and Mongols.
The Kalmuks settled in the steppes of Western Russia while Russia was still
forming its own culture. Horsemen and herders, they represented the
furthest extension of Buddhism into European territory. They interacted
with the Cossacks as friends and as foes. Intelligent European Russians
found Kalmuks colorful and amusing but ignorant Russians hated and

insulted them, which sometimes provoked drastic retaliation. Stalin destroyed their European enclaves on charges of complicity with the Germans during World War II and those Kalmuks who survived long boxcar rides without adequate food or water were dumped in Kazakstan. Substantial Kalmuk communities exist in Braunschweig, Germany, and around New Brunswick, New Jersey. The *Jungfrauverein*—"Virgins Union"—was a social club for Catholic girls of good character. *Jungfrau* refers to The Virgin Mary while *Jungfer* describes virgins in general. *Maruschka*—Polish for "Little Mary"—is an affectionate nickname for a Polish girl. The favorite fast dance of Europe at the time was the Polka, which means "Polish girl." The Germans and some neutral observers charged that some ethnic Polish Silesians who advocated remaining part of Germany were singled out not just for verbal intimidation but for outright murder. Ehrhardt uses the word *Prügel*, a non-commissioned officer's stick used to inflict mild beatings as part of recruit training before the Army was re-organized. The *Prügel* is equivalent to a humiliation by public spanking rather than a serious—and often fatal—beating inflicted with steel ramrods or with rifle butts. The *Dreikaiserecke*—"three emperor corner"—was the name for the corner where German, Austrian, and Russian Imperial territory had converged before the war had started. Mann, who was obviously not Jewish but not an organized anti-Semite, refers to Passover as 'Easter' but he fully understood the importance of the flour used for matzohs, unleavened bread, for the Jewish Seder. He grasped the implications not only for street agitation but for journalism and finance.)

Marshall, S. L. A. *The American Heritage History of the First World War.* New York: American Heritage Publishing, 1964.

Massie, Robert K. *Dreadnought: Britain, Germany, and the Coming of the Great War.* New York: Random House, 1991.

McAleer, Kevin. *Dueling: The Cult of Honor in Fin-de-Siècle Germany.* Princeton: Princeton University Press, 1994.

Manchester, William. *The Last Lion: Winston Spencer Churchill: Visions of Glory, 1874-1932.* New York: Bantam, 1984
(Ted Morgan [born a French nobleman!] covered the same problem in greater detail in *Young Man in a Hurry.*)

Marx, Rudolph, M.D. *The Health of the Presidents.* New York: G. P. Putnam's Sons, 1960.

Moore, Captain John RN. *Jane's Fighting Ships of World War I*. New York: Military Press, 1990.

(Details of Hermann Ehrhardt's half-flotilla torpedo boats, which the British source refers to as destroyers. The only double sinking of British destroyers in the first half of 1916 was the loss of HMS Nomad and HMS Nestor but these were listed as "sunk by gunfire of German Battle Fleet at the battle of Jutland, May 31." Since the German Navy contained a large quotient of leftists—the mutiny in the German Navy started the revolution of November 1918—Ehrhardt could hardly have pulled off a fraudulent double sinking without being exposed. The records could be wrong or British wartime propaganda may have transposed the date. HMS 'Amphion' was a light cruiser mounting 10 4-inch guns and was mined and sank in the Thames on August 6. V 27 and V 29 were logged in as having been sunk by fire from British battleships at the battle of Jutland, May 31, 1916. The L-10, a late-war design brought out in 1917-1918, was the only submarine of her class sunk. Jane's Fighting Ships of World War I, Page 5: L 10 sunk in action with German Destroyer S 33 of Texel, night of 3rd-4th October, 1918.)

Morgan, Ted. *Churchill: Young Man in a Hurry 1874–1915*. New York: Simon & Schuster, 1982.

Morley, James William. *The Japanese Thrust into Siberia, 1918*. New York: Columbia University Press, 1957.

Müller, Georg Alexander von and Walter Görlitz. *The Kaiser and His Court: The Diaries, Note Books and Letters of Admiral Georg Alexander von Müller, Chief of the Naval Cabinet 1914–1918*. New York: Harcourt, Brace & World, 1961.

(Walter Görlitz, a post-war German historian, bluntly describes the whole concept as "crazy." Austria-Hungary, however, became increasingly willing to give up some holdings to the new Polish state as the military situation worsened. The amount of land for the new Poland demanded by the Treaty of Versailles, however, stunned the Germans. Some of the most prominent German soldiers—Paul von Hindenburg—scientists—Fritz Haber—and artists—Emil Ludwig—had been born in German-speaking cities like Posen and Breslau.)

Nöhbauer, Hans F. *Bavaria*. Cologne: Ziethen-Panorama Verlag, 1989.

Payne, Robert. *The Life and Death of Adolf Hitler*. New York: Praeger Publishers, 1973.

Pakenham, Thomas. *The Boer War*. New York: Random House, Inc., 1979.

(Pakenham, a British nobleman from a family of eminent historians, is exceptionally objective and never fudges numbers.)

Pakenham, Thomas. *The Scramble for Africa; The White Man's Conquest of the Dark Continent from 1876 to 1912*. New York: Random House, Inc., 1991. (Objective general outline.)

Preston, Diana. *The Boxer Rebellion: The Dramatic Story of China's War on Foreigners that Shook the World in the Summer of 1900*. New York: Walker & Company, 2000.
(Preston's key German outrage is the looting of some astronomical instruments from Peking: She cites the Russians, the French, and British Indian troops for most of the rapes and, like Jack London, gives the Japanese, who brought their own contract prostitutes with them, a clean bill of health on rape and murder, though they too looted with abandon.)

Preston, Diana. *A Higher Form of Killing: Six Weeks in World War I that Forever Changed the Nature of Warfare*. New York: Bloomsbury Press, 2015.

Röhl, John. *1914: Delusion or Design?* London: St. Martin's Press, 1973.
(Text of Eulenberg's essay on the German Navy.)

Rosenbaum, Ron. *Explaining Hitler: The Search for the Origins of His Evil*. New York: Random House, 1998.

Roth, Joseph and Michael Hofmann. *What I Saw: Reports from Berlin 1920–1933*. New York: W. W. Norton & Company, 1996.

Sheean, Vincent. *Not Peace But a Sword*. New York: Doubleday, Doran & Company, 1939.

Shirer, William L. *The Rise and Fall of the Third Reich: A History of Nazi Germany*. New York: Simon & Schuster, 1960.

Simpson, Colin. *The Lusitania: Finally, the Startling Truth About One of the Most Fateful of All Disasters of the Sea*. Boston: Little, Brown & Company, 1972.
(Colin Simpson may have been premature in assigning total credit or blame to Curt Thummel, who certainly played a part.)

Snyder, Louis. *The Blood and Iron Chancellor: A Documentary Biography of Otto Von Bismarck*. Princeton: Van Nostrand Reinhold Inc., 1967

Sönke, Neitzel and Harald Welzer. *Soldaten: On Fighting, Killing, and Dying, The Secret WWII Transcripts of German POWs*. New York: Knopf, 2012.

Spengler, Oswald. *The Decline of the West*. New York: Modern Library, 1980.

Smith, Truman and Robert Hessen. *Berlin Alert: The Memoirs and Reports of Truman Smith*. Stanford: Hoover Institute Press, 1984.

Snyder, Louis L. *The Blood and Iron Chancellor: A Documentary-Biography of Otto von Bismarck*. Princeton: D. Van Nostrand Company, Inc., 1967.

Stern, Fritz. *Gold and Iron: Bismarck, Bleichröder and the Building of the German Empire*. New York: Alfred A. Knopf, 1977.
(A detailed overview of Bismarck's financial dealings with Gerson Bleichröder, his Jewish banker, a personal friend and a Prussian patriot.)

Strachan, Hew. *The First World War*. New York: Viking, 2001.

(Strachan, an Oxford don, could never be described as unduly pro-German and his description in an acclaimed mainstream history confirms what the Nye Commission in the U.S. found in the 1930s: a pattern in which manufacturers and investors promoted military actions of no benefit and possible great detriment to the average U.S. citizen. Strachan points out that the awful impact of the machine gun was so powerful that even British regulars sometimes had to be shot for cowardice to keep men in line, and that the French Army executed more of its own men in the last six months of 1914 than in any other year. Strachan omits to mention that Kaiser Wilhelm and some of his diplomats had been talking about an independent Poland from the first days of the war, though they expected Upper Silesia and the area around Danzig [now Gdańsk] to remain German.)

Strachan, Hew. *The First World War in Africa*. Oxford: Oxford University Press, 2004.

Swearingen, Ben E. *The Mystery of Hermann Göring's Suicide*. New York: Harcourt Brace Jovanovich, 1985.

Taylor, A. J. P. *History of World War I*. London: Octopus Books, 1974.

Taylor, A. J. P. *Bismarck: The Man and the Statesman*. New York: Alfred A. Knopf, 1955.

Thalmann, Rita, Emmanuel Feiner and Gilles Cremonesi. *Crystal Night: 9–10 November, 1938*. New York: Coward, McCann & Geoghegan, 1974.

Toland, John. *Adolf Hitler*. Garden City, NY: Doubleday & Company, 1976.

(The Ehrhardt Brigade is mentioned once, Hermann Ehrhardt not at all.)

Turner, Henry Ashby, Jr. *Hitler's Thirty Days to Power: January 1933*. Reddy, MA: Addison-Wesley, Inc., 1996.

(To follow the complicated intrigue, much of which took place behind closed doors but was confirmed afterwards, the whole book has to be read. Turner's key pages are those from Page 53 to the conclusion on Page 162. His summary offers the plausible thought that after a military take-over, the Prussian generals, minus Hitler, would never have done anything like Auschwitz and might have avoided war with Britain, France, and the United States, and possibly with the Soviet Union—though not with Poland. Lippe, humorously celebrated as the smallest of German states, was heavily Protestant with a population of small farmers and handicraft workers—not good political turf for the Catholic parties, the Social Democrats or the Communists. But the Nazis barely held their own despite massive efforts.)

Ullrich, Volker. *Die Revolution von 1918/19 (Beck'sche Reihe)*. Munich: C.H. Beck, 2009.

Valtin, Jan. *Out of the Night: The Memoir of Richard Julius Herman Krebs alias Jan Valtin*. New York: Alliance Book Corporation, 1941.

Waite, Robert G. L. *The Psychopathic God: Adolf Hitler*. New York: Basic Books, 1977.

(Two mentions of the Ehrhardt Brigade, none of Hermann Ehrhardt.)

Waite, Robert G. L. *Vanguard of Nazism: The Free Corps Movement in Postwar Germany 1918–1923*. New York: W. W. Norton & Company, 1952.

(Franz Ritter *(Knight)* von Epp was a *Freikorps* leader who helped put down the Soviet Republic of Bavaria in 1919. Robert Waite reports that Hitler's service with the *Freikorps* Epp has never been confirmed. Most sources suggest that Hitler was still serving in the *Reichswehr* at this time but took no active part in the fighting. Oddly, Waite's subsequent book portrayed Ehrhardt as a proto-Nazi. In *Vanguard of Nazism* [1952] Waite's description of Ehrhardt is factually responsible though some of his interpretation may be subject to debate. Waite says the Küstrin Putsch was completely bloodless but a more detailed analysis in the German-language *Paul Schulz* [1898 – 1963] reports one man killed. Buchrucker's career was ruined.)

Watt, Richard. *Dare Call it Treason*. New York: Dorset Press, 1969.

(French Army mutiny of 1917 with background.)

Wilcox, Marrion. *Harper's History of the War in the Philippines*. New York: Harper & Brothers, 1900.

(Contains a description of the genuinely heroic death of Edmund Smith.)

Wise, L. F. and W. W. Egan. *Kings, Rulers, and Statesmen*. New York: Bantam Books, 1967.

Whitehouse, Arch. *The Zeppelin Fighters*. New York: Ace Publishing, 1966.

Wilson, A. N. *Hitler: A Short Biography*. New York: Basic Books, 2012.

(Cites a medical report by Otfrid Foerster, a renowned neurosurgeon who examined Hitler's medical records in 1932 and gave it as his opinion that Hitler's late-war blindness after the 1918 gas attack was due to "hysterical amblyopia"—intense fear that caused a physical symptom with no physical trauma. Emil Ludwig, born Emil Ludwig Cohn in Breslau, a Hitler contemporary and no fan of Hitler, missed the shrapnel wound in his 1958 World Book Encyclopedia article, Volume 8, Page 3434, but concurred that the blindness was due to front-line hysteria rather than poison gas. The American popular historian Richard Hanser co-author of "Victory At Sea," suspected the same thing in *Putsch*.)

Ziegler, Jean and John Brownjohn. *The Swiss, the Gold, and the Dead: How Swiss Bankers Helped Finance the Nazi War Machine*. New York: Harcourt Brace & Company, 1998.

Zuber, Terence. *The Real German War Plan 1904–14.* Gloucestershire, UK: The
    History Press, 1911.
    (Zuber, a former U.S. Infantry and Counter-Intelligence Office, describes
    Germany's entry into the war as essentially an act not of aggression but of
    desperation due to the overwhelming odds. He argued that had the
    Germans practiced universal conscription rather than limiting conscription
    to 55 percent, they could possibly have overwhelmed both France and
    Russia in the first few months of the war.)

ARTICLES

Ackermeier, Jan. "Le Capitaine Hermann Ehrhardt: ennemi de la République de
    Weimar et combattant clandestin." *euro-synergies.hautetfort.com.* Euro-
    Synergies. February 20, 2012.
    (French-language site on Ehrhardt, based on Austrian sources in *"Die Zeit,"*
    2011.)
"Albert Leo Schlageter." *de.wikipedia.org.* Wikipedia (German). 2016
"Alexander von Falkenhausen (General)." *de.wikipedia.org.* Wikipedia (German).
    2016.
    (Falkenhausen later served as Chiang Kai-shek's military advisor against
    both the Chinese Communists and the Japanese in the 1930s. As German
    Military Governor of Belgium, Falkenhausen refused to force Jews to wear
    the Yellow Star and spared most of those Jews who were not Communists.
    He remained a hero in China and the Allies let him off easy because he had
    supported the June 1944 plot against Hitler.)
"Alexander von Falkenhausen: Was ich Dachte und Was ich Tat." *zeit.de.* Zeit
    Online. May 4, 1950.
    (German-language interview with Alexander von Falkenhausen)
"Altmann: Tote der Revolution 1918-1921." *deutsche-revolution.de.* Deutsche
    Revolution. 2016.
    (The information comes from court records examined by Emil Julius
    Gumbel, a statistician trained at the University of Munich who taught at
    Heidelberg and later at Columbia University in New York City. Gumbel's
    expertise was in "extreme values" and his book was extremely valuable to
    writers who wanted to link the monarchist Right to Hitler. Gumbel describe
    himself as a "pacifist" for immigration purposes but his actual affiliation was
    the the USPD, the Independent Socialists who were communists in
    everything but name and later jointed the KPD. Paraphrasing Gumbel. The
    Prussian Guard was made up of the tallest and sturdiest soldiers drawn from
    regional Prussian units who would otherwise have been based near their

home towns. The fact that three of the convicted assailants have Polish names indicates objective recruiting standards. The fact that the Poles were brought up on charges and that the officer was not charged suggests that the three Poles and Mueller may have been scapegoats. Regular amnesties probably had all four of them out of prison in a few months to a few years.)

"Anton Graf von Arco auf Valley." *en.wikipedia.org*. Wikipedia. 2016.

(Worth noting is that Oswald Spengler, the most famous German historian of the era—who quickly came to dislike Hitler but admired Herman Ehrhardt—including the Jews and Arabs, the Chinese and Japanese, and the Mayans and Aztecs as among the world's great cultures.)

"B. Traven." *en.wikipedia.org*. Wikipedia. 2016.

(Based on his writings B. Traven was more of an anarchist than a Marxist and opposed all forms of arbitrary government. He spent the last part of his life in Mexico and loved Mexican Indians. Stalin murdered vast numbers of Communists with convictions like these before and during the Blood Purge of 1938, along with many old Bolsheviks who themselves had had no objections to spilling blood in the previous years.)

Bernstein, Mark F. "Woodrow Wilson Revisited." *Princeton Alumni Weekly*. April 19, 2006.

(Summary of 21st Century scholarly opinions of Wilson's attempts and failures.)

Bigelow, William. "New Evidence From His Doctors Shows Hitler Was Gay." *Breitbart.com*. Breitbart. May 9, 2013.

(Dr. Theodore Morell's medical report sold at auction.)

Bildarchiv Preußischer Kulturbesitz. "The Accused in the Rathenau Trial (October 13, 1922)." *germanhistorydocs.ghi-dc.org*. German Historical Institute: German History in Documents and Images. 1922.

(Group photo of the suspects)

"Black Tom explosion." *en.wikipedia.org*. Wikipedia. 2016.

Blake, Heidi. "Hitler Had African and Jewish Roots, DNA Tests Show." *The Telegram*. August 24, 2010.

(Account of the two Belgian researchers who found that many Hitler relatives had E1b1b1 haplogroup genes, a founding lineage of the Jewish population unusual among Germanic or Slavic groups.)

Brown, Cyril. "Army Officers Held in Rathenau Murder." *New York Times*. June 27, 1922.

Bytwerk, Randall. "The German National Catechism." *research.calvin.edu*. German Propaganda Archive. 2003.

(Translations of Nazi propaganda from the early 1930s, assembled by Randall Bytwerk and translated into English in 2003.)

"Charles Fryatt." *en.wikipedia.org*. Wikipedia. 2016.

Chojnowski, Dr. Peter E. "1917: Democratic Jihad and the Pope's Peace." *drchojnowski.blogspot.com.* Articles by Dr. Peter Chojnowski. October 31, 2006.

"Clemens von Ketteler." *en.wikipedia.org.* Wikipedia. 2016.

Crowther, Bosley. "'The Sea Chase'; John Wayne Stars at the Paramount." *nytimes.com.* The New York Times. June 11, 1955.

Cohen, Jennie. "Study Suggests Adolf Hitler Had Jewish and African Ancestors." *History in the Headlines.* August 26, 2010.
(The 'Africans' referred to are Berbers from Morocco, not Bantu Africans, but the authors do not seem to know what a Berber is. Berbers are dark-skinned Caucasians.)

Dauvé, Gilles and Denis Authier. "The Communist Left in Germany 1918–1921: Chapter 7: The Confrontations: November 1918 to May 1919." *Marxists.org.* Marxists Internet Archive. 1976.

Dierckmann, E. L. "The Prophet of Mission Hills." *American History Illustrated.* November, 1976.
(Eugene P. Lyle and his "prophetic" 1919 article, "THE WAR OF 1938.")

Doder, Dusko and Peter Essick. "Experiment that Failed: The Bolshevik Revolution." *National Geographic.* October, 1992: Pages 111–130.
(A Russian former Communist sums up 70 years of the Soviet Union or an awrd-winning journalist: "It was all a big lie." Quite.)

Duffy, Michael. "Back To Bavaria" *firstworldwar.com.* FirstWorldWar.com: a Multimedia History of World War One. 2009.

Eisenbichler, Ernst. "Hintergrund die Bayerische Revolution 1918/1919: Vom Umsturz zum Absturz." *br.de.* Bayerischer Rundfunk. November 24, 2008.
(German-language political biography of Kurt Eisner.)

"Emil Eichhorn." *en.wikipedia.org.* Wikipedia. 2016.

Enigmajones. "Dead by Dawn: The Road to Revolution." *AlternateHistory.com.* Alternate History Forum. August 28, 2012: Posts 274–277 (Where Are They Now).
(This site has Hermann Ehrhardt heading the secret police for the late Weimar Republic. This mythical Ehrhardt escaped to Argentina after the fall of the Heydrich Government in 1968 [*!*]—The historical Reinhardt Heydrich was executed by Czech partisan assassins in 1942.—Not all sources are reliable. See Gumbel above.)

"Erich Mühsam." *en.wikipedia.org.* Wikipedia. 2016.

"Ernst Thälmann." en.wikipedia.org. Wikipedia. 2016.
(Ernst Thälmann was arrested by the Nazis, kept alive under harsh conditions for 11 years as a bargaining ploy, and ultimately executed in August of 1944.)

"Ernst von Oven." *de.wikipedia.org*. Wikipedia (German). 2016.

(Ernst von Oven was born in 1859 in Oldenburg, a semi-autonomous grand dukedom before November 1918 and a largely Catholic enclave surrounded by mostly Protestant Hanover, but seems to have realized that, located between France and Russia, Germany and Oldenburg needed unity more than particularism.)

"Ernst von Salomon." en.wikipedia.org. Wikipedia. 2016.

(Salomon also wrote a book denouncing the American occupation of West Germany and making fun of the American de-Nazification program in 1951. It became a best-seller.)

"Eugen Leviné." *de.wikipedia.org*. Wikipedia (German). 2016.

(The German-language articles are generally three to five times as long as the English-language Wikipedia articles and cite different sources.)

"Eugen Schiffer." *de.wikipedia.org*. Wikipedia (German). 2016.

(Dr. Eugen Schiffer was thrown out of the government by the Social Democrats but re-emerged and attempted to avoid the extremes of Socialism such as confiscation of major industries. His honesty and intelligence were widely respected by many Germans, but in 1943 the Nazi administration sent him to the Berlin Ghetto along with his daughter, simply because they were of Jewish ancestry. They both survived the war and Schiffer himself lived until 1954.)

Evans, Richard J. "Prophet in a Tuxedo." *LRB.co.uk*. London Review of Books. Vol. 34 No. 22 November 22, 2012: Pages 20–22.

(Analysis of Walther Rathenau, book review.)

Falk, Gerhard, Dr. "Albert Ballin & Walther Rathenau: Two German Patriots." *jBuff.com*. Jewish Buffalo on the Web. 2005.

"Friedrich Freksa." *de.wikipedia.org*. Wikipedia (German). 2016.

(Actually Kurt Franz Georg Friedrich-Freksa, 1882 – 1955, born in Berlin, volunteered and served briefly in World War I when he was already pushing 40, and wrote in many genres, including comedy, ancient history, Germanic folklore, and science fiction. The German-language articles are generally three to five times as long as the English-language Wikipedia articles and cite different sources.)

"Freiherrn Klemens von Ketteler." *deutsche-schutzgebiete.de*. Der Boxeraufstand 1900/01. 2000. (Diana Preston's account is substantially accurate. The Wikipedia account is full of mistakes. The German account is based on verbatim stories from contemporary newspapers.)

Garbrecht, Günter. "Die Bremer Räterepublik von 1918 1919." *www–user.uni-bremen.de*. Universität Bremen. 2016.

> (Karl Tillessen and his brother Heinrich had both served with Ehrhardt in the torpedo boat squadrons. Ehrhardt does not supply a first name but since Heinrich had been left behind watching the interned High Seas Fleet at Scapa Flow, this conversation would have been with Karl. Both were involved in *Freikorps* activities including terrorism and—unlike Ehrhardt—they later served in Hitler's navy during World War II.)

"Gelbe Gefahr." *de.wikipedia.org*. Wikipedia (German). December 21, 2016.

> (details of Kaiser Wilhelm's attempt to expose the Asian plans for a world take-over, with art.)

Gercke, Achim, Dr. "Solving the Jewish Question." *research.calvin.edu*. German Propaganda Archive. 2007.

> (Translations of Nazi propaganda from the early 1930s, assembled by Randall Bytwerk and translated into English in 2003.)

"German federal election, November 1932." *en.wikipedia.org*. Wikipedia. 2016.

"Gestorben: Hermann Ehrhardt." *spiegel.de*. Der Spiegel. October 11, 1971.

> (Brief on-line archive obituary.)

Gumbel, Emil Julius. "Tote Der Revolution 1918–1919." *DeutscheRevolution.de*. Deutsche Revolution. 1922.

> (German-language web site. The headline for the Saint Joseph Gesellschaft executions says 21 were killed, the text of the story says seven.)

"Gustav Landauer." *en.wikipedia.org*. Wikipedia. 2016.

"Hans Paasche." *en.wikipedia.org*. Wikipedia. 2016.

"Harden–Eulenburg affair." *en.wikipedia.org*. Wikipedia. 2016.

"Hartmut Plaas." *de.wikipedia.org*. Wikipedia (German). 2016.

"Harry Graf Kessler." *de.wikipedia.org*. Wikipedia (German). 2016.

> (The German-language articles are generally three to five times as long as the English-language Wikipedia articles and cite different sources.)

"Heinrich Schulz." *de.wikipedia.org*. Wikipedia (German). 2016.

"Heinrich Tillessen." *de.wikipedia.org*. Wikipedia (German). 2016.

"Herbert von Bose." *en.wikipedia.org*. Wikipedia. 2016.

> (The English site has details not on the German site but the same facts.)

"Herero People: The Fearless and War-Like African Tribe That Suffered The World's First Holocaust at the Hands of Germans." *kwekudee-tripdownmemorylane.blogspot.com*. Trip Down Memory Lane. October 26, 2012.

> (235 pages covering Herero culture and many historic and modern photographs, rich is detail and anecdote but somewhat lacking in objectivity. The late Brigitte Lau, an ethnic German who lived in Namibia, argued conversely that while the war was extremely brutal, no extermination was ever contemplated: the Herero death toll, mostly from diseases that also

killed hundreds of Germans, was vastly exaggerated by the British after World War I to cover their own tracks in the Boer War and the tracks of their Belgian allies in the Congo. The one tale of tossing a child on a bayonet was provided not by a proverbially honest Boer, but by a 'Cape Coloured' named Jan Cloete who probably said what he told to say. He said it after the Belgian atrocity stories at a time when the British wanted to take over Germany's African colonies.)

"Hermann Ehrhardt." *de.wikipedia.org*. Wikipedia (German). 2016.
(The German version, despite some questionable construements, contains six pages of facts and references. The English-language version is one page long. The French Wikipedia article is four full pages long.)

Herzberg, Gary. "The Anti-Semitism Behind the Assassination of Walther Rathenau." *Swarthmore.edu*. Swarthmore College. 2010.
(term paper from Swarthmore, 2010—excellent analysis. The student author concluded that Rathenau's murder was primarily political rather than racial or religious.)

Hoffstadt, Anke and Richard Kühl. "'Dead Man Walking': der 'Fememörder' Paul Schulz und seine 'Erschießung am 30. Juni 1934'." *ssoar.info*. Social Science Open Access Repository. 2009: Pages 273–285.

Hofmann, Ulrike Claudia. "Fememorde." *historisches-lexikon-bayerns.de*. Historisches Lexikon Bayerns. May 15, 2006.
(Accounts in German of individual political murders in the Munich area in 1920 – 1921.)

"Hohenlohe Princess arrested for aiding Putsch leader." *New York Times*. July 16, 1923.

Hunnicutt, Alex. "Eulenburg-Hertefeld, Philipp, Prince zu (1857-1921)." *glbtqarchive.com*. glbtq, Inc. 2004.

"Ich Lies Rosa Luxemburg Richten." *Der Spiegel*. 16: April 18, 1962.
(Pabst, at a time when the Berlin Wall was still standing, justifies his orders to kill Rosa Luxemburg and Karl Liebknecht.)

Institut für Zietgeschichte. "Paul Schulz." *ifz-muenchen.de*. Archiv, Findmittel online: Bestand: ED 438.
(details of his life and family with important dates and references.)

Jahn, Ralf G., Dr. "Hitler-Genealogie." *adel-genealogie.de*. Der Geschichtsdetektei. 2002.
(The last word on incest in the Hitler clan and the retardation it caused for many of them, along with Hitler's mixed Czech ancestry.)

Jones, Nigel. "The Assassination of Walther Rathenau: Nigel Jones on the Redemption Sought by the Assassin of Weimar Germany's Foreign Minister." *History Today*. Volume 63, Issue 7, July 2013.
(Nigel Jones accurately reports Ehrhardt as an anti-Nazi who was almost murdered in 1934.)

"Juye Incident." *en.wikipedia.org*. Wikipedia. 2016.

"Karl Liebknecht." *de.wikipedia.org*. Wikipedia (German). 2016.
    (The German-language articles are generally three to five times as long as
    the English-language Wikipedia articles and cite different sources.)

Kelley, Tina. "John Weitz, 79, Fashion Designer Turned Historian, Dies." *New
    York Times*. October 4, 2002.

"Kiautschou Bay concession." *en.wikipedia.org*. Wikipedia. 2016.

Kilgannon, Corey. "Three Quiet Brothers on Long Island, All of Them Related to
    Hitler." *New York Times*. April 24, 2006.

Kistenmacher, Olaf. "From 'Judas' to 'Jewish Capital': Antisemitic Forms of
    Thought in the German Communist Party (KPD) in the Weimar Republic,
    1918-1933." *engageonline.wordpress.com* Engage Journal. Issue 2: May 2006.
    (Shows that German Communists, like their Nazi adversaries, often used
    anti-Semitism to attract useful thugs. Prince Albert was intelligent,
    sensitive, somewhat neurotic, drank very moderately, loved music and hated
    hunting: this was all the evidence anti-Semites needed, but British authors
    note that his parents' marriage was stable until four years after his birth.
    John Houseman, whose own father was Jewish, included the legend in his
    play *Victoria Regina*, minus the explicit Jewish reference. The story today has
    very little credence.)

Koster, John. "The Scapegoats of Chancellorsville." *America's Civil War*. May, 2008.
    (Sociology of Prussian and other German officers in the Union Army
    1861-1865 explored through records from the U.S. National Archives and
    rare books.)

Lehnert, Erik, Dr. "Der Kaiser in der Kritik: Die sogenannte Hunnenrede (1900)."
    wilhelm-der-zweite.de. Wilhelm II: Deutschlands Letzter Kaiser 1888-1918.
    2016.
    (Scholarly website article by Dr. Erick Lehnert. Both the official [censored]
    version and the unofficial [reportedly verbatim] version mention not taking
    prisoners, but only the unofficial version mentions the Huns as role models.
    Wilhelm reportedly objected when the official press censors deleted the
    Huns.)

Leicht, Robert. "Matthias Erzberger: Patriot in der Gefahr." *zeit.de*. Zeit Online.
    August 18, 2011.
    (Sees Erzberger as a scapegoat-victim)

Leicht, Robert. "Weimarer Republik: Nachts bin ich Bolschewist." *zeit.de*. Zeit
    Online. June 21, 2012.
    (a sympathetic look at Walther Rathenau's deeply enigmatic stance on the
    politics of Socialism versus the need to maintain German sovereignty)

"List of the Pour le Mérite (military class) recipients." *en.wikipedia.org*. Wikipedia. 2016.

(Friedrich von Miaskowski, Pabst's friend and fellow General Staff Officer, was an *Oberst* (colonel) by the end of the war and received the Pour le Mérite, Prussia's highest decoration for valor and service, on April 21, 1918. A half-dozen other Prussian officers with Polish surnames received Prussia's and Germany's highest medal, as did Erwin Rommel, Manfred and Lothar von Richthofen, aviator Werner Voß—and Hermann Göring.)

"Leo Jogiches." *de.wikipedia.org*. Wikipedia (German). 2016.

(The German-language articles are generally three to five times as long as the English-language Wikipedia articles and cite different sources.)

"Lothar von Trotha." *en.wikipedia.org*. Wikipedia. 2016.

"Ludwig Ebermayer." *de.wikipedia.org*. Wikipedia (German). 2016.

Ludwig, Emily. "Biography of Adolf Hitler." *World Book Encyclopedia*. 1958.

Lundy, Darryl. "Person Page - 11261: Hermann Ehrhardt." *thepeerage.com*. The Peerage. January 13, 2003.

(genealogy of Hermann Ehrhardt's second marriage, his children with Princess Margrethe and their own upper-class marriages.)

Lyle, Eugene P. "The War of 1938." *Everybody's Magazine*. September, 1918

(Classic Germanophobia postulated on the Prussian bid to take over the planet at all costs....)

"Margarete of Hohenlohe convicted of aiding Ehrhardt." *New York Times*. July 24, 1923.

"Matthias Erzberger." de.wikipedia.org. Wikipedia (German). 2016.

"Matthias Erzberger." en.wikipedia.org. Wikipedia. 2016.

"Max Levien." de.wikipedia.org. Wikipedia (German). 2016.

McFadden, Robert D. "Nicholas Winton, Rescuer of 669 Children From Holocaust, Dies at 106." *New York Times*. July 1, 2015

(Describes the extensive degree to which the Gestapo accepted bribes from Holocaust rescuers.)

Migge, Torsten. "Chronik aller Attentate auf Hitler." *geschichtsthemen.de*. Geschichtsthemen. 2016.

(Lists all serious attempts to assassinate Hitler after the Munich Putsch, including the Mylius-Ehrhardt conspiracy of 1934 – 1935 and the Liedig-Heinz plot of 1938.)

Mischer, Olaf. "Die Mörder Rosa Luxemburgs." *geo.de*. Geo Epoche. Nr. 27: August, 2007.

"Murdering Hitler: the failed attacks on Hitler's life." *valkyrie.greyfalcon.us*. Target Hitler. February 21, 2005.

(detailed accounts of attempts to assassinate Hitler, including the Mylius-Ehrhardt conspiracy.)

"Night of the Long Knives." *en.wikipedia.org*. Wikipedia. 2016.

Office of Strategic Services, Research and Analysis Branch. "The Pattern of Illegal Anti-Democratic Activity in Germany After the Last War: The Free Corps." *lawcollections.library.cornell.edu*. Cornell University Law Library. Nuremberg, Germany: International Military Tribunal, Vol 106: October 13, 1944.

"Organisation Consul." *en.wikipedia.org*. Wikipedia. 2016.
    (The source is Robert Waite, *Vanguard of Nazism*, who quoted the original source, the *Münchener Post* of December 27, 1922. The *Münchener Post* was a Social Democratic newspaper with Leftist editorial policies and at the time the article appeared, Hermann Ehrhardt was already in prison.)

"Otto Gessler." *en.wikipedia.org*. Wikipedia. 2016.

"Otto Pittinger." *de.wikipedia.org*. Wikipedia (German). 2016.

"Paul Schulz (1898-1963) Oberleutnant a.D. – Teil I:Kurzbiographie." *oberleutnant-schulz.de*. Paul Schulz (1898-1963), Oberleutnant a.D. 2016.

"Paul Schulz (Politiker)." *de.wikipedia.org*. Wikipedia (German). 2016.

"Philipp Scheidemann." *de.wikipedia.org*. Wikipedia (German). 2016.

"Pope Benedict XV's Peace Proposal." *wwi.lib.byu.edu*. The World War I Document Archive. August 1, 1917.

"Preuß Denounces Demand of Allies." *New York Times*. September 14, 1919.
    (The Congress of Vienna was organized by the former opponents of Revolutionary and Napoleonic France and aimed at preventing any more new republics in Europe. France itself was forced to accept Louis XVIII as a legitimate monarch, followed by the even more unpopular Charles X. Some Prussians who had fought against Napoleon found the Prussian regulations fostered by the Congress of Vienna after the liberal reforms of 1813 to be too reactionary. Franz Lieber, later an American citizen and grandfather of the Hague Convention and the Geneva Convention on military law and treatment of prisoners, went to prison twice for his objections.)

Radek, Karl. "Leo Schlageter: The Wanderer into the Void" *marxists.org*. Marxists Internet Archive. June, 1923.

Rauscher, Richard. "Geschichte des Ortes Brunn am Wald, Gemeinde Lichtenau." *lichtenau.at*. Lichtenau im Waldviertel. November 15, 2007.
    (German-language site from Austria offers views and a brief history of what became the Ehrhardt estate from 1934 to 1976.)

"Reich Chancellery meeting of 12 December 1941." *en.wikipedia.org*. Wikipedia. 2016.

"Revolution in München." *jusos-muenchen.de*. Jusos München. 2016.
    (article on the Alois Lindner murders at the Bavarian Diet after the Eisner assassination.)

"Rosa Luxemburg." *de.wikipedia.org*. Wikipedia (German). 2016.
    (The German-language articles are generally three to five times as long as the English-language Wikipedia articles and cite different sources.)

Ross, Alex. "Diary of an Aesthete: Count Harry Kessler Met Everyone and Saw Everything." *The New Yorker*. April 23, 2012.

"Rudolf Berthold." *en.wikipedia.org*. Wikipedia 2016.

"Rudolf Berthold (Jagdflieger)." *de.wikipedia.org*. Wikipedia (German). 2016.

"Rudolf Egelhofer." *de.wikipedia.org*. Wikipedia (German). 2016.

(Dr. Menzi was apparently arrested and jailed but not executed, according to Erich Mühsam's diary on Project Gutenberg, which saw all the Reds, including the death squad organizer Rudolf Egelhofer, as martyred heroes. Mühsam survived because he was already in custody when the *Reichswehr* arrived in Munich. He himself was viciously tortured and murdered by the Nazis in 1934. Wikipedia shows both versions, apparently retouched photographs. The late Bernard Perlin produced a plausible likeness of Egelhofer with his clenched fist for the U.S. propaganda poster "Avenge Dec. 7." Perlin was born and raised in the United States. At the age of 92 he told the author that he had no conscious memory of hearing about Rudolf Egelhofer, who had killed about the time he was born, but that Egelhofer's free-floating rage mysteriously embodied what Perlin wanted to convey in the illustration.)

Ruggenberg, Rob. "Extremes in No Man's Land: Adolf Hitler and Erich Maria Remarque Fighting Together." *greatwar.nl*. The Heritage of the Great War. 2017.

(Somewhat cynical about Remarque but very cynical about Hitler's heroism.)

Samichaiban, L'espace. "Le Fils Français Cache d'Adolf Hitler." *Libre de Exclusif*. February 18, 2012.

(Hitler's fake French son discovered by Werner Masser was later exposed by DNA as a fake French son.)

Schmalzl, Markus. "Zweite Revolution, 1919." *historisches-lexikon-bayerns.de*. Historisches Lexikon Bayerns. June 8, 2009.

(provides full names and details of the murders at the Bavarian Diet after the Eisner assassination. Richard Hanser's names in *Putsch* are incomplete but his context is accurate.)

Schulz, E. H., Dr. and Dr. R. Frercks. "Why the Aryan Law? A Contribution to the Jewish Question." *research.calvin.edu*. German Propaganda Archive. 1999.

(Translations of Nazi propaganda from the early 1930s, assembled by Randall Bytwerk and translated into English in 2003.)

Schulz, Paul Alexander, Dr. "Paul Schulz (1898-1963) Oberleutnant a.D. – Teil I."
    oberleutnant-schulz.de. Dr. Paul Alexander Schulz. March, 2010.
    (compiled from remembrances and research by Paul Schulz's son, who
    obviously respected him, the 20-page biography offers the fullest picture of
    Schulz as a war hero, *Feme* assassin, and German patriot. The second
    installment contains affadavits from Jews and others whom Schulz rescued
    from death camps, sometimes at risk of his own life.)

Seipp, Bettina. "Keine Kinder für die Familie Hitler." *welt.de*. Welt. January 23,
    2007.
    (Discusses Dr. Ralf Jahn's genealogical survey of Hitler's extended family
    and the many retarded members and hereditary diseases probably due to
    incest.)

Sewell, Rob. "Germany: From Revolution to Counter-Revolution." *marxists.org*.
    Marxists Internet Archive. 1988.
    (Marxist from an essentially Trotskyite, but non-Bolshevik perspective)

Simkin, John. "Hermann Ehrhardt." *spartacus-educational.com*. Spartacus
    Educational. August 2014.

Simkin, John. "Adolf Hitler." *spartacus-educational.com*. Spartacus Educational.
    March 2016.
    (The Mend Protocol is also quoted in Lothar Machtan's book *The Hidden
    Hitler*.)

Simkin, John. "Leo Jogiches." *spartacus-educational.com*. Spartacus Educational.
    August 2014.

Simkin, John. "Walter Krivitsky." *spartacus-educational.com*. Spartacus Educational.
    August 2014.
    (Krivitsky later defected in the United States and was found dead in a
    Washington DC hotel room locked from the inside. The theory was that the
    NKVD told him that they would spare his family if he committed suicide
    immediately.)

Simkin, John. "Rosa Luxemburg." *spartacus-educational.com*. Spartacus Educational.
    February 2016.

Simkin, John. "Rose Leviné-Meyer." *spartacus-educational.com*. Spartacus
    Educational. August 2014.
    (Favorable biography with many quotes and biographical detail not available
    elsewhere)

Staas, Christian. "Nationalsozialismus: Hitlers willige Landser." *zeit.de*. Zeit
    Online. October 4, 2012.
    (a look at German soldiers' attitudes.)

"Stahlhelm: Ehr und Wehr." *Der Spiegel*. 42: October 9, 1967.
    *(retrospective article about the Nazi destruction of the front-line veterans group
    for not being sufficiently anti-Semitic or sufficiently obedient to Hitler)*

Thadeusz, Frank. "Mad King Ludwig?: Study Claims Bavarian Monarch Was Sane." *spiegel.de*. Spiegel International. January 31, 2014.

"Theobald von Bethmann-Hollweg." *en.wikipedia.org*. Wikipedia. 2016

Thery, Fabrice. "La terreur Zeppelin 1914-1916: comment le Zeppelin devient un bombardier stratégique, entre 1914 et 1915." *histoquiz-contemporain.com*. Histoquiz Contemporain. 2007.

(In French—detailed and objective study of Zeppelin campaign)

Thornton, Michael. "The Nazi Relative that the Royals Disowned." *Daily Mail*. December 1, 2007.

"Traugott von Jagow." *de.wikipedia.org*. Wikipedia (German). 2016.

"Uprising of 1953 in East Germany." *en.wikipedia.org*. Wikipedia. 2016

"Von Hunnen und Sachsen." *stumpfeldt.de*. Hamburger China-Notizen. August 13, 2000.

(Study of the *Hunnenbriefe* as probable forgeries.)

Vossische Zeitung. "Lynchmord in Dresden: Sturm auf das Kriegsministerium." *bommi2000.de*. Bommi 2000. April 12, 1919

(full account of the lynching of Social Democrat Gustav Neuring by angry veterans)

"Waldemar Pabst." *de.wikipedia.org*. Wikipedia (German). 2016.

(The German-language articles are generally three to five times as long as the English-language Wikipedia articles and cite different sources.)

Weyerer, Benedikt. "Münchner Räterepublik: Mord im Luitpold-Gymnasium." spiegel.de. Spiegel Online. September 25, 2007.

(Full details as far as they are known of the shooting of hostages in Munich. Complete names under *Der Mord*.)

Wilson, A. N. "Were Queen Victoria and Prince Albert Both Jewish?" *Daily Mail*. March 4, 2009.

Zenz, Helmut. "Hugo Preuß: Jurist, Politiker (1860-1925)." *helmut-zenz.de*. Helmut Zenz SDB im Netz. 2003.

(Detailed biography of Hugo Preuß with family and career history in German)

VIDEO DOCUMENTARIES

Title. Dir. First M. Last. Perf. First M. Last. Distributor, Year Published. Media
   Type.
Killing Hitler: The True Story of the Valkyrie Plot. Pegasus Entertainment, 2008.
Nazi Mega Weapons. Darrow Smithson Productions, 2013.
Poland. Megan McCormick. Globe Trekkers, 2014.
Why We Fight #1: Prelude to War. Frank Capra. U.S. War Department, 1942
World War II Behind Closed Doors: Stalin, the Nazis, and the West. Andrew Williams.
   BBC / PBS, 2008.
Nova: Zeppelin Terror Attack. Ian Duncan. PBS, 2014.

FEATURE FILMS

Hitler's Children. Edward Dmytryk and Irving Reis. RKO Radio Pictures, 1943.
The Sea Chase. John Farrow. Warner Bros., 1955
So Ends Our Night. John Cromwell. United Artists, 1941

NOVELS

Remarque, Erich Maria. *The Road Back* (A.A. Wheen Translation). Boston: Little,
   Brown and Company, 1931.
Remarque, Erich Maria. *Flotsam* (A.A. Wheen Translation). Boston: Little, Brown
   and Company, 1941.

THEATRICAL PLAY

Housman, Lawrence. "Victoria Regina." *Sixteen Famous British Plays*, compiled by
   Bennett Cerf and Van H. Cartmell. New York: The Modern Library, 1942.

AUDIO CD

German University Songs, Volumes I–IV (2000) [CD]. New York: Vanguard Classics.

Idle
Winter
Press